D60S44.08

Proceedings of the ISSS symposium on water and solute movement in heavy clay soils

W9-BUJ-064

P. M. C ...
May 1986.

Horizontal cross-section of a dry, cracked clay soil at 40 cm below surface. (Photo courtesy of the Netherlands Soil Survey Institute.)

Proceedings of the ISSS symposium on water and solute movement in heavy clay soils

Edited by J. Bouma and P.A.C. Raats

ILRI
International Institute for Land Reclamation and Improvement,
P.O. Box 45, 6700 AA Wageningen, The Netherlands 1984

The International Society of Soil Science / ISSS

International Symposium on
water and solute movement in heavy clay soils
Wageningen, The Netherlands. August 27–31, 1984
The Symposium was sponsored by Commissions I and V of ISSS

CIP-data Royal Library, The Hague

Proceedings

Proceedings of the ISSS symposium on water and solute movement in heavy clay soils/ed. by J. Bouma and P.A.C. Raats – Wageningen: International Institute for Land Reclamation and Improvement. – Ill. – (ILRI publication; 37)
With lit. references
ISBN 90-70260-97-2
SISO 631.2 UDC 631.4
Keyword: Soil Science

Printed in The Netherlands.

PREFACE

At the 12th International Congress of Soil Science, held in New Delhi, India in February 1982, a proposal by the Dutch Soil Science Society to host an international symposium on water movement in heavy clay soils was accepted. The symposium was to be sponsored by the International Society of Soil Science (ISSS). To further develop contacts between pedologists and soil physicists, it was decided that the symposium be organized by Commissions I (Soils Physics) and V (Soil Genesis, Classification, and Cartography) of the ISSS. Later, it was also agreed that the symposium be co-sponsored by the European Geophysical Society.

In the fall of 1982 an organizing committee was formed consisting of G.H. Bolt (Chairman), J. Bouma (Secretary), and W.A. Blokhuis, J.W. van Hoorn, P.A.C. Raats, K. Rijniersce, and G.P. Wind (members).
The ISSS was represented on an ad hoc basis by its Secretary General W.G. Sombroek. Secretarial assistance was provided by Mrs. M. Rijk of the Netherlands Soil Survey Institute.

At its first meeting the organizing committee decided to expand the scope of the symposium to cover not only water but also solute movement in heavy clay soils, thus including aspects of ISSS Commission II (Soil Chemistry).

The three-day symposium was held from 27 to 29 August 1984, at the International Agricultural Centre, Wageningen, The Netherlands. Organizational aspects were handled by Mrs. L. Hotke and Mrs. E. van de Wetering. The schedule included a reception given by the Dutch Ministry of Agriculture and Fisheries, a visit to the International Soil Reference and Information Centre (ISRIC) in Wageningen and, of course, a symposium dinner. Following the symposium were two day-long excursions. The first, organised by K. Rijniersce of the IJsselmeerpolders Development Authority, was to very young clay soils in a newly-reclaimed polder. The second, to young and old riverine clays near the Rhine, was organized by M. Kooistra, J.H.M. Wösten,

L.W. Dekker, and J. Bouma of the Netherlands Soil Survey Institute, and by R. Miedema and J. Versluys of the University of Agriculture, Wageningen. Excursion guides were prepared, but they have not been reproduced in these proceedings.

The following themes were covered in the symposium:

1. Development of structural patterns in swelling and shrinking clays;
2. Transport phenomena: water movement;
3. Transport phenomena: solute transport;
4. Measurement and simulation techniques.

The sessions in which these themes were covered were chaired by G.H. Bolt, W.A. Blokhuis, J.R. Philip, G.D. Towner, D.H. Yaalon, and L.P. Wilding.

Prior to the symposium, ten individuals with particular experience in one of the four themes were asked to submit a keynote paper of a maximum ten pages in length. Each paper was to be presented during a half hour period, at least fifteen minutes of which was to be devoted to discussion. This unusual timing was possible because of the fact that the participants had received the papers well before the symposium took place, and had been asked to submit their comments and questions in writing preferably before, but also during, the symposium.

Voluntary papers were also invited, to be submitted in the form of expanded abstracts with a minimum of 2 and a maximum of 4 pages. These were available to all participants during the symposium. To stimulate meaningful discussions, the authors were asked to include specific information, such as tables, graphs, and selected key references. They were encouraged to submit the complete papers elsewhere as well. Each expanded abstract was reviewed by a member of the organizing committee. Preprints of keynote papers and expanded abstracts were prepared by Messrs. T. Beekman and J. van Manen of ILRI.

The present book consists of the keynote papers, the expanded abstracts, and the discussions during the symposium, the latter only insofar as questions and answers were submitted in writing, to the symposium committee. Thus, not all exchanges are reflected in these Proceedings. All authors were given the opportunity to submit corrected versions of their papers.

Heavy clay soils occupy large areas of the world. Their complex and variable structure leads to intricate patterns of water and solute movement. This book shows that the soil surveyors, experimentalists, theorists, and the scientists who are building (computer) models are, individually and jointly, facing up to the challenges that these soils present. We hope that the symposium and these proceedings will prove to be a contribution towards a more mature assessment of the problems and potentials of heavy clay soils.

The editors.

CONTENTS

* keynote paper
** paper submitted but not presented

Development of structural and microfabric properties in shrinking and swelling clays

L. P. Wilding and C. T. Hallmark
Texas Agricultural Experiment
Station and Texas A&M University

Abstract

Soil structure in the pedological context is the physical constitution of soil material expressed by size, shape and arrangement of solid particles and voids into secondary polyhedral assemblages of primary particles. The secondary units (peds) are separated from adjoining cohesive aggregates by natural surfaces of weakness. Surfaces of weakness include simple or compound concentration coatings (cutans) of sesquioxides, clays, organic-clay complexes, carbonates, albic materials, and/or rearrangement of _in situ_ clay plasma by stress. In high shrink-swell clay systems, structural surfaces are commonly generated by microshear, macroshear (slickensides), or plastic deformation stresses. Microfabrics are lattisepic, vosepic, masepic, skelsepic and crystic. Several generations of structural formation and subsequent instability are evident from microfabric analysis. Size of structural units is generally smallest near the surface and increase with depth while strength of structural development is the converse. While cutanic surface features may comprise only a small proportion of the ped bulk volume, their impact on inter- and intra-ped solute and water transfer may be inordinately great. Structural units are stabilized by interparticle bonding associated with organic matter, amorphous inorganic compounds, and silicate clays. Bonding forces include polar and non-polar van der Waal forces, coulombic attractions, and organic chelation - complexation of polyvalent metals at silicate surfaces.

1 Introduction

Pedologically, structure may be defined as the arrangement of sand, silt and clay into aggregates and the arrangement of these aggregates (including pores) into a composite pattern (Baver et al., 1972). Soil Survey Staff (1975) defines soil structure as "the aggregation of primary soil particles into compound particles, or clusters of primary particles, which are separated from adjoining aggregates by surfaces of weakness". This is essentially the concept of "pedality" as defined by Brewer (1976) and is the definition of structure to be used herein. Field pedologists have long recognized structure and its spacial variability as important attribute that directly or indirectly determines soil porosity, transfer of liquids and solutes, plant rooting volumes, and the intensity, mode and mechanisms of pedogenic development. Pedologists qualitatively describe structure in terms of shape (type), size (class) and strength (grade) of development (Soil Survey Staff, 1975). Few advances have been made to further quantify structure for more definitive interpretations of soil management and hydrological interpretations (Bouma, 1983).

Processes responsible for structural development are physical, chemical, and biological. In cracking clayey soils, surface horizons are commonly granular, angular blocky, or subangular blocky. In subsoils prismatic, angular blocky, wedge-shaped aggregates, or compound prismatic-blocky structure occurs. Size increases with depth while grade decreases. Excellent reviews on this topic for Vertisols have been presented by Blokhuis (1982) and Ahmad (1983). Microfabrics of Vertisols and Vertic intergrades have been summarized by Nettleton et al. (1983). Baver et al. (1972) and Martin et al. (1955) have also provided extensive reviews on soil structure and aggregation. The purpose of this paper is to outline modes of structural aggregate formation, to consider means by which aggregates are stabilized and to present macro and micromorphical evidence that structure in clayey soils controls water and solute movement.

2 Formation of structural aggregates

Major processes in the formation of aggregates in soils are desiccation, shear failure, and biological activity.

2.1 Desiccation

Upon desiccation and dewatering of saturated sediments, open fabrics become denser and more compact. Edge to edge orientation of clays become face to face, and clays become oriented around sand and silt skeleton grains. Contractual forces exceed tensile strength, and polygonal fractures develop to form very coarse prismatic structure. Prisms subsequently part along bedding planes to form coarse angular or subangular blocky, and thick platy structure (Pons and Zonneveld, 1965; Mitchell, 1976; Pons and Van der Molen, 1973). Fabric reorientation is in response to contractive capillary water forces pulling particles closer together as water content decreases (Figure 1).

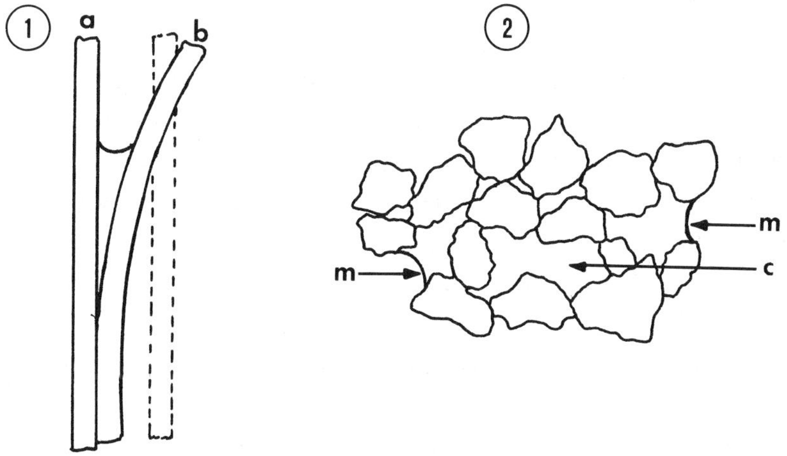

Figure 1. A schematic to illustrate the effect of drying on packing of moist powders. (1) capillary attraction drives particle b from initial position (discontinuous line) toward particle a; (2) contractible force of menisci m will bring adjacent particles closer together and may cause collapse of vault above c (modified from Biekerman, 1958, p. 36).

Grossman (1983) reports that once the new orientation has been assumed, wetting alone without accompanying mechanical disturbance does not restore the original fabric and its original high water content. This irreversible change in fabric organization would be expected to enhance upon increasing frequency of desiccation-rewetting cycles and with proximity to the soil surface; hence, smaller aggregates should be expected and are commonly found at or near the soil surface.

2.2 Shear Failure

Wedge-shaped aggregates in subsoils are formed by intersecting shear planes (slickensides) whenever the swelling pressures of a confined system exceed the shear strength of the soil (Yong and Warkentin, 1975; Yaalón and Kalmar, 1978; Blokhuis, 1982; Ahmad, 1983). The shear strength of a clayey soil is a function of cohesion which is dependent on bulk density, clay content, clay mineralogy and moisture content (McCormack and Wilding, 1979). Upon wetting and subsequent swelling of a dry soil, vertical and lateral stresses are generated (Figure 2A). In unconfined surface horizons, vertical stresses are relieved by upward movement. When the vertical stresses are confined and lateral stresses exceed the shear strength, failure occurs along a grooved shear plane approximately at 45° to the horizontal (Figure 2B), but in practice from 10 to 60° (Smart, 1970). In subsoils, overburden pressures confine vertical movement; upon crack closure, and when swelling pressures exceed the shear strength, the soil fails along diagonal shear planes. Often such mechanisms give rise to tilted lenticular (bicuneate) or larger wedge-shaped aggregates (Blokhuis, 1982; Ahmad, 1983). Several factors appear to determine the depth and character of such aggregates including: seasonal wetting-drying patterns, organic matter content, clay content, clay mineralogy, saturating cation species, and maximum difference in water content between wet and dry cycles. Shrinking and swelling is an equidimensional phenomenon (Yule and Ritchie, 1980). It's magnitude is positively correlated with the external surface area (size) of colloids, independent of mineralogy (Dixon, 1982; Wilding, 1984).

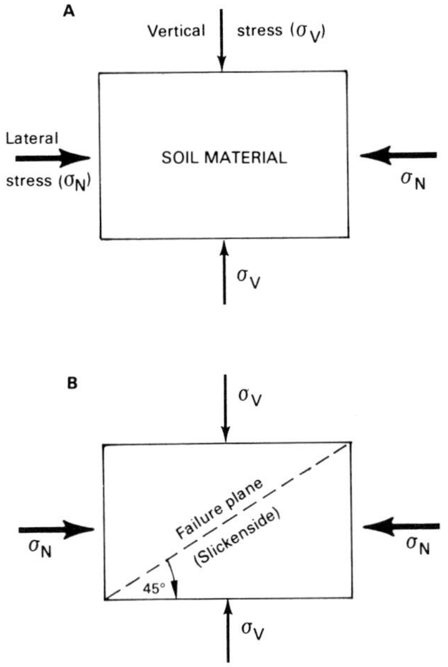

Fig. 2 Soil mechanics model of slickenside formed when confining stresses exceed soil shear strength:
(A) vertical and horizontal stresses acting on a soil ped; and
(B) orientation of shear plane at 45° to the principal stress.
(Size of vectors proportional to stress magnitude.)

2.3 Biological Activity

Granular aggregates in soils are commonly formed by the mixing of organic and mineral constituents by biota particularily arthropods and annelids (Kubiena, 1970; Ugolini and Edmonds, 1983). Earthworms have often been noted for their effectiveness in developing a granular sponge-like structure in surface horizons by forming coalescent worm droppings (Kubiena, 1970).

This promotes high infiltration, high water-holding capacity and good aeration. Fibrous root systems and root, fungal and bacterial metabolites also favor formation and stability of granular structure (Baver et al., 1972; Ugolini and Edmonds, 1983). The role of roots may be to separate larger aggregates into smaller granules or to desiccate soil around the root causing shrinkage and formation of fracture planes (Baver et al., 1972).

3 Stabilization of aggregates

The stability of aggregates is a function of two factors--the relative degree of cohesion within aggregates versus adhesion among aggregates. Water-particle inter-actions and cementation by organic matter, sesquioxides, silica, carbonates and clay impact cohesion. Adhesive forces are dependent on surfaces of weakness that separate abutting natural aggregates and the degree of surface accommodation among peds.

3. 1 Surfaces of weakness

3.1.1 Plasma concentrations

Translocation of mobile constituents (plasma) during pedogenesis often results in concentrations or coatings of plasma along ped surfaces (Figures 3A, B, C, D,). The two most commonly cited processes for formation of these features are illuviation and diffusion (Brewer, 1976). Illuviated clays (argillans) that occur along ped surfaces are indicative of ped stability requiring geologic periods of time (several thousand years) to form. Illuviated argillans exhibit optical birefringence (Figure 3A) and extinction pheonmena indicative of their crystallinity, continuity, thickness, orientation, packing density and distribution. They are commonly laminated indicating deposition in successive increments.

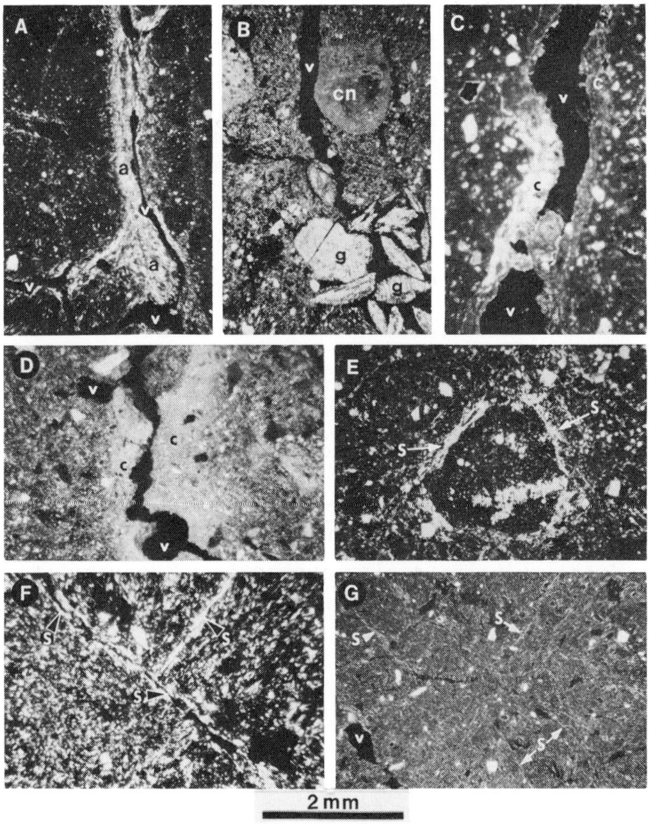

Figure 3. Thin section micrographs taken in cross-polarized light illustrating: (a) illuviated argillan along a channel void (Btg horizon, Fulton series, Aeric Ochraqualf, Ohio; (b) gypsan and carbonate nodule along a planar ped void (Btky horizon, Lufkin series, Vertic Albaqualf, Texas); (c) calcan along a channel void (Bk horizon, Algoa series, Aquic Calciustoll, Texas); (d) calcan along a channel void in oxidized calcareous glacial till (C horizon, Celina series, Aquic Hapludalf, Ohio) (e) stress-oriented plasma separation around a skeleton grain in a skelsepic plasmic fabric (Bt horizon, Rumple series, Udic Arguistoll, Texas); (f) stress-oriented plasma separations along planar ped voids in a vosepic plasmic fabric (Bg horizon, Toledo series, Mollic Haplaquept, Ohio); and (g) plamsa separations in masepic plasmic fabric indicating microshear failure (A horizon, El Carmen series, Typic Pellustert, El Salvador). Bar length is equal to 2 mm. (a = argillan, c = calcan, cn = carbonate nodule, g = gypsan, s = plasma separation and v = void.

Ped instability is indicated by embedded illuviation argillans and stress-oriented plasma separations within the s-matrix (Smeck et al., 1968; Smith and Wilding, 1972; Rostad et al.,1976; Ritchie et al., 1974). Embedded argillans often outline former structural units. Ped instability appears greatest in soils that have high shrink-swell potential, high clay content, and expandable 2:1 layer clay minerals (Ritchie, et al., 1974). Shrink-swell activity and ped instability may become so great in some soils as to deter formation of illuviation cutans or destroy preexisting cutans (Nettleton et al., 1969 and Smeck et al., 1981).

Other cutanic or subcutanic plasma concentrations that occur at ped surfaces include: carbonates (calcans, Figure 3C, D), gypsum (gypsans, Figure 3B)), silica (silans), iron oxides (ferrans), manganese oxides (mangans), sesquioxides (sesquans) and albic materials (skeletans or albans). These cutanic features also represent surfaces of weakness because of their differential composition, texture and particle orientation with the host ped s-matrix.

3.1.2 Plasma separations (stress reorientation)

Cutans in clayey soils can also be formed by differential swelling pressure causing macro and microshear planes with translational or plastic flow deformation (Crampton, 1974; McCormack and Wilding, 1974; Wilding, 1984). Grooved and polished slickensides result from translational shear along the failure zone while plastic deformation causes reorientation as pressure faces that often lack the shiny, grooved appearance (Blokhuis, 1982).

In thin sections, microshear planes are identified by linear oriented patterns of striated clay aggregates that exhibit preferential plate orientation and optical birefrengence (plasma separations, Figure 3E, F, G). Figure 4 schematically illustrates postulated stages in developing stress-oriented plasmic fabrics in high shrink-swell clayey soils.

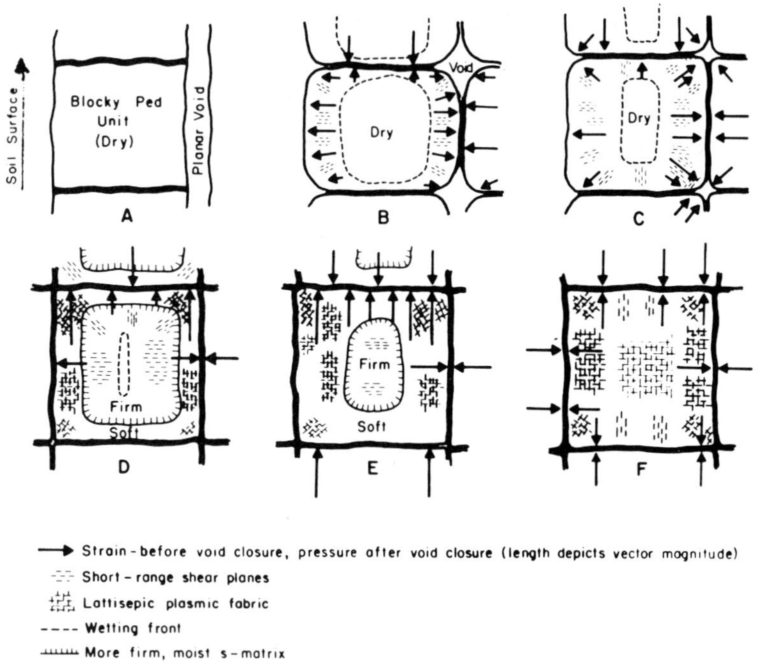

Figure 4. Schematic stages of wetting as water front advances into blocky ped, showing postulated direction of strain and shear planes commonly found within structural units of high shrink-swell soils. Taken from Figure 5, McCormack and Wilding, 1973.

The process involves wetting a dry soil with strongly expressed prismatic-blocky structure. Differential wetting, swelling and development of microshear, first in the horizontal and then in the vertical axis, are proposed. Strong anisotropic wetting and swelling patterns are in response to the size and patterns of the structural voids (McCormack and Wilding, 1974; Ahmad, 1983). In shrinking and swelling clayey soil systems with moderate to high shrink-swell potential, microfabrics reflecting microshear and ped instability yield skelsepic (Figure 3E), vosepic (Figure 3F), lattisepic, masepic (Figure 3G), and combinations of these plasmic fabrics (Holzhey et al., 1974; Nettleton et al., 1969, 1983; Smeck et al., 1981; Blokhuis, 1982).

Crystic is also a common plasmic fabric in Vertisols (Wilding, 1984). The degree of stress-oriented soil fabric is inversely related to the occurrence of illuviation argillans. Stress oriented plasmic fabric can also be generated by root pressure, mass movement, and ice crystal growth.

3.2 Cementation effects: Bonding of soil separates in aggregates

Bonding between particles is possible only when particles are sufficiently close for bonding mechanisms to become effective. Therefore, physical rearrangement to increase the proximity of particles may be necessary. This is accomplished by processes including flocculation, hydration-dehydration, shrinking-swelling, freezing-thawing, movement by gravity, and forces exerted by roots, earthworms, and other biota. For significant structure formation to endure, bonding must be sufficiently strong to withstand destruction from these same physical forces.

Bonding among particles is primarily attributed to organic matter, silicate clays, and amorphous oxy-hydroxy compounds of Al, Fe, and Si although cementation and engulfment of soil particles by relatively large quantities of calcite, amorphous silica (and opal), gypsum, crystalline Fe-oxide, and illuviated organic-metal compounds are known (Soil Survey Staff, 1975). Temporal to long-term bonding of soil separates is imparted by organic matter. Temporary bonding is primarily from physical entanglement of soil separates by filamentous structures of the microflora, principally fungi and actinomycetes (Swaby, 1949). Such structures are short-lived as the microbial bodies are substrate for subsequent decomposition by microbes. The bonding by polysaccharides (long chain, flexible polymers of varying content of alcohol, amino, carboxyl and phenolic groups) is more long-term than filamentous structures, but eventually they too are decomposed by microbes (Fehrmann and Weaver, 1978; Harris et al., 1963).

Evidence suggests that long-term bonding by organic matter is due primarily to microbial resistant fractions, most notably the humic acids.

The bonding of soil particles by both crystalline and amorphous Fe, Al, and Si forms is easily recognized in extreme cases (Soil Survey Staff, 1975). However, in less acute concentrations, bonding by these compounds is not readily obvious. Deshpande and co-workers (1964, 1968) found oxy-hydroxy Al compounds to be more important to aggregate stability than Fe oxides when levels were sufficiently low to preclude engulfment of the soil separates. Recent work on loamy fragipans indicates amorphous Si, Al, and Fe compounds even in low quantities are important in imparting brittleness and strength to the fragipan (Hallmark and Smeck, 1979; Steinhardt and Franzmeier, 1979).

The importance of silicate clays in bonding soil separates together has been long recognized because stable aggregate formation seldom takes place in sand or silt in the absence of these colloids. Preferential orientation of silicate clays in structural units occurs and has been observed as links between adjacent sand-silt grains (Wang et al., 1974). In fine-textured soils, the clay may occur as a dense groundmass surrounding sand and silt separates. Hydration state, exchangeable cation and soil solution salt content affect the expansion and orientation of silicate clays and undoubtedly their ability to bond within structural units.

Although organic matter, amorphous inorganic compounds and silicate clays are generally recognized as instrumental in structure development, the mechanisms of bonding are poorly understood. Greenland (1965) recognized that coulombic attractions and van der Waals forces (both polar and non-polar) were involved in the interaction of clays and organic compounds. The multiplicity of soil organic matter functional groups (i.e., carboxyl, amines, hydroxyls, carbonyls, etc.) provides for chelation-complexation of multivalent cations on crystalline silicate edges or on amorphous coatings of soil minerals.

Although the intercrystalline forces between adjacent clay particles may be sufficiently strong to account for binding in aggregate formation (Martin et al., 1955), recent work showing the amorphous nature of outer silicate surfaces suggests that bonding or polymerization across mineral grains through an amorphous phase is likely (Ribault, 1971). Such a mechanism would allow coherence of clay particles to quartz sand grains as well as between silicate clay particles.

4 Macro-and micromorphological evidence of preferential water and solute movement at ped interfaces

The importance of macrovoids on saturated water movement and dissolved solutes has steadily gained the attention of soil physicists and chemists (Blake et al., 1973; Bouma, 1983; Ritchie et al., 1972; Kissel et al., 1973; and Thomas, 1970). There is also substantial evidence from a pedological perspective that water moves preferentially along ped interfaces. The evidence is based on the following observations: (1) immobilization of suspended clay that is carried in an advancing water front along structural surfaces and other macrovoids (Figure 3A); (2) ped interfaces that have higher moisture contents and lower strengths than ped interiors (McCormack and Wilding, 1979); (3) occurrence of calcans (Figure 3D), ferrans, mangans,and argillans along vertical fissures and prisms in oxidized and unoxidized sedimentary deposits 4 to 8 m below the surface (Smeck et al., 1968; Smith and Wilding, 1972; Ritchie et al., 1974); (4) occurrence of roots preferentially distributed along structural surfaces (Miller et al., 1971; Ritchie et al., 1974) (5) occurrence of albans and skeletans (albic materials) along ped surfaces (Vepraskas et al., 1974; Vepraskas and Wilding 1983a, 1984b); (6) occurrence of soluble salts (gypsans Figure 3B), and silica (silans) along structural conductive voids (Brewer, 1976); and (7) the preferential flow of water along ped surfaces and other macrovoids upon pit excavation in a saturated soil.

5 Impact of peds on solute and water movement

Soils are not homogeneous media either on a macro or micro scale. They have vertical and lateral anisotropic properties with both systematic and random spacial dependence (Wilding and Drees, 1983). At the level of a ped, zonation of plasma and skeleton grains is common. Although ped cutans may comprise less than 1 or 2% of the total soil volume, they may impart a disproportionally large environmental influence on the soil as a medium for plant growth (Miller and Wilding, 1972). The pattern and orientation of slickensides in Vertisols and associated gilgai topographic relief give rise to cyclic subsurface horizonation, with strikingly different leaching potentials between microlows and microhighs (Ritchie et al., 1972; Wilding, 1984). Ped argillans have been demonstrated to markedly reduce rates of diffusion and mass flow from the ped surface to the s-matrix (Gerber et al., 1974). This impact on water movement is due primarily to increased tortuosity, but chemical interactions may be important with solute transfer depending on the mineralogical composition of the cutan and species involved. Khalifa and Buol (1968) and Miller and Wilding (1972) conclude that argillans have deleterious effects on fine root penetration into a ped and upon nutrient uptake.

Losses of soluble nutrients from the root zone will be enhanced by a water front that moves preferentially along structural voids and short circuts the s-matrix during the wetting phase in cracking clay soils (Bouma, 1983; Kissel et al., 1973; Thomas, 1970). This likewise has important implications on the suitability of soils as a media for disposal for toxic and non-toxic wastes. Conversely, leaching, weathering and nutrient transfer out of the s-matrix will be retarded by short circuit water movement through cracks of clayey soils. This is evidenced by ped interiors that are calcareous while superjacent ped exteriors have been leached of carbonates.

6 Summary

In clayey, high shrink-swell soil systems, ped interfaces and Biological macrovoids represent the major avenues of water movement. Strongly anisotropic wetting, swelling and shear patterns in these soils govern structural type, stability and preferential water and solute movement

REFERENCES

Ahmad, N. 1983. Vertisols. In L. P. Wilding, N. E. Smeck and G. F. Hall (eds.). Pedogenesis and Soil Taxonomy II. Soil Orders. Developments in Soil Science 11B. 91-123 pp.

Baver, L. D., W. H. Gardner, and W. R. Gardner. 1972. Soil physics (4th Ed.). John Wiley and Sons Inc., New York, NY 130-177 pp.

Bickerman, J. J. 1958. Surface chemistry. Academic Press, New York, NY., 501 pp.

Blake, G., E. Schlichting and E. Zimmerman. 1973. Water re-charge in a soil with shrinkage cracks. Soil Sci. Soc. Am. Proc. 37:669-672.

Blokhuis, W. A. 1982. Morphology and genesis of vertisols. In Vertisols and Rice Soils of the Tropics, Symposia Papers II. 12th. ICSS. New Delhi, India. 23-45 pp.

Bouma, J. 1983. Hydrology and soil genesis of soils with aquic moisture regimes. In L. P. Wilding, N. E. Smeck and G. F. Hall (eds.). Pedogenesis and Soil Taxonomy I. Concepts and Interactions. Developments in Soil Science 11B. 253-284 pp.

Brewer, R. 1976. Fabric and mineral analysis of soils. Krieger Pub. Co., Huntington, NY. 205-233 pp.

Crampton, C. B. 1974. Microshear-fabrics in soils of the Canadian north. In G. K. Rutherford (ed.). Soil Microscopy. 4th Int. Working Meetings on Soil Micromorphology. The Limestone Press, Kingston, Ontario. 655-664 pp.

Deshpande, T. L., D. J. Greenland, and J. P. Quirk. 1964. Role of iron oxides in the bonding of soil particles. Nature 201:107-108.

Deshpande, T. L. D. J. Greenland, and J. P. Quirk. 1968. Changes in soil properties associated with the removal of iron and aluminum oxides. J. Soil Sci. 19:108-122.

Dixon, J. B. 1982. Mineralogy of vertisols. In Vertisols and Rice Soils of the Tropics, Symposia Papers II. 12th. ICSS. New Delhi, India. 48-59 pp.

Fehrmann, R. C. and R. W. Weaver. 1978. Scanning electron microscopy of *Rhizobium* spp. adhering to fine silt particles. Soil Sci. Soc. Am. J. 42:279-281.

Gerber, T. D., L. P. Wilding, and R. E. Franklin. 1974. Ion diffusion across cutans: A methodology study. In G. K. Rutherford (ed.). Soil Microscopy. 4th Int. Working Meetings on Soil Micromorphology. The Limestone Press, Kingston, Ontario. 730-746 pp.

Greenland, D. J. 1965. Interaction between clays and organic compounds in soils. Part I. Mechanisms of interaction between clays and defined organic compounds. Soils and Fert. 28:415-425.

Grossman, R. B. 1983. Entisols. In L. P. Wilding, N. E. Smeck, and G. F. Hall (eds). Pedogenesis and Soil Taxonomy II. Soil Orders. Developments in Soil Science 11B. Elsevier Pub., Co. Amsterdam. 55-90 pp.

Hallmark, C. T. and N. E. Smeck. 1979. The effect of extractable aluminum, iron, and silicon on strength and bonding of fragipans of northeastern Ohio. Soil Sci. Soc. Am. J. 43:145-150.

Harris, R. F., O. N. Allen, G. Chesters, and O. J. Altoe. 1963. Evaluation of microbial activity in soil aggregate stability and degradation by the use of artifical aggregates. Soil Sci. Soc. Am. Proc. 27:542-545.

Holzhey, C. S., R. D. Yeck, and W. D. Nettleton. 1974. Microfabric of some agrillic horizons in udic, xeric and torric soil environments of the United States. In G. K. Rutherford (ed.). Soil Microscopy. 4th Int. Working Meetings on Soil Micromorphology. The Limestone Press, Kingston, Ontario. 747-759 pp.

Khalifa, E. M. and S. W. Buol. 1968. Studies of clay skins in a Cecil (Typic Hapludult) soil: I. Composition and genesis. Soil Sci. Soc. Am. Proc. 32:857-861.

Kissel, D. E., J. T. Ritchie and E. Burnett. 1973. Chloride movement in undisturbed swelling clay soil. Soil Sci. Soc. Am. Proc. 37:21-24.

Kubiena, W. L. 1970. Micromorphological features of soil geography. Rutgers University Press. Brunswick, NJ. 254 pp.

Martin, James P., W. P. Martin, J. B. Page, W. A. Raney and J. D. DeMent. 1955. Soil aggregation. Advances in Agron. 7:1-37.

McCormack, D. E. and L. P. Wilding. 1974. Proposed origin of lattisepic fabric. In G. K. Rutherford (ed.). Soil Microscopy. 4th Int. Working Meetings on Soil Micromorphology. The Limestone Press, Kingston, Ontario. 761-771 pp.

McCormack, D. E. and L. P. Wilding. 1979. Soil properties influencing strength of Canfield and Geeburg soils. Soil Sci. Soc. Am. J. 43:167-173.

Miller, F. P., L. P. Wilding and N. Holowaychuk. 1971. Canfield silt loam, a Fradiudalf: II. Micromorphological, physical and chemical properties. Soil Sci. Soc. Am. Proc. 35:324-329.

Miller, M. H. and L. P. Wilding. 1972. Microfabric studies in relation to the root-soil interface. In R. Protz (ed.). Proceedings of a symposium on Microfabrics of Soil and Sedimentary deposits. Dept. of Land Resource Sci. Univ. of Guelph, Ontario. C.R.D. Pub. No. 69. 75-110 pp.

Mitchell, J. K. 1976. Fundamentals of soil behavior. John Wiley, New York, NY. 422 pp.

Nettleton, W. D., K. W. Flach and B. R. Brasher. 1969. Argillic horizons without clay skins. Soil Sci. Soc. Am. Proc. 33:121-125.

Nettleton, W. D., F. F. Peterson, and G. Borst. 1983. Micromorhological evidence of turbation in vertisols and soils in vertic subgroups. In P. Bullock and C. P. Murphy (eds.). Soil Micromorphology. Vol. 2 Soil Genesis. AB Academic Pub. Berkhamsted, Herts. 441-458 pp.

Pons, L. J. and I. S. Zonneveld. 1965. Soil ripening and soil classification. Int. Inst. Land Reclamation and Improvement, Publ. 13. 128 pp.

Pons, L. J. and Van der Molen. 1973. Soil genesis under dewatering regimes during 1000 years of polder development. Soil Sci. 116:228-235.

Ribault, L. L. 1971. Presence d'une pellicule de silice amorphe a la surface de cristaux de quartz des formations sableuses. C. R. Acad.

Sci. Paris. 272, Serie D: 1933-1936.

Ritchie, J. T., D. E. Kissel, and E. Burnett. 1972. Water movement in undisturbed swelling clay soil. Soil Sci. Soc. Am. Proc. 36:874-879.

Ritchie, A., L. P. Wilding, G. F. Hall and C. R. Stahnke. 1974. Genetic implications of B horizons in Aqualfs of northeastern Ohio. Soil Sci. Soc. Am. Proc. 38:351-358.

Rostad, H. P. W., N. E. Smeck, and L. P. Wilding. 1976. Genesis of argillic horizons in soils derived from coarse-textured calcareous gravels. Soil Sci. Soc. Am. J. 40:739-744.

Smart, P. 1970. Residual shear strength. Proc. Am. Soc. Civil Engrs., Soil Mech. and Foundations Div. 96:2181-2183.

Smeck, N. E., L. P. Wilding, and N. Holowaychuk. 1968. Genesis of argillic horizons in Celina and Morley soils of western Ohio. Soil Sci. Soc. Am. Proc. 32:550-556.

Smeck, N. E., A. Ritchie, L. P. Wilding, and L. R. Drees. 1981. Clay accumulation in sola of poorly drained soils of western Ohio. Soil Sci. Soc. Am. J. 45:95-102.

Smith, H. and L. P. Wilding. 1972. Genesis of argillic horizons in Ochraqualfs derived from fine textured till deposits of northwestern Ohio and southwestern Michigan. Soil Sci. Soc. Am. Proc. 36:808-815.

Soil Survey Staff. 1975. Soil Taxonomy. Agric. handbook no. 436. USDA-Soil Conservation Serv., Washington, D. C.

Swaby, R. J. 1949. The influence of humus on soil aggregation. J. Soil Sci 1:182-193.

Steinhardt, G. C. and D. P. Franzmeier. 1979. Chemical and mineralogical properties of the fragipans of the Cincinnati catena. Soil Sci. Soc. Am. J. 43:1008-1013.

Thomas, Grant W. 1970. Soil and climatic factors which affect nurtient mobility. In O. P. Englestad (ed.). Nutrient mobility in soils: Accumulation and losses. Soil Sci. Soc. Am. Special Publ. 4, Madison, WS. pp 1-20.

Ugolini, F. C. and R. L. Edmonds. 1983. Soil biology. In L. P. Wilding, N. E. Smeck and G. F. Hall (eds.). Pedogenesis and Soil Taxonomy I. Concepts and Interactions. Developments in Soil Science IIB. Elsevier Pub., Amsterdam. 193-231 pp.

Vepraskas, M. J., F. G. Baker and J. Bouma. 1974. Soil mottling and drainage in a Mollic Hapludalf as related to suitability for septic tank construction. Soil Sci. Soc. Am. Proc., 38:497-501.

Vepraskas, M. J. and L. P. Wilding. 1983a. Aquic moisture regimes in soils with and without low chroma colors. Soil Sci. Soc. Am. J. 47:280-285.

Vepraskas, M. J. and L. P. Wilding. 1983b. Albic neoskeletans in argillic horizons as indices of saturation and iron reduction. Soil Sci. Soc. Am. J. 47:1202-1208.

Wang, C., J. L. Nowland and H. Kodama. 1974. Properties of two fragipan soils in Nova Scotia including scanning electron micrographs. Can. J. Soil Sci. 54-159-170.

Wilding L. P., and L. R. Drees. 1983. Spacial variability and pedology. In L. P. Wilding, N. E. Smeck and G. F. Hall (eds.). Pedogenesis and Soil Taxonomy I. Concepts and Interactions. Developments of Soil Science. 11B. 83-116.

Wilding, L. P., M. H. Milford and M. J. Vepraskas. 1983. Micromorphology of deeply weathered soils in the Texas Coastal Plains. In P. Bullock and C. P. Murphy (eds.). Soil Micromorphology. Vol. 2 Soil Genesis. AB Academic Pub. Berkhamsted, Herts. 567-574 pp.

Wilding, L. P. 1984. Genesis of vertisols. Proc. of Fifth Int. Soil Classification Workshop. Sudan. 1982. (In Press).

Yaalon, D. H. and D. Kalmar. 1978. Dynamics of cracking and swelling clay soils: Displacement of skeleton grains, optimum depth of slickensides, and rate of intro-pedonic turbation. Earth Surface Processes 3:31-42.

Yong, R. N. and B. P. Warkentin. 1975. Soil properties and soil Behavior. Elsevier Publ. Co., Amsterdam. 449 pp.

Yule, D. R. and J. T. Ritchie. 1980. Soil shrinkage relationships of Texas vertisols: I. Small cores. Soil Sci. Soc. Am. J. 44:1285-1291.

Discussion

T.M. Addiscott:

I was interested in your comment on the effects of cutans, etc. on intraped diffusion. Are such effects likely to differ much between peds from uncultivated horizons and those at the surface, where fresh surfaces are produced by breakage during cultivation?

Author:

Most of the cutans, wich consist of coatings with modified physical, chemical, or biological properties on ped surfaces are in subsoils and would only be important in cultivated surface horizons if these subsoil materials were incorporated into the surface by plowing eroded soil areas or where soils had thin surface horizons naturally. However, cutans resulting from rearrangement with preferred orientation parallel to the surface of the agregates are common in many clayey surface horizons as a consequence of both natural and tillage-induced plastic deformation and perhaps desiccation.

J. Bouma:

You list six types of observations from which evidence is deduced that water moves preferentially along ped faces. You do not mention staining tests which have been applied by several investigators, who found stains exclusively on ped faces. Is there a particular reason for not mentioning this approach?

Author:

No. It is really an oversight in the paper and should be added to the list. Certainly dyes and solutions of $CaSO_4 . 1/2H_2O$ (plaster of Paris) are direct means to demonstrate water movement along ped faces and via biological macrovoids. We have used dye methods to monitor the movement of organic toxic wastes through clay liners which have failed. In our paper we have put the emphasis on observations related to natural pedogenic transfers.

H.F.M. ten Berge:

Associated with diurnal temperature waves in the topsoil are 'waves' of moisture content. A damping depth for moisture can be defined as

PD/π, where P is the period (one day) and D is the soil water diffusivity. Do you expect that a causal relationship exists between, on the one hand,this damping depth and the amplitude of the water content and,on the other hand,the size of the peds?

Author:

I would expect a causal relationship of this kind to exist, but not induced by diurnal temperature waves. I believe the time scale involved is too short to result in maximum fluctuation in water content necessary to result in tensile stress failure. The rate of water transmission from the matrix of a clayey soil is slow. This, coupled with the break in capillary pore continuity associated with the 'dust mulch' or 'self mulching' effect near the soil surface, would make the rate of desiccation a long-term event of the order of weeks or months.

H.F.M. ten Berge:

Under the climatic conditions where Vertisols may occur, diurnal fluctuations in topsoil temperature may cover 50°C or more. Do you expect that these diurnal temperature waves play an important role in crack formation?

Author:

No. Our experience, and that reported in the literature, indicate that crack formation upon entering the drying cycle requires weeks or months to initiate. It is, again, a long-term process relative to diurnal changes.

D.H. Yaalon:

Which micromorphological feature would you consider as best evidence of turbation in Vertisols? Have you observed in thin sections indications of surface material falling into cracks?

Author:

As I reported on the genesis of Vertisols at the International Soil Classification Workshop in the Sudan two years ago, I am not a strong proponent of the pedoturbation model for these soils.
In fact, based on organic carbon, carbonate, and salinity profiles, I do not believe they strongly churn. This suggests to me that most of

the physical activity in these soils is upward or downward thrust along the slickenside planes that act as fault zones and avenues for preferential root growth, and water and solute movement. Thus no unique micromorphic features can be attributed to Vertisols that are not also recognized in other orders of soils that shrink-swell and crack to a more limited extent. Occasionally, microscopically and macroscopically, one observes partial crack infilling that looks similar in thin section to other tubular infillings.
Again, these features are not unique to Vertisols. Evidence of sepic fabrics with plasma separations are most indicative of high shrink-swell and microshear properties.

D.H. Yaalon:
Please note that on Figure 2 the shear plane angle is not 45°, while theoretically it should be 45° minus the angle of internal friction/2, which essentially accounts for differences in particle arrangements and their mineralogy.
Author:
Your comment is well taken and, in fact, true. However, in this model I had assumed the angle of internal friction was quite small because the skeleton grains are sufficiently widely spaced that they do not interlock, and the effect of clay mineralogy was not considered.

G.P. Wind:
It has been observed in the Netherlands that in heavy clays under grassland high-quality drainage (deeper drains, spaced closer together) causes smaller clods and smaller cracks than low-quality drainage. Can you explain this?
Author:
I believe the explanation for this phenomenon lies in the fact that deeper drains with closer spacing lower the water table more uniformly across these areas. This favors larger and more frequent fluctations of the water content near the soil surface. Both of the latter processes should enhance the development of smaller structural units with greater stability, as noted in our paper.

T.M. Addiscott:

How thick are the cutans on ped surfaces, and how far does moisture penetrate into the ped matrix?

Author:

Cutan thickness varies with the pedogenic processes under which a soil has developed, the depth in a soil, and whether one is considering a plasma concentration (coating) or plasma separation (stress orientation) feature. Plasma concentrations may range from 10 μm or less to several millimeters, while plasma separations are invariably thin (5-10 μm). In the latter case, plasma separates (micelles) may assume parallel orientation with the planar void surface at 100 μm or slightly deeper into the s-matrix.
No suitable answer can be given to the second part of your question, except that obviously the ped surfaces are the most dynamic zones, while ped interiors are the most static regions to water and solute movement.

Comment by F.F.R. Koenigs at end of symposium

Wilding's concept of the origin of slickensides is based on lateral compression and subsequent failure of the soil mass by the expansion on wetting of dry material fallen into vertical cracks. But according to Yaalon, no substantial amounts of material are falling into cracks. So another approach is needed and, for that, one might start from the wet end. As far as I have seen, the best lenses (lentiform, 2 m wide units delineated by slickensides) are found in drying lake deposits of montmorillonite clay. Of these, those with a known history and especially those which have been reclaimed fairly recently are of most interest. In case of the deposits of the Tisza River (in Hungary and Rumania, reclaimed 200 years ago) only slickensides and lenses are found and no vertical cracks or prisms. The volume loss after drainage will be of the order of 70%. The resulting horizontal tensile stresses can cause failure planes (see the following paper by Raats). The lenses delineated by these planes slump and slide over each other thus causing the slickensides.

I close with three comments on related matters: 1. Philip pointed out that stress relief near a crack is accompanied by a rise of the soil along the edges of the crack. This has often been observed. 2. Hartge demonstrated the importance of friction with underlying layers. If the friction is small, then the soil can travel considerable distances without cracking. I have observed that filterpaper may be torn, and that a sintered glass plate may break. 3. The tensile strength is the crucial parameter for crack formation. It is recommended that measurements of this parameter are made in the near future.

Mechanics of cracking soils

P.A.C. Raats
Institute for Soil Fertility, P.O. Box 30003, 9750 RA Haren, The Netherlands

Abstract

The deformation gradient tensor of the solid phase is the central concept in the description of swelling and shrinkage of soils. The appearance of slip surfaces and cracks is governed by relationships among the stress tensor and parameters characterizing the strength of the soil. An analysis of the perturbation of stress induced by a crack gives some insight in spacings, angles of intersection, and depths of cracks.

1 Deformation of the solid phase

A description of the motion of the solid phase of a soil gives the places $\underset{\sim}{x}$ occupied by any parcel $\underset{\sim}{\chi}_s$ of the solid phase in the course of time t (Raats, 1984a):

$$\underset{\sim}{x} = \underset{\sim}{x}\left[\underset{\sim}{\chi}_s, t\right] \tag{1}$$

As labels for the parcels χ_s one may use their locations $\underset{\sim}{x}_o$ in the reference configuration at some reference time t_o. Differentiation of (1) with respect to t and $\underset{\sim}{\chi}_s$ gives, respectively, the velocity vector $\underset{\sim}{v}_s$ and the deformation gradient tensor $\underset{\sim}{F}_s$:

$$\underset{\sim}{v}_s = \partial \underset{\sim}{x} / \partial t|_{\underset{\sim}{\chi}_s}, \qquad \underset{\sim}{F}_s = \partial \underset{\sim}{x} / \partial \underset{\sim}{\chi}_s|_t \tag{2}$$

The determinant of $\underset{\sim}{F}_s$ for a parcel χ_s compares the volume currently occupied by the parcel to the volume occupied in the reference configuration. It can be related to a concept more familiar in soil science by considering the isotropic, homogeneous *swelling* of a cube with current edge 1 and edge 1_o in the reference configuration. Then

$$\det \underset{\sim}{F}_s = (1/1_o)^3, \tag{3}$$

and the coefficient of linear extensibility, COLE, can be defined by:

$$\text{COLE} \equiv \frac{1 - 1_o}{1_o} = (\det \underset{\sim}{F}_s)^{1/3} - 1 \tag{4}$$

For isotropic, homogeneous *shrinkage* of a cube, the negative of COLE is defined as the coefficient of linear contractibility, COLC.

The measures of deformation COLE and COLC are mainly useful to describe the volumetric aspects of isotropic homogeneous deformations. By contrast, the deformation gradient tensor $\underset{\sim}{F}_s$ can be used to describe all aspects of any continuous deformation. Not only COLE and COLC, but numerous other concepts describing various aspects of deformations can be derived from $\underset{\sim}{F}_s$ (Truesdell and Toupin, 1960). In particular, it can be shown that the deformation at any point may be regarded as resulting from a translation, a rigid rotation of the principal axis of strain, and stretches along these axis.

The deformation gradient tensor $\underset{\sim}{F}_s$ plays a key role in the description of movement of water in deforming soils (Raats, 1984a, b). Disregarding the influence of gravity, so that the gradient of the pressure head h is the only driving force, and using the solid phase as a reference, the product of the volumetric water content θ and $\det \underset{\sim}{F}_s$ can be shown to satisfy a nonlinear, inhomogeneous diffusion equation (Raats, 1984b):

$$\partial\, (\theta \det \underset{\sim}{F}_s)/\partial t\big|_{\underset{\sim}{\chi}_s} = \partial\, \{\mathfrak{D}\, \partial(\theta \det \underset{\sim}{F}_s)/\partial \underset{\sim}{\chi}_s\}/\partial \underset{\sim}{\chi}_s \tag{5}$$

with the transformed diffusivity tensor $\mathfrak{D}$ given by

$$\mathcal{D} = (\det \underset{\sim}{F}_s)\, \underset{\sim}{F}_s^{-1}\, (\underset{\sim}{F}_s^{T})^{-1}\, D \tag{6}$$

and in turn D related to the hydraulic conductivity k by:

$$D = k\, d\,(\theta \det \underset{\sim}{F}_s)/dh \tag{7}$$

As a special case, it can further be shown that for flow in the axial direction of a swelling and shrinking, thin, porous rod (6) reduces to (Raats, 1969):

$$\mathcal{D} = (\det \underset{\sim}{F}_s^{-1})^{2n-1}\, D \tag{8}$$

Figure 1 shows plots of (8) for purely axial deformation (n=1), purely lateral deformation (n=0), deformation for which the effects of the axial and lateral deformations upon the diffusivity cancel each other ($n=\frac{1}{2}$), isotropic deformation ($n=\frac{1}{3}$), and also for $n=\frac{2}{3}$. The value $n=\frac{1}{2}$ separates two opposite trends. For $n < \frac{1}{2}$ swelling causes $\mathcal{D}/D$ to increase and shrinking causes $\mathcal{D}/D$ to decrease. For $n > \frac{1}{2}$ swelling causes $\mathcal{D}/D$ to decrease and shrinking causes $\mathcal{D}/D$ to increase.

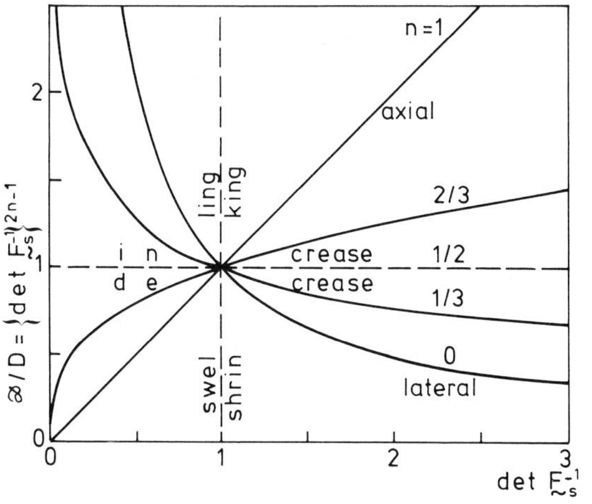

Figure 1. Transformed diffusivity tensor for flow of water in the axial direction of a swelling and shrinking, thin porous rod

The case of purely axial deformation has been studied in great detail, yielding valuable results with regard to equilibrium of water, steady upward and downward flows of water, adsorption and infiltration of water, and sedimentation and filtration of slurries (Philip and Smiles, 1982). But in many cases there is a need for a 2- or 3- dimensional stress-strain theory (Miller, 1975). Even splitting of parcels χ_s must be dealt with: shrinkage may lead to cracks (Corte and Higashi, 1960; Neal et al., 1968; Blokhuis, 1982) and swelling may lead to slip surfaces (Krishna and Perumal, 1948; De Vos et al., 1969). Tension cracks may also be induced by penetrating roots (Barley et al., 1965) and the passage of wetting fronts (Dexter, 1983; Parlange and Sawhney, 1976).

In the next Section two crack-slip failure criteria are discussed briefly. In Section 3 a theory for stress perturbation due to the presence of cracks is summarized.

2 Two crack-slip failure criteria

At any point within the soil the force upon any surface through that point is determined uniquely by the normal to that surface. By using the balances of momentum and moment of momentum, it can be shown that the forces upon all surfaces through any point are determined fully by the symmetric stress tensor field $\underset{\sim}{T}$. If ρ is the bulk density, $\underset{\sim}{g}$ the gravitational force, and $\underset{\sim}{a}$ is the acceleration, then

$$\operatorname{div} \underset{\sim}{T} + \rho \underset{\sim}{g} = \rho \underset{\sim}{a} \tag{9}$$

In dealing with soils, force balances can be written for each of the phases and for the soil as a whole. The symmetry of $\underset{\sim}{T}$ implies that, at any point, $\underset{\sim}{T}$ is completely determined by three principal stresses τ_1, τ_2, and τ_3 associated with the principal directions $\underset{\sim}{n}_1$, $\underset{\sim}{n}_2$, and $\underset{\sim}{n}_3$. Given (τ_1, τ_2, τ_3) and $(\underset{\sim}{n}_1, \underset{\sim}{n}_2, \underset{\sim}{n}_3)$ at a point the normal stress τ_n and the shear stress τ_t on any plane through that point can be calculated or determined graphically by Mohr's circle diagram.

By analogy with friction between separate bodies, in 1773 Coulomb formulated a simple criterion for slip failure of a granular material (Jaeger and Cook, 1979). According to Coulomb's criterion slip failure will occur if on any plane the shear stress τ_t, the normal stress τ_n, the angle of internal friction ψ, and the shear strength σ_s satisfy

$$|\tau_t| - (\tan\psi)\,\tau_n \geqslant \sigma_s \tag{10}$$

If the soil is partially saturated and τ_n is not an effective but a total stress, then σ_s includes a cohesive component due to capillarity. Coulomb's criterion can also be expressed in terms of the major principal stress τ_1 and the minor principal stress τ_3:

$$\tau_1 - \nu\,\tau_3 \geqslant \sigma_c \tag{11}$$

The flow value ν and the compressive strength σ_c appearing in (11) are given by

$$\nu = \tan^2(\pi/4 + \psi/2) \tag{12}$$

$$\sigma_c = 2\sigma_s \tan(\pi/4 + \psi/2) \tag{13}$$

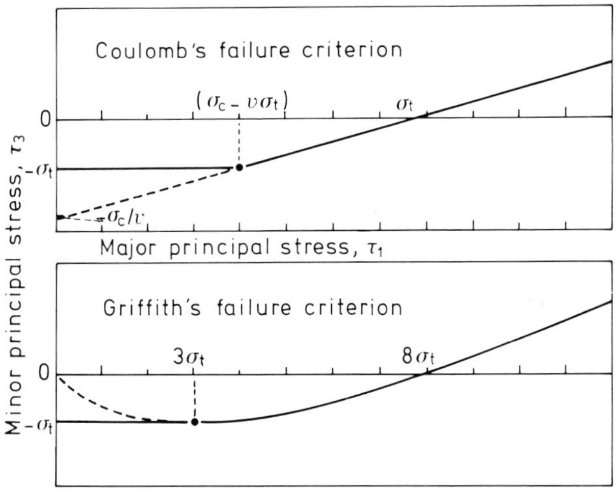

Figure 2. Two failure criteria

A plot of (11) is shown in Figure 2A. It can be shown that, according to Coulomb's criterion, slip failure may occur in two planes passing through the principal direction n_2 and making angles of $(\pi/4 - \psi/2)$ with the principal direction n_1.

In Fig. 2A the intercept on the τ_1-axis is the compressive strength σ_c. The intercept $-\sigma_c/\nu$ on the τ_3-axis could be interpreted as the tensile strength, were it not that cracks normal to the n_3 - direction occur at some value $\tau_3 > -\sigma_c/\nu$. To account for this, in 1961 Paul proposed to introduce the tensile strength σ_t as an additional parameter and replace the slip failure criterion (11) by the combined crack/slip failure criterion (Jaeger and Cook, 1979; see also Figure 2A):

$$\tau_3 \leqslant -\sigma_t \qquad \text{if } \tau_1 \leqslant \sigma_c - \nu\sigma_t \tag{14}$$

$$\tau_3 \leqslant (\tau_1 - \sigma_c)/\nu \qquad \text{if } \tau_1 \geqslant \sigma_c - \nu\sigma_t \tag{15}$$

At $\tau_1 = \sigma_c - \nu\sigma_t$ the Coulomb/Paul (CP) criterion predicts simultaneous occurrence of cracks normal to the principal direction n_3 and slip surfaces passing through the principal direction n_2, making angles of $(\pi/4 - \psi/2)$ with the principal direction n_1. Nadai (1931, pp. 330-331) described an experiment of W. Riedel in which tension cracks and shearing slip surfaces indeed occurred simultaneously. Hartge and Rahte (1983) attempt to distinguish cracks and slip surfaces on the basis of angles between faces of structure elements.

An alternative theory, relating the tensile strength to the growth of pre-existing, minute cracks, was introduced by Griffith in 1921 and further developed by Irwin and by Orowan in the fourties and fifties (Irwin, 1958). On the basis of an analysis of the stress distribution near tips of such cracks, Griffith derived the failure criterion (Figure 2B).

$$\tau_3 \leqslant -\sigma_t \qquad \text{if } \tau_1 \leqslant 3\sigma_t \tag{16}$$

$$\tau_3 \leqslant \{\tau_1/\sigma_t + 4 - 4\{1 + \tau_1/\sigma_t\}\sigma_t \qquad \text{if } \tau_1 \geqslant 3\sigma_t \tag{17}$$

The tensile strength can be shown to be a function of the energy G associated with the creation of new crack surfaces (Griffith) and/or plastic deformation near the tips of the cracks (Irwin and Orowan), the length 2λ of the cracks, Young's modulus Y and Poisson's ratio ν of the soil, and a factor α of order unity depending on the geometry and boundary conditions.

$$\sigma_t = \alpha \{ (G/\lambda) Y/(1 - \nu^2) \}^{\frac{1}{2}} \tag{18}$$

For $\tau_1 \leqslant 3\sigma_t$, the Griffith/Irwin/Orowon (GIO) criterion predicts cracks normal to the n_3 direction. At $\tau_1 = 3\sigma_t$ a *gradual* change of the orientation of the failure surfaces sets in. By contrast the CP criterion predicts an *abrupt* change at $\tau_1 = \sigma_c - \nu\sigma_t$.

The GIO criterion implies a value 8 for the ratio of compressive and tensile strengths. According to Farrell et al. (1967) measurements of this ratio range from 3 to 13. They attribute this wide variation to anisotropy, sample preparation, and changes in geometrical configuration under different test conditions. Recently Hettiaratchi and O'Callaghan (1980) presented a mechanics of unsaturated soils incorporating both the GIO failure criterion and the so-called critical state theory. They emphasize the dependence of the mechanical properties upon the water content.

3 Stress perturbation due to presence of cracks

The GIO failure criterion is concerned with the initiation of macro-cracks by growth of randomly oriented pre-existing minute cracks. To understand the spacing, width and depth of macro-cracks, an analysis is required of the perturbation of the original tensile stress field due to the presence of macro-cracks. Lachenbruch (1961, 1962) made such an analysis and applied it to cooling joints in lava and ice-wedge polygons in permafrost. The analysis is based on the linear theory of elasticity. Lachenbruch himself suggested that a drying mud, despite its plasticity, can be expected to conform to the theory in at least a qualitative way.

Such use of the theory of elasticity is rather common in soil mechanics (Koolen and Kuipers, 1983).

It is worthwhile to point out that the application of Lachenbruch's theory to ice-wedge polygons is of interest to soil scientists in its own right. In soils that were once subject to permafrost conditions, relicts of ice-wedges occur widely as wedges filled with foreign material, usually finer than the host material (Christensen, 1974, 1978; de Gans, 1983). Their presence is sometimes revealed by polygonal plant growth and ripening patterns, most clearly in cereal during dry summers. Christensen found that the available water capacity and rootability are the primary causes of the differences in crop growth.

Lachenbruch (1961, 1962) derived two exact solutions for stress fields near cracks in infinite media with, respectively, a step function and a linear function initial stress distribution. From these exact solutions he developed approximate solutions for tension-cracks at the surface of semi-infinite media. The approximate solutions are based on an iterative procedure by which, in the solutions for infinite media, the normal stresses upon the planes of symmetry normal to the cracks are eliminated and at the same time the walls of the cracks are kept stress free.

For the step function initial tensile stress distribution, Figures 3A and B show the relief of the principal stress in the soil surface and normal to the crack as a function of the distance from the crack. The stress relief is expressed in units of the initial tensile stress P and, not surprisingly, in these units varies from full relief of -1.0 at the crack to vanishing relief of 0 at large distance from the crack. The parameter on the curves is the ratio of stress depth a and crack depth b. Figure 3A shows that, for a given crack depth b, the width of the zone of stress relief is strongly dependent upon the depth distribution of the initial tensile stress. Figure 3B shows that, for a given stress depth a, the width of the zone of stress relief increases as the crack depth b increases up to a certain limit. Multiplication of the stress relief normal to the crack by Poisson's ratio (< 1) gives the stress relief parallel to the crack. The nature of the anisotropy causes a

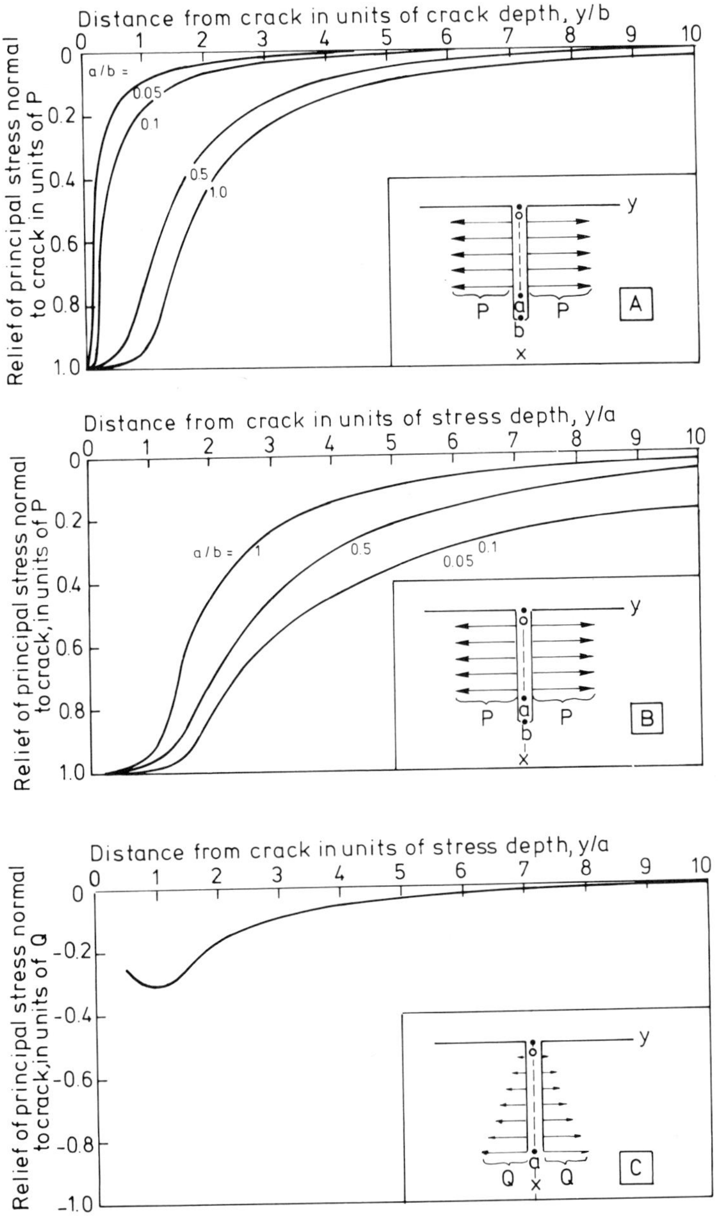

Figure 3. Stress relief at the soil surface due to a crack (inserts show the geometries and boundary conditions)

growing crack to orient itself normal to a pre-existing crack. Figure 4 shows a typical pattern of cracks with predominantly orthogonal intersections. The arrows indicate the directions in which the cracks were growing. These directions can be inferred not only from time-series of photographs but also from characteristic markings on the walls of the cracks. Euler's theorem, relating the numbers of vertices,edges,

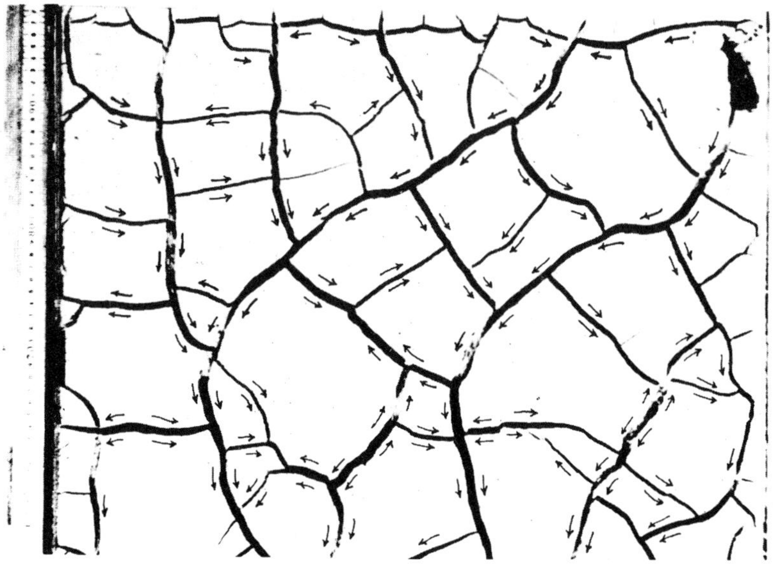

Figure 4. Cracking pattern (from Corte and Higashi, 1960)

and faces of any convex polyhedron, implies some simple properties of averages of parameters characterizing random crack patterns (Gray et al., 1976; Raats, 1984c). For instance if all junctions of edges are of the types  and , then the average number of edges and vertices per polygon is 6.

Figure 3C shows the result for a linear stress distribution. It can be used to account for the influence of the self-weight of the soil.

According to both the CP and GIO theories a crack will be initiated at the soil surface as soon as $\tau_3 \leqslant -\sigma_t$. At the ('mathematical') tip of a crack the stress will be infinite. At any finite distance from the tip the maximum stress occurs in the plane of the crack. This stress can be expressed in terms of the radial distance r from the crack tip and the crack-edge stress intensity factor κ by:

$$\tau_{max} = \kappa \, (2\,r)^{-\frac{1}{2}} \tag{19}$$

For the step function and the linear function initial tensile stress distribution the crack-edge stress intensity factor κ is of the form

$$\kappa = \gamma \, [a/b] \, \sqrt{b} \, P \text{ or } Q \tag{20}$$

Figure 5 shows the normalized crack-edge stress intensity factor $\gamma = \kappa / \{\sqrt{b}\, P \text{ or } Q\}$ as a function of the stress/crack depth ratio a/b. The x for an infinite body are exact while the • for a semi-infinite body are approximate. The latter are for the step function at a/b = 1 in good agreement with exact results of Wigglesworth and Irwin (Barenblatt,1962). Lachenbruch (1961, 1962) shows how Equations (19) and (20) and Figure 5 in combination with the tensile strength given by equation (18) can be used to estimate the depths of unstable propagation and of arrest of cracks.

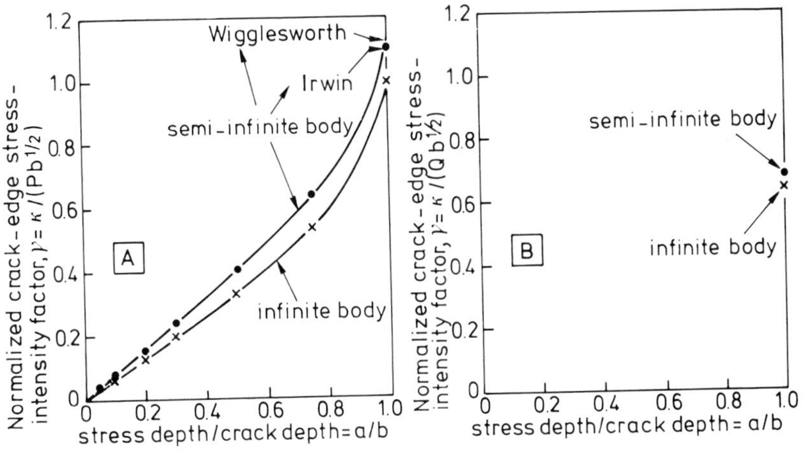

Figure 5. Crack-edge stress intensity factor (part A corresponding to Figure 3A, B and part B corresponding to Figure 3C)

4 Concluding remarks

Building upon experience with rigid unsaturated soils, the last two decades brought some progress with movement of water in deforming soils. Separately, classical soil and rock mechanics has recently inspired some progress with the mechanics of the solid phase of natural soils. Further progress might be expected from appropriate combinations of the separate disciplines soil physics, soil and rock mechanics, and, last but not least, modern continuum mechanics.

5 Acknowledgement

I wish to thank G.H. Bolt and C. Dirksen for their constructive criticism.

References

Barenblatt, G.I. 1962. The mathematical theory of equilibrium cracks in brittle fracture. Appl. Mech. 7: 55-129.

Barley, K.P., D.A. Farrell and E.L. Greacen 1965. The influence of soil strength on the penetration of a loam by plant roots. Aust. J. Soil Res. 3: 69-79.

Blokhuis, W.A. 1982. Morphology and genesis of vertisols. Trans. 12th. Intern. Congr. Soil Sci., New Delhi, India 3: 23-47.

Christensen, L. 1974. Crop-marks revealing large-scale patterned ground structures in cultivated areas, southwestern Jutland. Boreas 3: 153-180.

Christensen, L. 1978. Waterstress conditions in cereals used in recognizing fossil ice-wedge patterns in Denmark and Northern Germany. In: Proc. Third Intern. Conf. on Permafrost, 10-13 July 1978, Edmonton, Alberta, Canada 1: 254-261.

Corte, A. and A. Higashi 1960. Experimental research on desiccation cracks in soil. U.S. Army Snow, Ice and Permafrost Research Establishment, Corps of Engineers, Research Report 66.

De Vos, T.N.C., J.H. Virgo, and K.J. Virgo 1969. Soil structure in vertisols of the Blue Nile clay plains, Sudan. J. Soil Sci. 20: 189-206.

Dexter, A. 1983. Two types of soil anisotropy induced by the passage of a wetting front. Soil Sci. Soc. Am. J. 47: 1060-1061.

Farrell, D.A., W.E. Larson, and E.L. Greacen 1967. A model study of the effect of soil variability on tensile strength and fracture. Presented at 1967 Meeting, American Society of Agronomy.

Gans, W. de 1983. Fossiele permafrostverschijnselen in Nederland (Permafrost relicts in The Netherlands). Grondboor en Hamer 6: 175-184.

Gray, N.H., J.B. Anderson, J.D. Devine and J.M. Kwasnik 1976. Topological properties of random crack networks Math. Geology 8: 617-626.

Hartge, K.H. and I. Rahte 1983. Schrumpf- und Scherrisse-Labormessungen. Geoderma 31: 325-336.

Hettiaratchi, D.R.P. and J.R. O'Callaghan 1980. Mechanical behaviour of agricultural soils. J. Agric. Eng. Res. 25: 239-259.

Irwin, G.R. 1958. Fracture. In: Encyclopedia of Physics. Springer, Berlin. VI: 551-590.

Jaeger, J.C. and N.G.W. Cook 1979. Fundamentals of rock mechanics. Chapman and Hall, London, 593 pp.

Koolen, A.J. and H. Kuipers 1983. Agricultural soil mechanics. Springer, Berlin, 241 pp.

Krishna, P.G. and S. Perumal 1948. Structure in black cotton soils of the Nizamsagar project area, Hyderabad State, India. Soil Sci. 66: 29-38.

Lachenbruch, A.H. 1961. Depth and spacing of tension cracks. J. Geophysical Res. 66: 4273-4292.

Lachenbruch, A.H. 1962. Mechanics of thermal contraction cracks and ice-wedge polygons in permafrost. Geol. Soc. Am. Spec. Pap. 70, 69 pp.

Miller, E.E. 1975. Physics of swelling and cracking soils. J. Colloid Interface Sci. 52: 434-443.

Nadai, A. 1931. Plasticity. A mechanics of the plastic state of matter. McGraw-Hill, New York, 343 pp.

Neal, J.T., A.M. Langer and P.F. Kerr 1968. Giant dessication polygons of Great Basin Playas. Geol. Soc. Am. Bulletin 79: 69-90.

Parlange, J.Y., and B.L. Sawhney 1976. Anisotropic effects in swelling soils. In: M. Kutilek and J. Sutor (Eds.) Water in Heavy Soils. The Czechoslovak Scientific Technical Society, Prague, Czechoslovakia, I, pp. 42-49.

Philip, J.R. and D.E. Smiles 1982. Macroscopic analysis of the behaviour of colloidal suspensions. Adv. Coll. and Interface Sci. 17: 83-103.

Raats, P.A.C. 1969. Axial fluid flow in swelling and shrinking porous rods. Abstracts 40th Annual Meeting The Society of Rheology, Inc., October 20-22, 1969, p. 13.

Raats, P.A.C. 1984a. Applications of the theory of mixtures in soil physics. In: C. Truesdell (Ed.), Rational Thermodynamics, 2nd ed. Springer, Berlin.

Raats, P.A.C. 1984b. Multidimensional transport processes in deformable soils. (In preparation)

Raats, P.A.C. 1984c. Topology of cracking patterns. (In preparation)

Truesdell, C. and R.A. Toupin 1960, The classical field theories. In: Encyclopedia of Physics. Springer, Berlin. III/1:226-794.

Discussion

D.H. Yaalon:

Could you explain the meaning of the arrows parallel to the cracks in Figure 4? Do they indicate propagation (growth) from a linear or point source? In a random tensional stress field there will be primary and secondary cracks, the latter tending to form at right angles to the primary cracks (cf. Hartge and Rahte, 1983). Hence, the mean number of edges per polygon is 4 to 5, as the figure clearly shows. Hexagons are an exception.

Author:

The meaning of the arrows is given in the text. The directions of crack growth were inferred from successive photographs. Among the sites where cracks originate are the edges of the sample and air bubbles in the mud. If collinear edges of polygons are counted as single edges, then, as the fraction of orthogonal intersections approaches unity, the number of edges per polygon approaches 4 (cf., Gray et al.,1976). However, if one regards the edges issuing from an orthogonal intersection as separate edges for each of the three adjoining polygons, then the average number of edges per polygon is 6.

S. van der Zee:

Numerical values of the parameters found in the failure criteria are obtained by loading a soil sample in specific ways, implying the application of an external force. However, cracking seems to be more of a pulling apart mechanism than of a compression mechanism. For sand, the moment of failure of a core will occur sooner when pulled apart than in the case of compression. Possibly this is also the case for clayey material. Should not, then, the apparent cohesion (cohesion plus water suction) be a better criterion for cracking of soil, in which case the normal stress due to compression is excluded?

Author:

I agree that, when cracks form, the material is being pulled apart. Figure 2 shows how, in a sense, crack failure overrules slip failure. Just as the shear strength σ_s (and thus the compressive strength σ_c), the tensile strength σ_t includes a component due to capillarity. The Griffith/Irwin/Orowan criterion even implies a definite value 8 for the ratio of compressive and tensile strengths.

S. van der Zee:

In Griffith's model, breakage of soil is characterized by elastic behaviour until breakage occurs. Afterwards fractures are found and separation of soil occurs, but no plastic flow. Is this rigid-matrix model applicable to clayey soil with a higher ability to deform, thus showing more plastic behaviour? Is clay, on the verge of cracking, dried to an extent that it does show rather elastic behaviour?

Author:

Griffith's original theory is concerned only with elastic behaviour. Irwin and Orowan extended the theory to include plastic deformation near the tip of the crack.

H.P. Blume:

What influence do variations in soil temperature have upon forming cracks? You showed soils with sand-filled cracks from Germany and Denmark. In our opinion these cracks are not formed by variations in water content, but by temperature variations during summer and winter times in the ice age. If 0.1-mm-thick cracks of a frozen soil will be filled by wind-blown sand, and this will be repeated for hundreds of years, sand-filled cracks with a diameter of 20-30 cm can be formed. Perhaps the same process will work in soils of the hot deserts by changes in soil temperature between day and night, summer and winter.

Author:

The cited papers by Christensen and Lachenbruch are indeed mainly concerned with the process of thermal contraction accompanying freezing.

Mechanics of colloidal suspensions with application to stress transmission, volume change, and cracking in clay soils

J. R. Philip, J. H. Knight,
CSIRO Division of Environmental Mechanics, Canberra, Australia
and J. J. Mahony
Department of Mathematics, University of Western Australia, Perth, Australia

1 Introduction

We report work in progress on the mechanics of colloidal suspensions. An important motivation is that the work promises to furnish new insights into various elements of the mechanics of heavy clay soils. We have in mind aspects such as: the state of stress of the water in heavy clay soils; stress tensors and stress trajectories on both microscopic and macroscopic scales; one-dimensional, three-dimensional, and other modes of soil volume change and deformation; the energetics of soil cracking.

Although a heavy clay soil is not simply a dense colloidal suspension, the latter offers a more physically meaningful point of departure for the study of the mechanics of such soils than does the classical elastic body, which is the (albeit now remote) starting point of conventional soil mechanics.

The viewpoint of this work has close affinities to that of Philip (1970a). It is also related to the unified approach to soil suspensions and to swelling soils recognized in Philip (1970b) and elaborated in Philip and Smiles (1982). One of us (J.R.P.) acknowledges gratefully the stimulus toward the present ideas of discussions with Professor G. H. Bolt in Canberra in 1969.

The work is essentially an investigation of electrical double-layer interactions in homogeneous swarms (Philip, 1970a) of charged particles, based on the Poisson-Boltzmann equation. We recognize the simplifications, omissions, and limitations of the model, but believe this exploratory work addresses the essential elements of the problems we seek to elucidate. Many limitations will be removable by later, more elaborate, calculations.

2 The Poisson-Boltzmann Equation in Homogeneous Swarms

We begin with the Poisson-Boltzmann equation applied to the diffuse double layer:

$$\nabla^2\psi = -(4\pi\varepsilon/D)\sum z_i n_i(0)\exp(-z_i\varepsilon\psi/kT) . \quad (1)$$

∇^2 is the Laplacian in physical space coordinates; ψ is the electrostatic potential; D is the dielectric constant of the solution; $n_i(0)$ is the number per unit volume of ions of species i in regions of the solution at potential 0; z_i is the signed valency; ε is the protonic charge; k is the Boltzmann constant, and T is the absolute temperature. The total number of ions per unit volume in regions at potential ψ,

$$N(\psi) = \sum n_i(0)\exp(-z_i\varepsilon\psi/kT) . \quad (2)$$

Minimum energy considerations require that the equilibrium configuration be a regular particle array. Each particle occupies its own basic cell, bounded by a surface on which the normal component of the electrostatic field strength, $\partial\psi/\partial\nu$, vanishes. All cells are identical, so the problem reduces to solving (1) subject to the conditions

$$\psi = \psi_0 \text{ on } A_0 ; \quad \partial\psi/\partial\nu = 0 \text{ on } A_1 . \quad (3)$$

Here A_0 is the particle surface and A_1 the surface of its basic cell.

3 Gibbs Free Energy, Variational Principle, and Microscopic and Macroscopic Stress Tensors

The Gibbs free energy of interaction per particle is

$$G = - \int_V \frac{D}{8\pi}[\nabla\psi]^2 + kT[N(\psi) - N(0)] \, dv \, , \qquad (4)$$

where the integral is taken over the volume of the cell external to the particle. Levine (1951) gave an elaborate derivation of (4); but (4) follows immediately from recognition that (1) yields a variational principle governing ψ, namely that $-G$ must be minimized.
The interaction produces on the water a local *microscopic stress tensor* which is the sum of an anisotropic electrostatic part and an isotropic "osmotic" one (Coolidge and Juda, 1946). Its principal components are

$$\mp \frac{D}{8\pi}[\nabla\psi]^2 + kT[N(\psi) - N(0)] \, , \qquad (5)$$

where the minus sign holds in the direction of the electric field. Tensile stress is positive.
We are concerned also with the *macroscopic stress tensor*. With the cell a cuboid with sides 2a, 2b, 2c in the principal directions of the particle array, the principal components (averaged over the relevant cross-sections) are

$$- V^{-1} \, \partial G/\partial \ln a \, , \quad - V^{-1} \, \partial G/\partial \ln b \, , \quad - V^{-1} \, \partial G/\partial \ln c \, . \qquad (6)$$

with cell volume $V = 8abc$.
To map the microscopic tensor we solve (1) subject to (3) at all points of the field. Evaluation of the macroscopic tensor (i.e. of $G(a,b,c)$ and its derivatives) is effected more economically. In the Debye-Hückel approximation (4) reduces to

$$G = - \frac{D}{8\pi\kappa} \int_V [(\nabla\psi)^2 + \psi^2] \, dv \, . \qquad (7)$$

Here and in (8) gradients, volumes, and surface areas are in dimensionless coordinates obtained by normalizing space coordinates with respect to the Debye length κ^{-1}:

$$\kappa = [4\pi\varepsilon^2 \sum_i n_i(0) z_i^2 / DkT]^{\frac{1}{2}} .$$

Applying the divergence theorem and using (3) reduces (7) to

$$G = \frac{D\psi_0}{8\pi\kappa} \int_{A_0} \frac{\partial\psi}{\partial\nu} \, dA_0 , \qquad (8)$$

with the normal derivative taken outward. This remarkable result also gives an excellent approximation to G for the full Poisson-Boltzmann equation in many circumstances.

A program of solving (1) to find ψ and the partial derivatives of G for various cell volumes and anisotropies is in train. Initial work is for two-dimensional systems and in the Debye-Hückel approximation. We plan to present to the symposium results and interpretations bearing on the various aspects listed in our opening paragraph.

References

Coolidge, A.S. and W. Juda 1946. The Poisson-Boltzmann equation derived from the transfer of momentum. J. Amer. Chem. Soc., 68:608-611.

Levine, S. 1951. The free energy of the double layer of a colloidal particle. Proc. Phys. Soc. A, 64:781-790.

Philip, J.R. 1970a. Diffuse double-layer interactions in one-, two-, and three-dimensional particle swarms. J. Chem. Phys. 52:1387-1396.

Philip, J.R. 1970b. Hydrostatics in swelling soils and soil suspensions: unification of concepts. Soil Sci. 109:294-298.

Philip, J.R., and D.E. Smiles 1982. Macroscopic analysis of the behavior of colloidal suspensions. Adv. Colloid Interface Sci. 17:83-103.

Discussion

L.P. Wilding:

1. What is the relevance of your diffuse double-layer theory with regard to volume changes due to shrinking and swelling, considering

the fact that, under field conditions, most volume changes that take place are at moisture states, where clay mineral interlayers are fully solvated?

2. How does clay particle orientation affect the energetics of your model, considering the fact that in most soil systems clays occur as booklets of clay particles (micelles) rather than completely dispersed systems?
3. How does cation saturation and electrolyte concentration affect this model?

Author:

1. At its present level of development, the analysis is, in fact, primarily applicable to swelling between, rather than within, particles.
2. As I explained, the results presented are for a highly schematic geometry. Any irreversible element in the geometry will, of course, lead to a corresponding irreversibility in the energetics. Results will vary quantitatively for different geometries, but their qualitative character should be preserved.
3. The chemistry enters the model through Ψ_o, the particle surface potential, and κ^{-1}, the Debye length.

G.H. Bolt:

Recognizing that your treatment at present refers to a sub-subset of charged-particle arrangements (i.e. uniform cylindrical rods situated at equal distances with a relative size of $\kappa r \simeq 1$, where r is the radius of the rod-shaped cell), the question may be posed whether one could hope (c.q. whether you have information) that more involved arrays, e.g. 'domains' of plates with perhaps positively charged edges, etc. would follow in principle the same or a very similar distribution pattern of stresses.

Author:

There seems to be no obvious reason why one would not obtain qualitatively similar results. But, as I said in my presentation, the work suggests that the constitutive equations for colloid particles of the type we consider should include one specifying the non-linear relation between the macroscopic strain tensor and the macroscopic stress tensor'. We have used a simple form of double-layer theory to make a first estimate of the character of that relation. Of course, the realistic goal is to use analyses of this type to suggest the basic form of the phenomenological macroscopic theory, and then go on to make appropriate measurements. Then, perhaps, we will have the beginning of the relevent mechanics.

J.C.W. Keng:
Your presentation was based on a simplified double-layer model which involves two clay particles. Would you please elaborate on the applicability of the electrical double-layer theory to macroscopic soil shrinkage problems, which involve hundreds and thousands of clay particles in a random structure-arrangement?

Author:
My presentation was based on a three-dimensional array of an effectively infinite array of particles. Because we have a regular array we need only the Poisson-Boltzmann equation within the basic cell - with the electrical field strengh zero on the cell boundary surface.

Extent and dynamics of cracking in a heavy clay soil with xeric moisture regime

D.H. Yaalon
Institute of Earth Sciences
Hebrew University, Jerusalem
D. Kalmar
Institute of Soils and Water
Volcani Center, A.R.O.

Cracks increase the rate at which a pedon can accept and absorb moisture during the wetting process. Their extent and dynamics are hence of interest in structure development and in water transport studies. We have measured for two seasons the development of cracking on a bare, weakly calcareous, alluvial Grumusol (Typic Pelloxerert), in the Zevulon valley, Israel. Measurements of surface cracking were made along a tape 20m by 20m and their depth was measured with a thin wire probe (ø 1.5 mm), essentially following the procedure of Zein el Abedine and Robinson (1971). Measurements were made monthly up to 200 days after the last winter rain. Linear extent, depth and volume of cracks were calculated and plotted.

First shrinkage cracks appeared 30 days after the last rain and some reached a maximum width of 9.5 cm. Width of most cracks increased with time to a mean width of 3.5 cm. Median distance between cracks decreased from over 50 cm to less than 20 cm, indicating that pattern development took about 150 days. Thereafter no new cracks are formed, remaining at a mean density of 3.3 cracks per linear meter. Maximum depth reached by the cracks measured was 80 cm., but most cracks did not extend beyond 40 cm. Volume of cracks increased steadily for the 200 days, reaching a total volume of 3% (Figure 1). During the same time the pedons contracted vertically by 5.7 cm. (Yaalon and Kalmar, 1972). In a cultivated orchard, the same volume of cracks was reached after 42 days of drying.

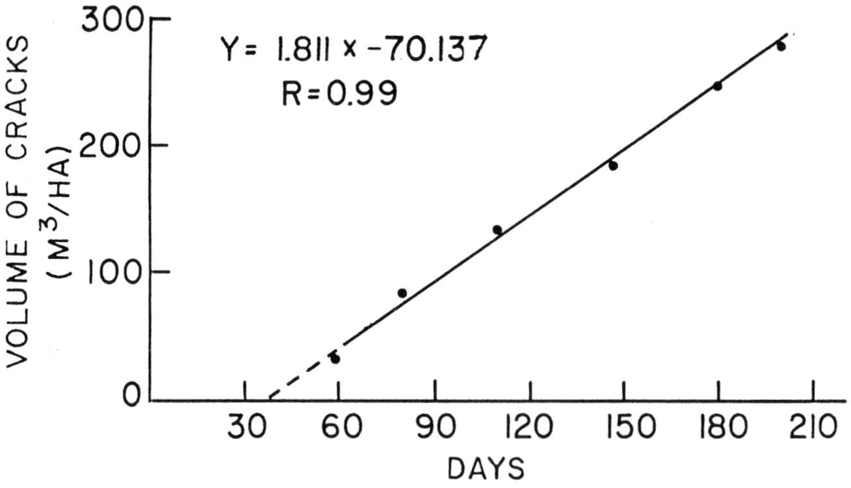

Figure 1. Mean volume of cracks during drying

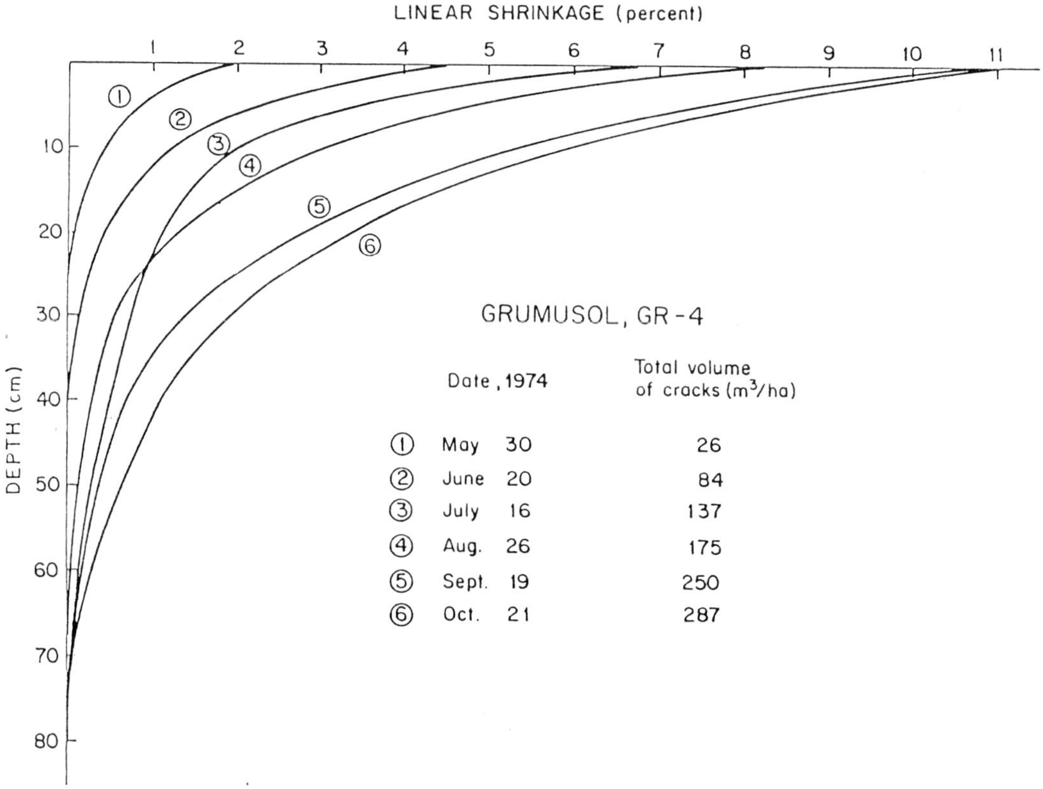

Figure 2. Distribution of shrinkage cracks with depth and season

The regression of Fig. 1 indicates a daily volume increment of cracks of 1.8 m^3/ha, or a mean depth of 0.18 mm/day. To this we add the vertical contraction of ∼0.3 mm/day. Together these soil-volume changes represent 0.3 of the loss of moisture (measured by neutron scattering). COLE measurements on undisturbed peds between field capacity and wilting moisture (Kalmar, 1979) produced a coefficient of volume change of 0.36, in good agreement with the above, but perhaps indicating that field measurements missed about 10% of the cracks or that structural changes in the field environment are somewhat lower than in laboratory measurements.

Spatial distribution of the cracks with time indicates that only once did upper soil material fall into deeper layers as delineated by the crossover lines on Figure 2. Its volume was only 0.05% of the total soil volume; in the preceding season the volume was 0.1%. Unless much material falls into the cracks during the wetting process, the turnover time in this case is over a thousand years. This is supported by other evidence (Yaalon and Kalmar, 1978).

It was not possible to ascertain whether surface cracks re-establish themselves at a similar pattern during the next drying cycle (Virgo, 1981) but micromorphological evidence from deeper layers suggests that to some extent it does develop the same rupture faces.

Deep seasonal cracking is important for the development of strong and stable soil structure. Measurements by others indicate that in well structured Israeli Grumusols (Vertisols) the steady state infiltration rate is five or more mm/h after 40-50 mm of rainfall. It decreases significantly on a disturbed or irrigated soil not allowed to dry out.

References

Kalmar, D. 1979. The dynamics of structure in heavy clay soils, Ph.D. thesis, Hebrew University, Jerusalem, 109 pp. (in Hebrew).

Virgo, K.J. 1981. Observations of cracking in Somali vertisols. Soil Sci., 131: 60-61.

Yaalon, D.H. and Kalmar, D. 1972. Vertical movement in undisturbed soil: continuous measurement of swelling and shrinkage with a sensitive apparatus. Geoderma, 8: 231-240.

Yaalon, D.H. and Kalmar, D. 1978. Dynamics of cracking and swelling clay soils: displacement of skeletal grains, optimum depth of slickensides and rate of intrapedonic turbation. Earth Surface Processes, 3: 31-42.

Zein el Abedine, A. and Robinson, G.H. 1971. A study on cracking in some vertisols of Sudan. Geoderma, 5: 229-241.

Discussion

P. Bullock:

Do you have any evidence to suggest that the fissures re-open in the same position year after year? The presence of slickensides may suggest that they do, but we do not know how long it takes for them to form. Perhaps they could form in a year. The presence of zones of preferred orientation of the clay on the matrix of some clayey soils may represent former slickensides which have not re-opened.

Author:

Our thin-section measurements show that the thickness of the oriented slickensiding zone increases with depth, thus suggesting that shearing takes place essentially in the same direction and position. But we could not ascertain whether the shrinkage cracks higher up in the profile also re-open at the same position year after year. Observations on saline muds indicate that several generations of shrinkage cracks are sometimes present.

Evolution of crack networks during shrinkage of a clay soil under grass and winter wheat crops

V. Hallaire
Institut National de la Recherche Agronomique
84140 Montfavet, France

1. Introduction

Pore space characteristics of swelling clay soils are mainly influenced by the soil-water content. A dynamic recording of cracks size and spacing (i.e. their network) is required to characterize hydrodynamic properties of these soils. In this paper, a study on the evolution of crack networks in a drying clay soil will be presented for two different crops (grass and winter wheat).

2. Material and methods

2.1. The soil

The investigated soil (Les Vignères) is a clayey soil, with a clay fraction (52 to 56 %) consisting mainly of montmorillonite and chlorite. The dry bulk density of the aggregates varies from 1.45 to 1.90 g.cm^{-3}, corresponding with field capacity (gravimetric water content w = 30 %) and shrinkage limit (w = 10 %). Three plots were investigated : with winter wheat, with grass (fescue) and without any vegetation.

2.2. Field studies

a) Soil moisture profiles : samples were augered weekly at 20 cm intervals, from 20 to 120 cm depth, from the beginning of March to the end of August.
b) At the same time, the thicknesses of three horizons (25-50 cm, 50-80 cm, 80-120 cm) were measured with reference marks fixed into the ground (Selig and Reinig, 1982).

c) Observations of cracks were made by photographying vertical walls and horizontal surfaces of soil.

d) The dry bulk densities were measured :

- by gamma-ray transmission equipment having a twin probe separation of 20 cm (Stengel et al., 1984).
- by cores sampling (1.2 dm^3).
- by Archimedes thrust on clods and on aggregates (Stengel, 1982).

Assuming that interaggregate pore spaces are cracks, the "cracks void ratio", $\mathbf{e}_f$, is given by :

$$\mathbf{e}_f = \mathbf{e}_T - \mathbf{e}_\tau$$

where :

$\mathbf{e}_T$ is the total void ratio

$\mathbf{e}_\tau$ is the aggregates void ratio.

2.3. Laboratory studies

Saturated undisturbed core samples (Ø 15 cm, h 7.2 cm) were collected during winter (w = 29 %). $\mathbf{e}_f$ was found to be negligible. The samples, kept in their cylinder, were subjected to different rates of evaporation. Vertical shrinkage, measured with displacement transducer, allowed the calculation of the total void ratio $\mathbf{e}_T$; diametral shrinkage, measured by filling up the space between core and sample by glass beeds (200 µm), allowed the calculation of the core void ratio $\mathbf{e}_c$ (annular crack excepted).

3. Results and discussion

3.1. Evolution of crack patterns observed on a horizontal plane

In winter, while the soil was saturated, a horizontal plane (50 cm x 50 cm) was prepared at a depth of 60 cm. Cracks were counted on successive photographs which were taken during evaporation. The images revealed two stages in the shrinkage process : at first, thin cracks (less than 5 mm wide) appeared, with about 3 cm spacing. Then, some of these cracks opened wider (to more than 1 cm), with about 20 cm spacing, while the remaining cracks were partially or even totaly closed (Figure 1).

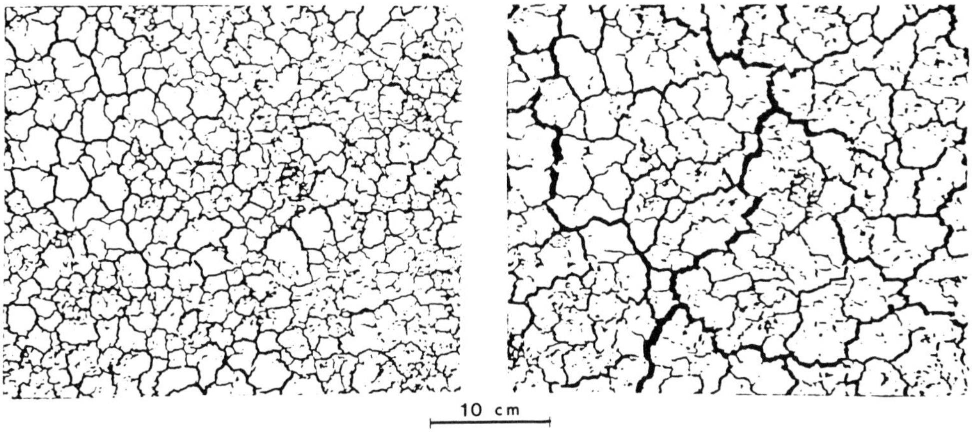

Figure 1 : Cracking process : first stage (a), second stage (b).

3.2. Evolution of cracking under crops

The study of mean values and standard deviations of $\mathbf{e}_f$, combined with field observation, showed that :

(1) At the end of the winter, the soil was at field capacity (w = about 30 %). The volume of the cracks was negligible.

(2) For 30 % > w > 23 %, the vertical shrinkage of the whole profile was not important (less than 1 %). Despite of the presence of thin (less than 2 mm) or even invisible cracks, they were detected by bulk density measurements ($\mathbf{e}_f$ reached 0.15). Because of the small standard deviation of $\mathbf{e}_f$, it is assumed that the crack network was homogeneous and that crack spacing was small compared with the twin probe separation.

(3) For 23 % > w > 18 %, vertical shrinkage was larger (up to 4 %) ; the mean value of $\mathbf{e}_f$ reached 0.25 under wheat, and 0.30 under grass. The large standard deviation appeared to indicate a wider crack network. This assumption was confirmed by observations from surface down to 1 m : cracks were big (often greater than 1 cm), and the distance between two cracks varied from 20 cm to 1 m.

The cracking process seems to be divided into two stages which are defined by water content. The second stage follows from the evaporative demand : in May for winter wheat, in June for grass. The start of the second stage is, however, associated with the same water content, approximatively 23 %.

3.3. Shrinkage of undisturbed core samples

Shrinkage curves are shown in Figure 2 (full line : intraaggregate void ratio $\mathbf{e}_\tau$; dashed line : total void ratio $\mathbf{e}_T$, including annular crack ; dotted line : core sample void ratio $\mathbf{e}_c$, annular crack excepted). The variation of $\mathbf{e}_c - \mathbf{e}_\tau$ with water content (Figure 3) showed a maximum for w = 23 %. Those measurements are illustrated by photographs (Figure 4) which clearly show the two stages of the cracking process.

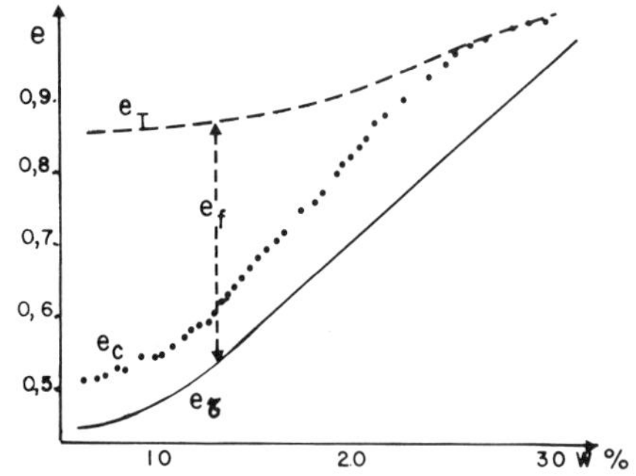

Figure 2 Shrinkage curves

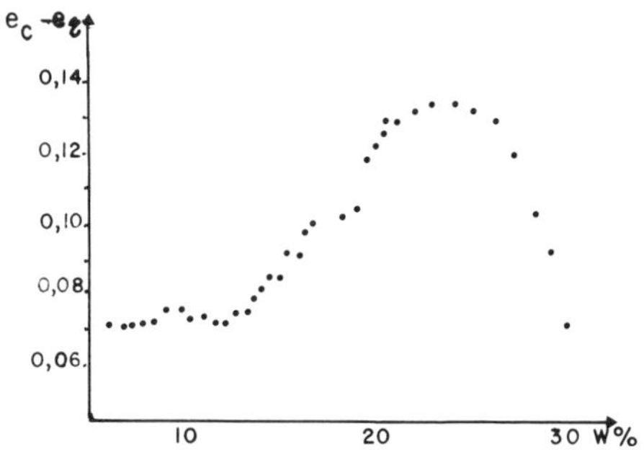

Figure 3 Variation of $\mathbf{e}_c$-$\mathbf{e}_\tau$ with w

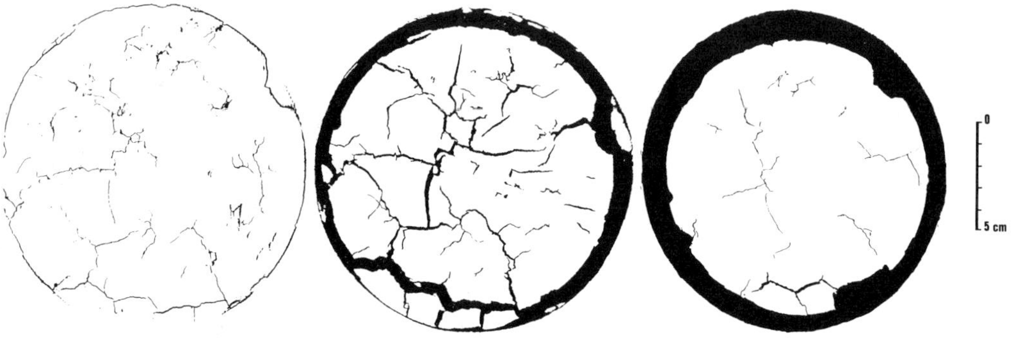

Figure 4 Shrinkage of an undisturbed sample

On Figure 5, analytical results obtained on core samples ($\mathbf{e}_f$ and $\mathbf{e}_c - \mathbf{e}_\tau$) and measurements for grass and winter wheat (mean value and standard deviation of $\mathbf{e}_f$) are compared. There is a good agreement between laboratory methods and in situ measurements and observations. A description of the cracking process in two stages as a function of the water content appears therefore to be useful for characterizing crack networks.

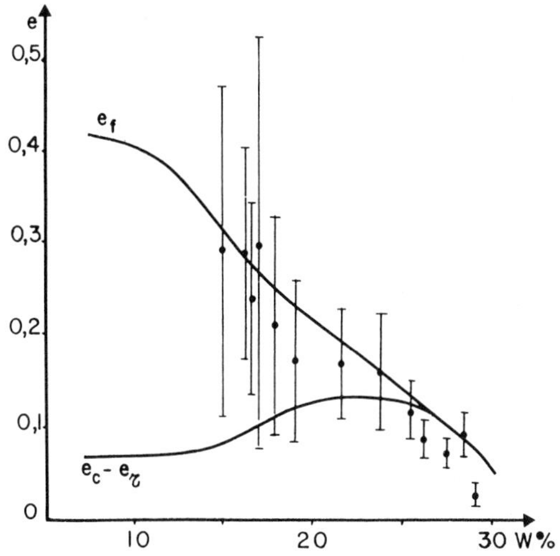

Figure 5 In situ results (grass and wheat) compared with laboratory results (core samples)

References

Selig E.T. and I.G. Reinig (1982). Vertical Soil Extensometer. Geotechnical Testing Journal, 5 (3/4) : 76-84

Stengel P. (1982). Swelling potential of soil as a criterium of permanent direct drilling suitability. 9th Conference of the International Soil Tillage Research Organization. Yugoslavia OSIJEK : 131-136

Stengel P., Y. Gabilly and J.C. Gaudu (1984). La double sonde gamma LPC-INRA : utilisation à des fins agronomiques. Bull. du GFHN (in press).

Discussion

P.A.C. Raats:

Figure 1 shows that, as drying progressed, the pattern of cracking became coarser. C.R.B. Lister ('On the penetration of water into hot rock'. Geophys. J.R. Astr. Soc. 39: 465-509, 1974) coined the name 'obdivision' for such a change. As the cracks deepen, the region of potential stress relief of each crack widens, and the original fine pattern tends to become unstable.

Cracking patterns in soils caused by shrinking and swelling

K.H. Hartge, Institut f.
Bodenkunde d. Univ. Hannover
Fed. Rep. Germany

1. Introduction

Angles formed on drying and wetting of clay slurries in the laboratory showed markedly different patterns for different soils. Here observations on in situ aggregates are reported and compared with the laboratory results.

2. Theory

In the mechanics of failure two kinds of fractures are known (Hahn, 1970): failures caused by tensile stresses, oriented rectangularly against direction of stress, and failures due to shearing stresses, oriented at an angle ($45^o - 1/2\varphi$) from the direction of the resultant first principal stress.

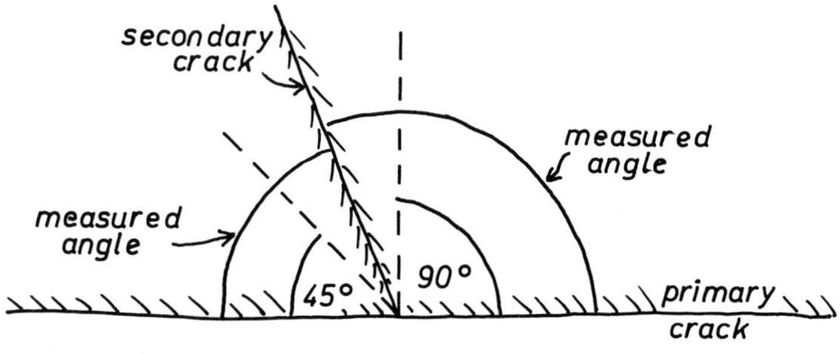

Figure 1. Angles between primary and secondary cracks

If now a distribution of values of angles between cracks is observed where besides 90°-angles there are two other peaks, then it is interesting to find out what is the distribution of the angles. For this purpose the values of the measured angles were devided into three groups in such a way that all of them are as symmetric as possible, while the central group retains its peak at 90°. This was done by trial and error and it turned out that, if both conditions were strictly followed, there was only little choice in allotting the values to the tree distributions. The mean values for the two distributions other than 90° were found to deviate equally from 90°, their sum being 180° (Figure 1).
This means, that the group of nonrectangular angles can be said to belong to just one direction of cracks which can be expressed by $135^{\circ} - \frac{\varphi}{2}$ or $45^{\circ} + \frac{\varphi}{2}$. If this assumption is correct, the corresponding failures must be associated with forces perpandicular to the original, older crack.

3. Materials and methods

Angles between two adjacent planes were measured on aggregates from 11 soils with clay contents between 5 and 69 %. The detailed description of the soils is given by Hartge and Rahte (1983).

4. Results on natural aggregates

The distributions of the angles of all the samples had irregular shapes. A prevailing frequency-maximum at 90° was not always distinguishable. As an example, the distribution for a chernosem from loess (boroll US-Tax) is given in fig. 2 above). The lower part of the figure shows how the values were allotted to the distributions following the rules given before. The peak of the 90°-distribution was placed as close as possible to 90°, keeping the distribution symmetrical to the same extent as the two distributions made up by the rest of values. The sum of the calculated means of these distributions gave 174°. The data for the other soils were treated in the same way. The result showed that:

(1) The fraction of angles allottable to the 90°-distribution varied.

(2) The deviations of the mean values of the two "nonrectangular" distributions from 90° were nearly equal.

(3) The sum of the mean values of these two distributions was close to 180° (8 out of 11 were between $177,5^{o}$ and $182,5^{o}$)

(4) The portion of 90°-angles from the total of all values of a sample was the higher, the lower the deviation of the rest of them was from 45° (see fig. 3).

(5) If this deviation was considered to represent $45^{o}+\frac{\varphi}{2}$ or $135^{o}-\frac{\varphi}{2}$ $(=90+45-\frac{\varphi}{2})$, then the rank of numeric values for φ corresponded with aggregate stability against water impact.

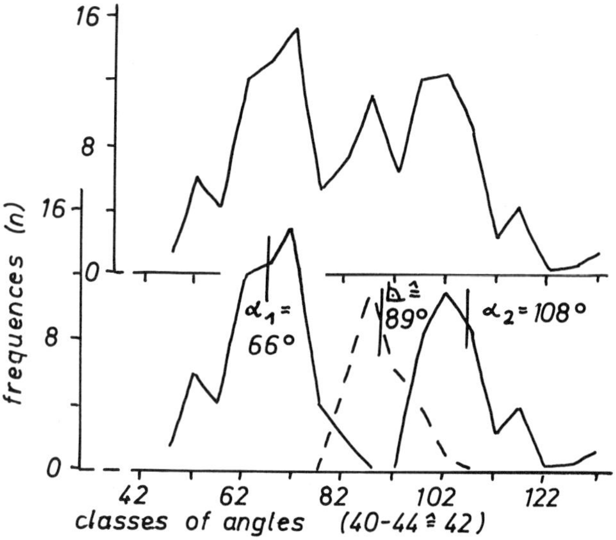

Figure 2 Distribution of angles of natural aggregates of the loess-boroll (top: frequency distribution of all values; lower: values split up to give three distributions)

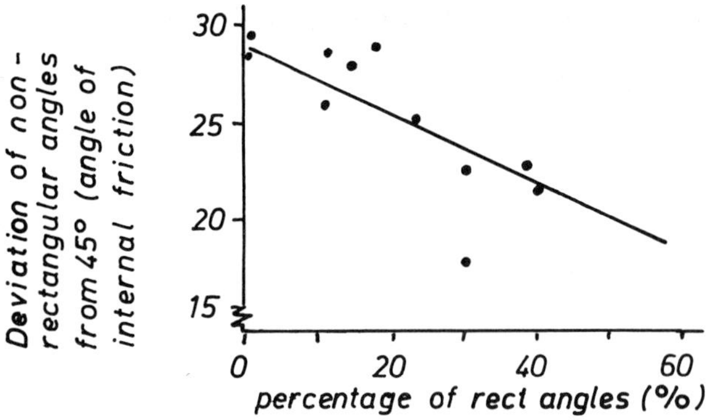

Figure 3 Percentage of right angles as related to divergence of nonrectangular angles from 45°

From the cracking pattern given in fig. 1 it can be seen, that the nonrectangular angles originate from forces normal to the direction of the elder primary cracks. If the forces were active tangential to the primarily developed crack, then the smaller angles should have distributions with peaks at <45°.

5. References

Hahn, H.G. 1970. Spannungsverteilungen in Rissen in festen Körpern, VDI-Forschungsschrift Nr. 542.

Hartge, K.H. and I. Rahte 1983. Schrumpf- und Scherrisse - Labormessungen. Geoderma 31, 325-336.

Crack formation in newly reclaimed sediments in the IJsselmeer polders

CRACK FORMATION IN NEWLY RECLAIMED
SEDIMENTS IN THE IJSSELMEERPOLDERS
K. Rijniersce
IJsselmeerpolders Development Authority
Lelystad, The Netherlands

1. Introduction

In the former ZuyderZee -nowadays called IJsselmeer- four polders have been reclaimed. Just after reclamation the clayey sediments in these polders can be characterised as very soft and wet and unsuited for agricultural production. Although the pore volume is high, there are no larger pores and so the soil is virtually impermeable, making drainage impossible. By a process, called soil ripening, the sediments are transformed in normal soils, very suitable for agricultural use.
The physical part of this process starts as soon as the water is pumped out of the polder. Due to the evaporation surplus in summer under the dutch climatic conditions, the watercontent decreases and the compaction of the soil leads to subsidence and crack formation. By the newly developed cracks the permeability of the soil increases so that water can be drained away. The compaction of the soil is almost completely irreversible.
Recently a simulation model for this process has been developed (RIJNIERSCE, 1983). A specific problem during this development was the simulation of the crack formation. In this abstract this part of the model will be described.

2. General description of the model for soil ripening

In the simulation model for ripening, the soil moisture suction Ψ is chosen as the "mastervariable". In this model the Ψ not only defines the stored amount of water and the conductivity, but also the rate of compaction. In the model the soil profile is divided into layers, in which the content of solid parts remains the same during the simulation. The soil in a layer is divided into "solid soil" including the solid parts, the water and evt. air and the so-called "big cracks". All the soil properties, like pF-curves, bulk densities etc. are related to the "solid soil".

3. Submodel for subsidence and crack formation

In soil mechanics subsidence is mostly calculated using Terzaghi's formula. In its most simple form this formula is:

$$\Delta z/z=(1/c).\ \ln(p_2/p_1) \qquad (1)$$

where $\Delta z/z$ is the relative subsidence, c is a consolidation constant and p_2 and p_1 are the grain pressures after and before loading. The c-value depends on soil properties. A relation between the pore space and this value is given by DE GLOPPER (1977).
This Terzaghi formula can give incorrect results. If p_2 is sufficiently greater than p_1, $\Delta z/z$ will be greater than 1, which is impermissible. Besides, the results obtained with this formula are affected by the magnitude of the steps in the increase of the load.
In the model for ripening it is assumed that p_2 is equal to the suction, so very high values can be reached. For the above mentioned reasons it was necessary to devclop a ncw compaction formula.
A formula which satisfies the condition that Δv(=relative compaction) $\rightarrow \varepsilon$ (ε=porosity before loading) if $p_2 \rightarrow \infty$ and which gives results independant of the magnitude of the load steps is:

$$\Delta v = \varepsilon.\{1/(\ln(p_2/p_1) + (1-\varepsilon)/K_2\ .\varepsilon)\}\ .\ \ln(p_2/p_1) \qquad (2)$$

where K_2 is a constant. By calibration it was found that this constant has a value of o.192.
The increase in bulk density during ripening can now be calculated using:

$\Delta v = \Delta\rho/\rho$ (3) and thus $\rho_2 = \rho_1/(1-\Delta v)$ (4)

where ρ_2 and ρ_1 are the bulk densities after and before loading.

An increasing bulk density can result in subsidence (shrinkage in the vertical direction) and in crack formation (shrinkage in the horizontal direction). If the compaction occurs only in the vertical direction, the soil will only subside and will not crak. The soil can also shrink equally in all three directions. Rather than shrinking entirely in one direction or uniformly in all three directions, the soil may show a distribution of cracking and subsidence which lies somewhere between these two extremes. If the distribution factor for crack formation and subsidence is designated as r_s, so that $r_s=3$ if the soil shrinks equally in all directions and $r_s=1$ if solely subsidence occurs, then it may be shown that:

$$d_2 = d_1 \cdot (1-\Delta v)^{\frac{1}{r_s}} \qquad (5)$$

and

$$\mu_2 = 1 - \{(1-\mu 1)(1-\Delta v)\}/(1-\Delta v)^{\frac{1}{r_s}} \qquad (6)$$

where d is the thickness of the layer and μ is volume fraction of the "big cracks".

On research spots on ripening soils in the IJsselmeerpolders RIJNIERSCE (1976) found that in layers, characterised as soft, the increase in bulk desity in a dry summer occured entirely in the vertical direction, whereas in layers regarded as firm the shrinkage took place in three directions. This observation may be explained as follows: a soil that is still soft is not coherent enough to crack and continue to bear the load by the overlaying layers. If any cracks would be developed, the soil immediately flows in and closes them. Crack formation can only occur if the soil is firm enough to bear the load without flowing. The limit at which cracking can occur was determined by plotting the load by the overlying layer against the waterfactor (n) of the soil. The waterfactor is the number of grams of water bound to 1 gram of clay (<2 μm). This relationship is shown in figure 1, from which it may be seen that as the load increases, the n-factor must became smaller in order to permit cracking, as one would expect from the above explanation.

The simulation model assumes a relationship represented by a line parallel to the first mentioned line above which the factor r_s has a value of 3. This line is also given in the figure. Using these relationships a good agreement was achieved between the results of the simulation and the values measured in the field.

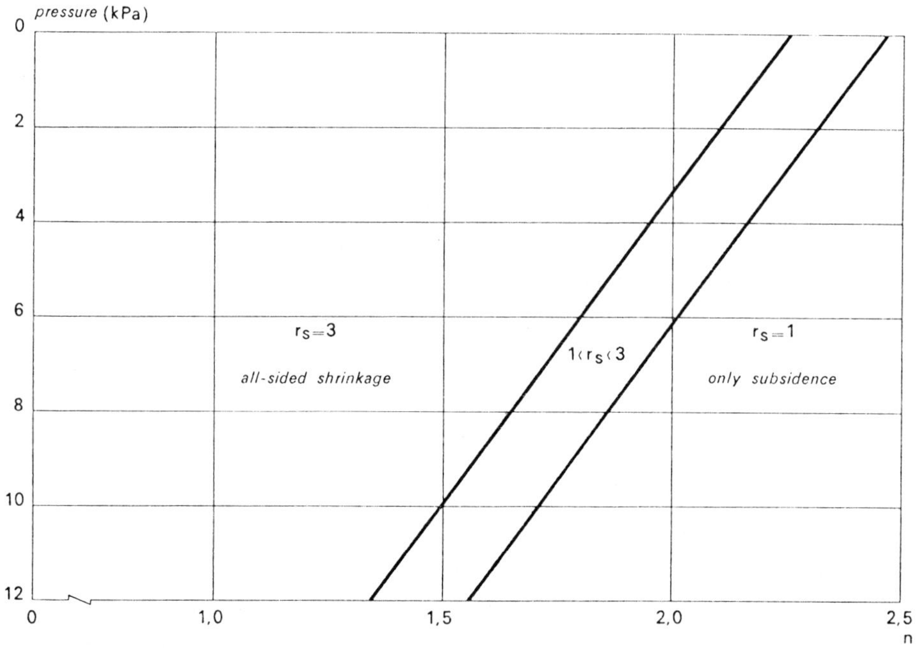

LITERATURE

Glopper, R.J. de, 1977. The application of consolidation constants derived from the pore space, in subsidence calculations. In: Proc. Land Subsidence Symposium, Washington.

Rijniersce, K., 1976. Het verband tussen het A-cyfer en het volumegewicht in jonge gronden na de droge zomer van 1976. R.IJ.P.-rapport 1976-34 Bbw, Lelystad.

Rijniersce, K., 1983. A simulation model for physical soil ripening in the IJsselmeerpolders. Flevobericht no. 203, Lelystad.

A technique for the description of the crack pattern and for predicting the hydraulic efficiency of heavy soils

F. Doležal[1], Š. Hrín[2], R. Mati[2], J. Harmoci[3], M. Kutílek[4]

Abstract

Introduction

Dessication cracks are only a part of the complex macrostructure of heavy soils. Their hydrological role has been recognized and continues to be a subject for attempts at quantitative simulations. The authors which described the cracks in soils quantitatively tried to do it by the direct measurement of the length of the cracks in a section. The assumption about the vertical direction of cracks was often accepted. No attempt has been made at the quantification of the anisotropy of the pattern. It is the purpose of this contribution to present a technique for the description of cracks based upon the Saltykov's method of directed secants, together with a simple geometric and hydraulic model, allowing the approximate calculation of hydraulic conductivity, as well as of other transport properties, under arbitrary boundary conditions.

---

[1]Design and Construction Institute of the Czechoslovak Ceramic Industries, Prague, Czechoslovakia
[2]Complex Agricultural Research Station, Michalovce, Czechoslovakia
[3]Research Institute for Soil Improvement, Experimental Station, Sobrance, Czechoslovakia
[4]Czech Technical University, Faculty of Civil Engineering, Department of Irrigation and Drainage, Prague, Czechoslovakia

Theory

Horizontal and vertical sections through the soil make it possible to count the intersections of cracks with lines (secants) in the direction of coordinate axes x, y, z. The number of intersections per unit length of the secants is m_x^p, m_y^p, m_z^p, for the p-th class of the widths of cracks. When ordered, these quantities can be designated as m_{max}^p, m_{med}^p, m_{min}^p. A model of three mutually perpendicular systems of parallel cracks (the normal vectors of which have the directions of axes x, y, z) is now used to substitute the real crack pattern. The one-sided specific surfaces of these crack systems are:

$$S_{min}^p = 2/3 \cdot m_{min}^p \qquad (1)$$

$$S_{med}^p = \pi/4 \cdot m_{med}^p + (2/3 - \pi/4) \cdot m_{min}^p \qquad (2)$$

$$S_{max}^p = m_{max}^p + (\pi/4 - 1) \cdot m_{med}^p + (2/3 - \pi/4) \cdot m_{min}^p \qquad (3)$$

where max, med, min is the same permutation of symbols x, y, z for both S and m. The p-th class of cracks contributes to the total hydraulic resistances by the amounts:

$$r_i^p = (w^p)^{-3} \cdot S_i^p \,, \; i = x, y, z \qquad (4)$$

where w^p is the representative width of the p-th class. Cracks of different widths are assumed to be ordered in series along the flow direction, the flow regime is laminar (for other transport processes, it suffice to replace the third power in (4) by other function of w^p). The main components of the hydraulic efficiency (or hydraulic conductivity) tensor are:

$$E = \begin{bmatrix} E_y + E_z, & 0 & , & 0 \\ 0 & , E_z + E_x, & 0 \\ 0 & , & 0 & , E_x + E_y \end{bmatrix} \qquad (5)$$

where:

$$E_i = \alpha \cdot (\sum_{p=m}^{n} S_i^p)^2 \,/\, \sum_{p=m}^{n} r_i^p \,, \; i = x, y, z \qquad (6)$$

and

$$\alpha \approx 628\,800 \text{ m}^{-1} \cdot \text{s}^{-1} \text{ (at } 10^\circ \text{ C)}.$$

Measurements

Sections with areas about 1 m^2 were made in two field agricultural soils in the Eastern Slovakia Lowland. The real pattern of cracks in the sections was discovered thoroughly with a spatula, and drawn by hand on a millimeter paper. True perpendicular widths of the cracks were measured, too. This technique is the simplest possible. Its obvious drawbacks are awkwardness and some amount of subjectivity. The narrowest cracks registered were about 0,3 mm in width. The images of cracks were processed as described above. Vertical component $E_x + E_y$ of the hydraulic efficiency tensor was compared with the initial infiltration rate from double-ring infiltrometers in different depths.

Results

1) Gleyic Vertisol, Plešany, about 90 % below 10 microns, montmorillonite prevails. Measured in August - September 1973 (dry period, cracks well developed, up to 8 mm width in the depth discussed). $E_x + E_y = 8{,}56 \cdot 10^{-3}\ m \cdot s^{-1}$, in the depth 50 - 70 cm. Initial infiltration rate (from 3 replicates): $5{,}82 \cdot 10^{-3}\ m \cdot s^{-1}$.

2) Gleyic Fluvisol, Milhostov, about 60 % below 10 microns, montmorillonite prevails. Measured in August - September 1979 (the soil was relatively moist, surface layer 0 - 5 cm remoistened by recent precipitations; crack narrower, max. width 3 mm in the whole profile). The results are given in the Fig. 1. Infiltration measurements were without replication. Hydraulic efficiency is given for different m in (6), i. e. for different minimum widths of cracks taken into account. This minimum width must be optimized by experiments. It probably lies in the vicinity of 1 mm. Since the space for water accumulation and lateral flow is smaller, the larger the depth, it is not surprising that the infiltration rates decrease with increasing depth. This is, however, rather the pronounce of boundary conditions than of the hydraulic conductivity in a given point.

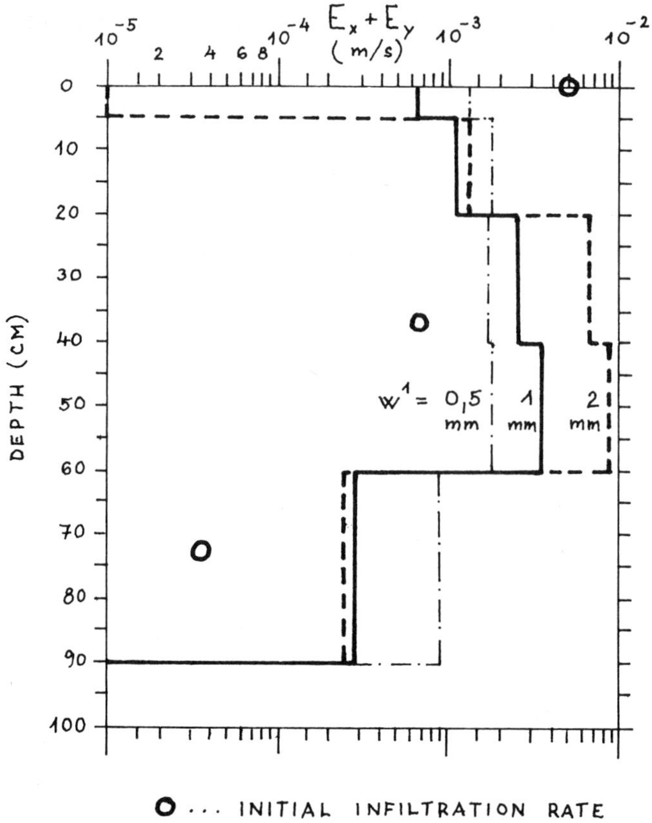

FIG. 1.

Conclusions

Preliminary experimental studies show that the method described above can give reasonable results, if it is supplied with reliable data and if the results are slightly calibrated for individual pedons. It is suitable for the assesment of the hydrological behaviour of large cracks, visible with the naked eye, where the problem of the connectivity of macropores usually does not occur. The model is open to refinements, to applications for other transport processes and for the prediction of hydraulic conductivity changes with changing soil moisture content.

Variations in hydraulic conductivity under different wetting regimes

P.B. Leeds-Harrison
Silsoe College, Silsoe, Bedford MK45 4DT
C.J.P. Shipway
Silsoe College, Silsoe, Bedford MK45 4DT

The importance of soil fissures to the effective drainage of clay soils has been demonstrated by Leeds-Harrison et al (1982). It is recognised however, that soil fissures change their dimensions with the swelling and shrinkage of the soil peds. This results in a change in the hydraulic characteristics of the soil. This paper makes a quantitative assessment of this change and relates it to the water movement in a swelling soil.

Soil monoliths measuring 330 mm diameter and 270 mm in height were taken from a clay loam soil having a 30% clay content and a CEC of 0.47 mequ/g. These were placed on a sand table and equilibrated with a water table at 800 mm below the soil surface. Some evaporation was then allowed from the soil surface and typical cracking patterns as shown in Fig. 1 developed. Using the method described by Youngs (1982) to measure the hydraulic conductivity profile, the variation in hydraulic conductivity with depth has been determined following two wetting treatments. In the first treatment the soil was rapidly wetted by bringing the water table to the surface within 1 hour. This was immediately followed by the determination of hydraulic conductivity over a period of six hours.

In the second treatment the water table was held at the soil surface for a period of 21 days before the hydraulic conductivity was measured.

The results of these measurements are shown in Fig. 2. Significant differences in the hydraulic conductivity are found between the two treatments. This is attributed to the visible closure of the shrinkage cracks with the long wetting of the second treatment. On treatment 1

closure of cracks due to shrinkage was only partial. The residual effect of a plough layer to 200 mm depth is seen in treatment 1, with a rapid fall in permeability occuring below the plough layer.
Measurement of the specific yield of the monoliths for the two treatments over the height of the sample also show significant differences. The results are shown in Fig. 3. Again a large increase in specific yield is noticed at 200 mm in treatment 1. Fig. 4 shows the moisture release characteristic for a small sample of the soil and this is taken to represent the inter-pedal retention characteristic. From this data it is noted that the yield of water observed from the monolith over an equivalent tension range is related to the macroporosity only.
Fig. 5 shows that there is a positive correlation between the hydraulic conductivity and specific yield. The relationship for the two treatments taken separately is not strong.
Solutions to the Stokes-Navier equation for laminar flow are quoted by Childs (1969) for cylindrical tubes and for planar voids. Using these equations and taking as the range of equivalent diameters of a cylindrical tube or the width of the macropore, values of 1 mm and 0.1 mm, the macroporosity, required to give the measured hydraulic conductivity, has been calculated. Calculated values of macroporosity are generally two orders of magnitude smaller than the measured values and this finding supports the observations of Bouma and Dekker (1978) who noted that the number of conducting macro-pores appears to be small compared to the total number of macropores.
This has implications for the reclamation of saline soils as the predominant leaching process will only occur in the macropores in which there is moving water. In these situations rapid fluctuations of the water table would be preferable to longer ponding of water as the peds would have a short opportunity time for swelling, the hydraulic conductivity would not fall and a greater movement of water within the profile would be possible with consequent increase in leaching.
In mole-drainage studies the hydraulic conductivity and specific yield can be used with the appropriate boundary conditions to predict the drain flow hydrograph. This has been done with the data presented here and the characteristically peaked hydrograph is predicted as observed by Childs (1943) and Leeds-Harrison et al (1982) in field studies. As expected higher flow rates are predicted with the rapidly wetted soil

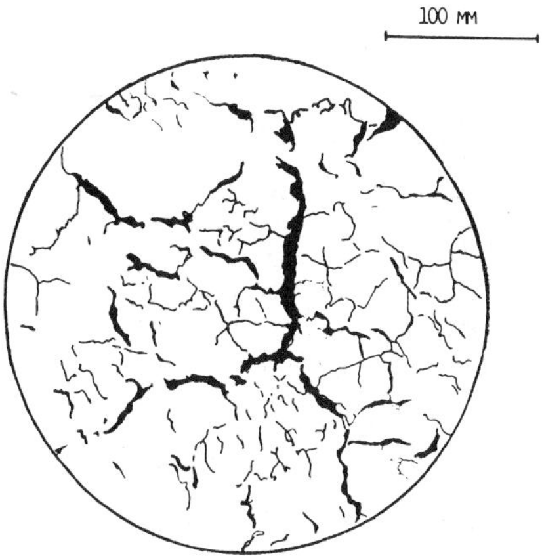

FIG. 1 CRACKING PATTERN ON SOIL MONOLITH

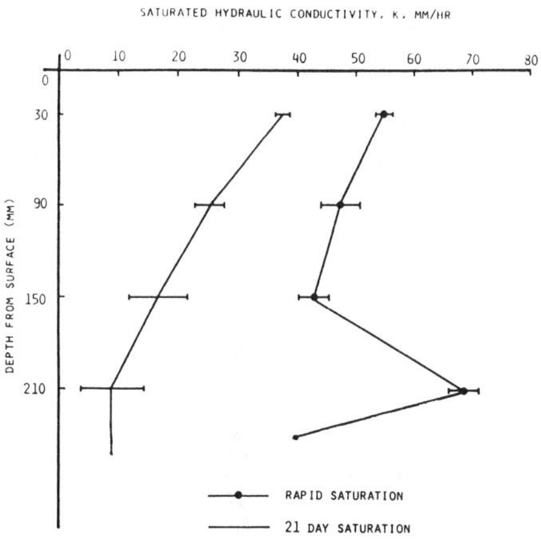

FIG. 2 VARIATION OF HYDRAULIC CONDUCTIVITY WITH DEPTH

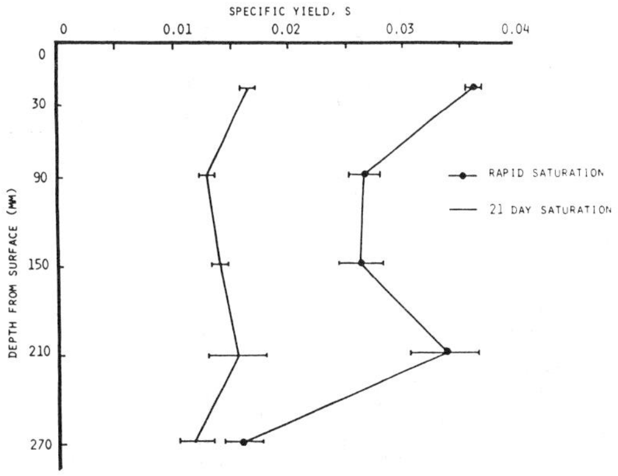

FIG. 3 VARIATION OF SPECIFIC YIELD WITH DEPTH

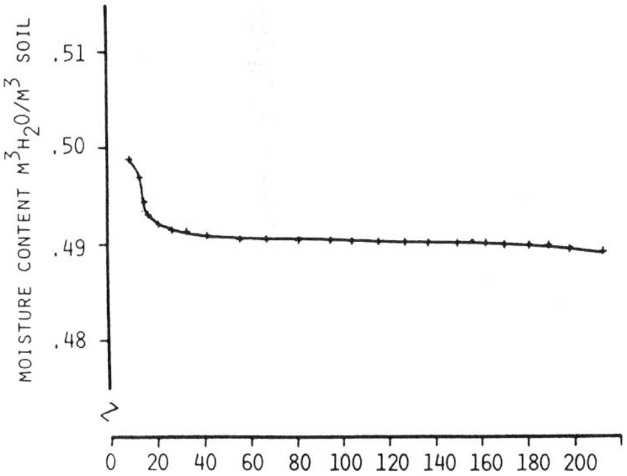

FIG. 4 SOIL MOISTURE CHARACTERISTIC FOR A SMALL SAMPLE

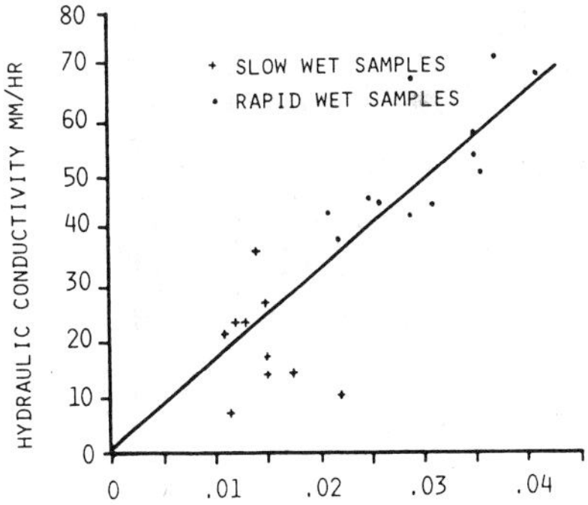

FIG. 5 CORRELATION BETWEEN HYDRAULIC CONDUCTIVITY AND SPECIFIC YIELD

data than with the slowly wetted soil data.

References

Bouma, J. and L.W. Dekker 1978. A Case Study on Infiltration into Dry Clay Soil. I. Morphological Observations. Geoderma 20, 27-40.

Childs, E.C. 1943. Studies in Mole Draining, Interim Report on an Experimental Drainage Field. J. Agric. Sci. 33, 136-146.

Childs, E.C. 1969. An Introduction to the Physical Basis of Soil Water Phenomena. Pub. J. Wiley and Sons, London.

Leeds-Harrison, P.B., G. Spoor, and R.J. Godwin. 1982. Water Flow to Mole Drains. J. Agric. Engng. Res. 27, 81-91.

Youngs, E.G. 1982. The Measurement of the Variation with depth of the Hydraulic Conductivity of Saturated Monoliths. J. Soil Sci. 33, 3-12.

Structural changes in two clay soils under contrasting systems of management

Mackie L. A.*, Mullins C. E. &
FitzPatrick, E.A.
Soil Science Department, Aberdeen University,
Meston Walk, Aberdeen AB9 2UE
* Now at The Macaulay Institute for Soil Research
Craigiebuckler, Aberdeen

The structure of clay soils can have great importance for infiltration, drainage and root growth. Because structure changes throughout the growing season (Andersson and Håkansson, 1966), can vary from year to year and is influenced by cultivations and the type of crop, it is difficult to study and characterise.

In this study (Mackie, 1983) an attempt was made to obtain a composite picture by combining many different methods. Two contrasting locations, Cruden Bay (Aberdeenshire) in North East Scotland and Compton Beauchamp (Oxfordshire) in the South of England were chosen.

Three sites were chosen at Cruden Bay, at different stages of a 7 year rotation, and sampled throughout two growing seasons. Although all three sites had the same parent material, belonged to the same soil series (Tipperty) and had similar clay mineralogies (mica, kaolinite, interstratified mica/ vermiculite and vermiculite), there were considerable differences in their textural characteristics. In the topsoil, both sites A and B were clay loams, whereas site C was a clay. In the subsoil sites B and C were clays, whereas site A was a sandy clay loam.

The structural effects of ploughing and direct drilling were compared with permanent grazing on the Denchworth series at Compton Beauchamp during one growing season. The clay mineralogies of the direct drilled and ploughed sites were very similar with the main minerals present as smectite, mica, kaolinite and interstratified smectite/chlorite with a trace of chlorite. Both sites had greater than 45% clay at all depths, the only difference between sites being a slight increase in clay content with depth in the direct drilled site.

Plaster of Paris ($CaSO_4 \frac{1}{2} H_2O$) was used to preserve the seedbed for sampling and to identify field cracks and earthworm channels. Blocks impregnated by the method of FitzPatrick and Gudmundsson (1978) using a fluorescent dye were used to study structure. Figure 1 shows a block sampled at the soil surface with the aid of plaster of Paris and subsequently impregnated with resin.

Figure 1. Soil block impregnated with plaster of Paris and resin

Macroporosity (pores $>350\ \mu m$) was measured by an Optomax image analyser. Water content was measured in the field with a neutron probe and infiltration measurements were made at Cruden Bay.

At Cruden Bay the grass site showed a different pattern of drying out than the spring barley sites. After a period of low rainfall in April/May 1980 the grass site dried out to 50 cm depth whereas the barley sites only dried out to 30 cm depth (Figure 2).

The early soil water deficit in the grass site could be explained by the minimal water use of the newly sown barley.

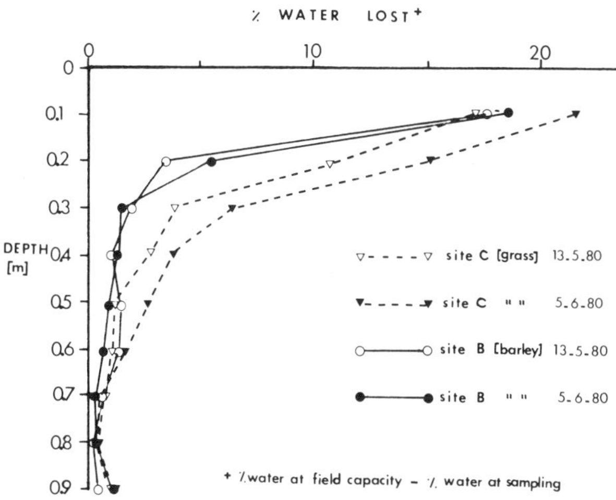

Figure 2. Soil moisture deficits at site B (barley) and site C (grass), Cruden Bay, in spring 1980

Infiltration with plaster of Paris outlines the hexagonal cracking pattern seen in the grass site at this time (Figure 3). The plaster of Paris also shows the continuity of the earthworm channels into the subsoil at this site.

Figure 3. Hexagonal cracking pattern in the grass site at Cruden Bay in the spring of 1980

Table 1 shows the change in crack porosity down this profile as measured on a series of plan photographs using an Optomax Image Analyser.

Table 1. Crack porosity Changes down the profile in site C (grass)

Depth (M)	Crack porosity, %	Average width of major cracks (mm)	% volumetric water content
0.01	8.62	11.3	
0.05	5.50	8.8	
0.10	5.12	8.0	30.3
0.15	9.75	6.4	
0.20	6.02	4.0	34.0
0.25	*	1.4	
0.30	*	-	36.9

* Below level of resolution

Photographs of profiles infiltrated with plaster of Paris in the field showed a number of contrasting structural features at Compton Beauchamp.

The direct drilled and grass sites had many continuous L. terrestris earthworm channels whereas the ploughed site did not. The ploughed site frequently had a plough sole at 20 cm created by the action of the plough forming a discontinuity between the topsoil and the subsoil. In July 1981 the soil had dried out to 90 cm depth and had only 20% volumetric water content at the surface. At this time a deep cracking pattern was seen in the direct drilled and ploughed sites whereas in the permanent grass site a finer structure had formed (Figure 4).

Results from the impregnated soil blocks showed that there were more large irregular macropores in the 0 to 10 cm zone of the ploughed site than in the direct drilled site, especially in October, just after cultivation. Boone et al. (1976) noted the same effect, where in the period just prior to the sowing of maize (March) the ploughed soil in the spring had many vughs (irregular pores) and interconnected pores, whereas the direct drilled samples collected at the same time of year had few vughs or interconnected vughs. Pagliai et al. (1983) also found that total porosity <30 μm was significantly higher in topsoils of conventionally tilled plots than in those of no-tilled plots, irrespective of sampling time.

Figure 4. Plaster of Paris impregnation in July 1981, Letcombe

Although there were many differences between the two locations there were also many similarities in structural change. In all ploughed sites the structure of the surface 0 to 10 cm zone was composite subangular blocky/granular just after seedbed preparation and settled continually until it was recultivated. In both locations, planar, irregular, and circular macropores developed on shrinkage and blocky and wedge structures were formed. On swelling, most of the macropores closed. This can be exemplified in the 0 to 10 cm zone under grass at Cruden Bay (Figure 5).

Changes in the macroporosity throughout the sampling period were often closely related to changes in volumetric water content (Figure 6), especially in the non-cultivated sites.

The effects of earthworm activity on the soil structure were also cyclic in nature corresponding to their pattern of annual activity. At Compton Beauchamp all sites had many vertical, sinuous and crescentic macropores and earthworm channels in July 1981, probably formed after April but

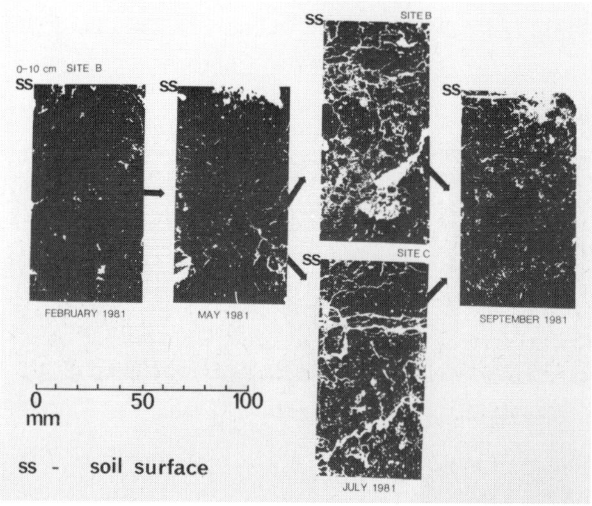

Figure 5. Ultra violet photographs of blocks at site B, Cruden Bay, showing summer shrinkage and winter swelling

before the severe drying period in July. The effect of earthworms on the soil structure was also evident on the sites at Cruden Bay, especially in one spring barley site in September 1980 where a surge in earthworm activity corresponded to a recent surface spreading of manure.

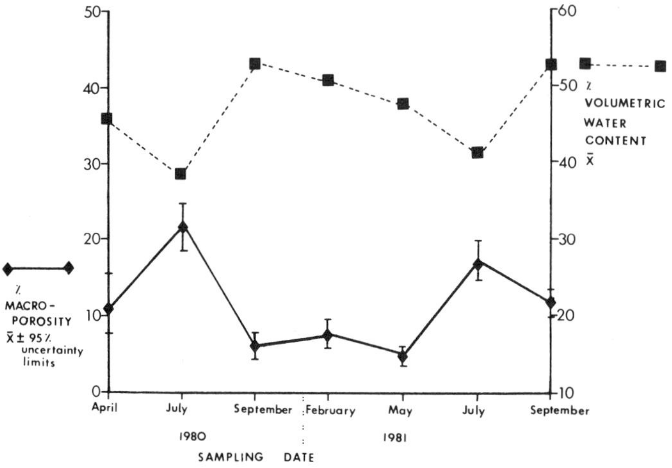

Figure 6. Macroporosity and volumetric water content changes in the 10 to 20 cm zone in the grass site C, Cruden Bay

These results illustrate the dynamic nature of clay soils. A sequence of processes both artificial and anthropogenic have resulted in a corresponding sequence of structural states. Depending on the intensity of the process operating, various structural states between fully developed and a massive structure were formed.

Acknowledgements

The authors wish to thank the following:
The Department of Agriculture and Fisheries for Scotland for providing financial support. The Department of Microbiology, The Macaulay Institute for Soil Research, for permission to use the Optomax Image Analyser, with special thanks to Dr. J. Darbyshire. The AFRC Letcombe Laboratory for use of their experimental field site, with special thanks to Dr. M. Goss and Mr. J. Douglas.

References

Andersson, S. and I. Håkansson. 1966. Markfysikaliska undersökningar i odlad jord. Grundförbättring., 3: 191-228.

Boone, F.R., S. Slager, R. Miedema, and R. Eleveld. 1976. Some influences of zero-tillage on the structure and stability of a fine textured river levee soil. Neth. J. of Agric. Sci., 24: 105-119.

FitzPatrick, E.A. and T. Gudmundsson. 1978. The impregnation of wet peat for the production of thin sections. J. of Soil Sci., 29: 585-587.

Mackie, L.A. 1983. Changes in the soil structure of two clay soils under contrasting systems of management. Ph.D. thesis. University of Aberdeen, 598 pp.

Pagliai, M., M. La Marca, and G. Lucamante. 1983. Micromorphometric and micromorphological investigations of a clay soil in viticulture under zero and conventional tillage. J. of Soil Sci., 34: 391-403.

Effect of Al-hydroxyde on the stability and swelling of soil (clay) aggregates

A.Muranyi

Research Institute for Soil Science and Agricultural Chemistry of the Hungarian Academy of Sciences, Budapest.

M.G.M. Bruggenwert

Department of Soil Science and Plant Nutrition, Agricultural University of Wageningen, the Netherlands.

Introduction

Hydraulic conductivity of soils is a function of structure. Structural stability as well as pore size distribution is, in turn related to chemical characteristics. In this context soil solution Electrical Conductivity (EC), composition of the adsorption complex (ESP) and the presence of cementing agents are important parameters. The role of the latter has received less attention than EC and ESP. This research was therefore directed towards contributing to the characterization of Al-hydroxyde influence on structure stability and swelling properties of soil (clay) aggregates.

Materials and methods

For this study aggregates from Winsum soil were used. This is a heavy clay soil from Northern Netherlands and contains 45% illite, 5% smectite, < 1% O.M., <1% $CaCO_3$. In the first experiment 2-4 mm aggregates were saturated with Al^{3+} at pH3 followed by equilibration at pH5 using HAc/HCl buffer. Treated and untreated aggregates were placed in columns and percolated with solutions containing varying EC- and SAR-values. Intrinsic permeability, K_i, was measured and related to Al-enrichment, EC and ESP. In each column this percolation starts at high salt concentration. When K_i becomes approximately constant, the concentration of the percolate was decreased and its composition adjusted to maintain the ESP. A new equilibrium was established and the process was repeated for each salt level used.

To investigate the effect of the degree of Al-interlayering on aggregate stability, a second experiment was initiated. In this case 1-2 mm aggregates were satured with Al^{3+} followed by partial neutralization with pH7 NaAc/NaOH buffer. In parallel systems, this treatment was repeated for a total of three and ten cycles. After these treatments total Al retained and Na and Al on the adsorption complex was determined. The influence of these Al-treatments on structure stability was measured as in the first experiment.

Results and discussion

As expected K_i-values decrease with decreasing EC and increasing ESP (Figure 1). For comparitive purposes, $K_{i,rel}$ is used and expressed as a percentage of K_i measured at the same ESP at an EC of 100 mS/cm. As shown (Figure 1) the Al-treatment limits the decrease of $K_{i,rel}$, with respect to the untreated soil. Those EC-ESP combinations at which liquid flow is effectively blocked ($K_{i,rel} < 1\%$) are of special interest. At this point the content of the soil columns become a homogeneous dispersed system. For the untreated soil this situation is achieved at ESP values of 40, 60 and 100% at EC values of 0.5, 1 and 3 mS/cm respectively. In the Al-treated soil, however, total structural dispersion only occurs at the lowest EC used in this first experiment (0.5 mS/cm), and than at approximately ESP = 100% (Figure 1).

In presenting the results of the second study, the Al-treatments are expressed in terms of a mutual coagulation factor (MCF) which is defined as percent blockage of the CEC. The three Al-treatments resulted in MCF's of approximately 10, 40 and 65%.

It is evident that the stability of the aggregates is improved with increasing MCF (Figures 2 and 3). Here $K_{i,rel}$ is expressed as a percentage of K_i meacured at the same ESP and at an EC of 10 mS/cm. As is shown in figure 2 the rate of decrease in $K_{i,rel}$ with increasing ESP decreases with increasing MCF. This phenomenon can be explained by the decrease of the (specific) surface area that is subject to swelling forces upon increasing MCF. The same behaviour occurs at higher salt

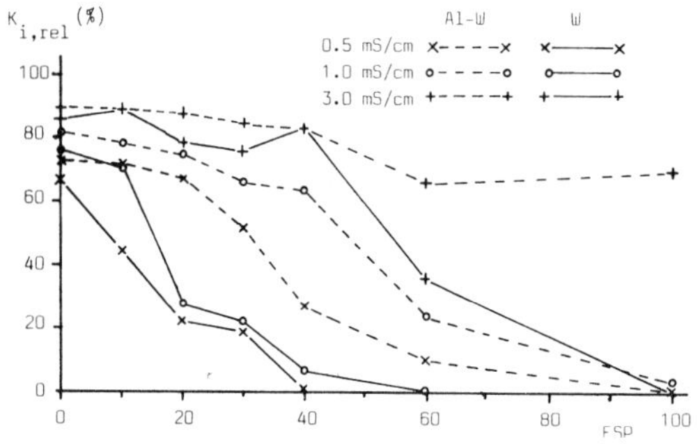

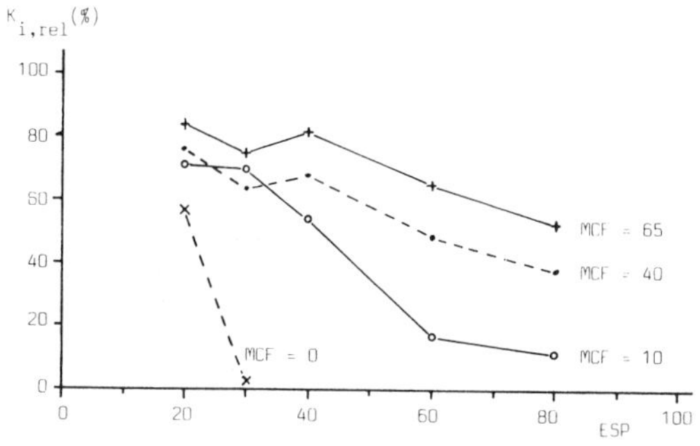

Figure 1. $K_{i,rel}$ as function of ESP at three salt levels (EC = 0.5, 1 and 3 mS/cm) for Winsum soil (W) and Al-treated Winsum soil Al-W)

Figure 2. $K_{i,rel}$ of Winsum soil as function of ESP and Al-treatment at constant salt level (EC = 0.5 mS/cm)

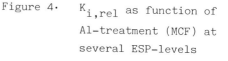

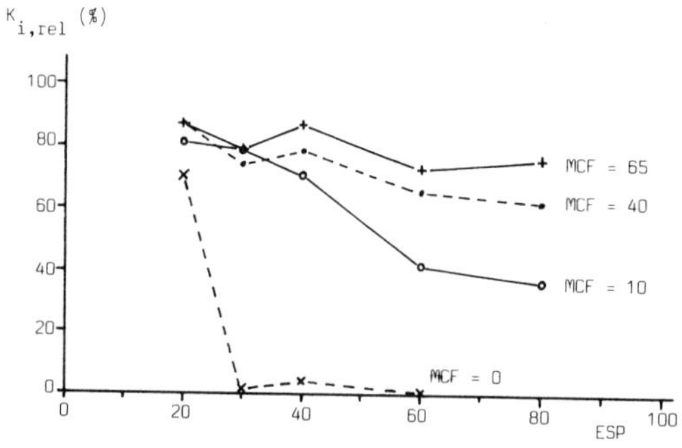

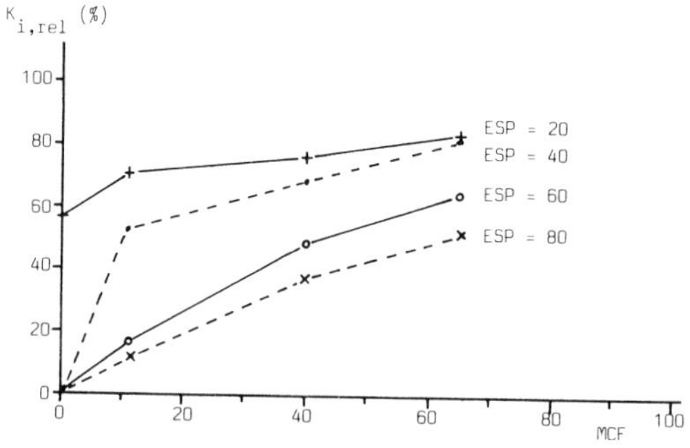

Figure 3. $K_{i,rel}$ of Winsum soil as function of ESP and Al-treatment at constant salt level (EC = 1mS/cm)

Figure 4. $K_{i,rel}$ as function of Al-treatment (MCF) at several ESP-levels

concentrations but with less intensity as a consequence of the decrease of the repulsion forces; compare figure 3 with figure 2. In other words the relative impact of the MCF on $K_{i,rel}$ is greatest the first 10% blockage of the CEC (Figure 4) and particular at low ESP-values (<40%) which is representative for agricultural soils.

For practical use it is important to focus the attention on the influence of EC-ESP combinations on the $K_{i,rel}$ as is done in figure 5. The original Winsum soil (Figure 5a) shows a sharp drop of permeability between ESP values of 20 and 40 when the EC is within the range normally occurring in arable soils. This behaviour of an unstable soil is already remarkably changed when the Al-hydroxide level is increased at a MCF of 10 (Figure 5b). Not only in the range of ESP at which permeability is blocked widened, but the point of complete blockage is pushed to the extreme case of very high ESP (>60) and low EC (< 0.1 mS/cm), a combination which is uncommon in soil. When MCF is increased to higher values it becomes even more difficult to block permeability almost completely (Figures 5c and 5d).

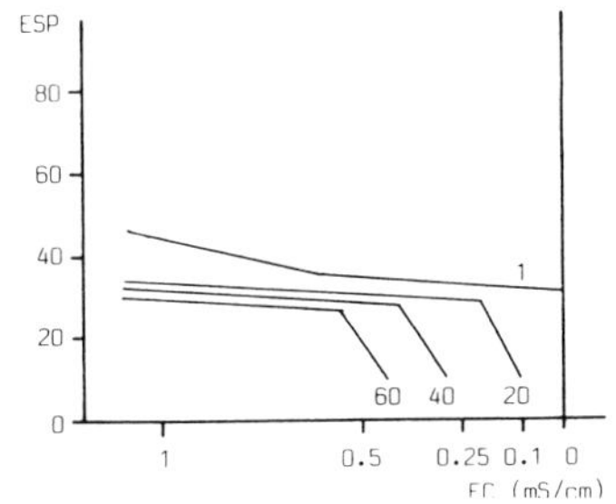

Figure 5a. Iso-permeability lines (iso-perms) as function of ESP and EC for untreated Winsum soil

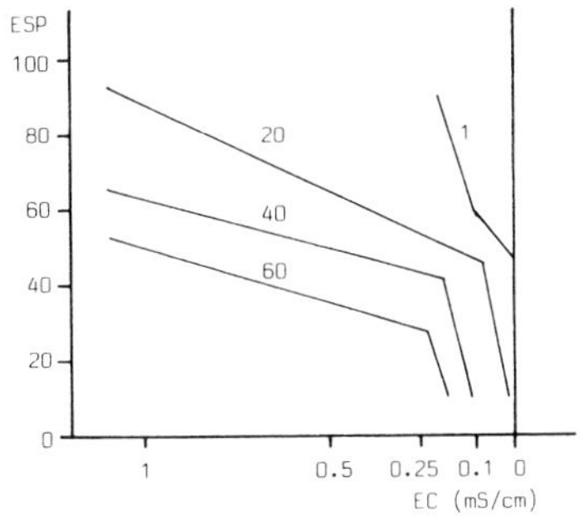

Figure 5b. Iso-perms as function of ESP and EC for Al-treated Winsum soil (MCF=10)

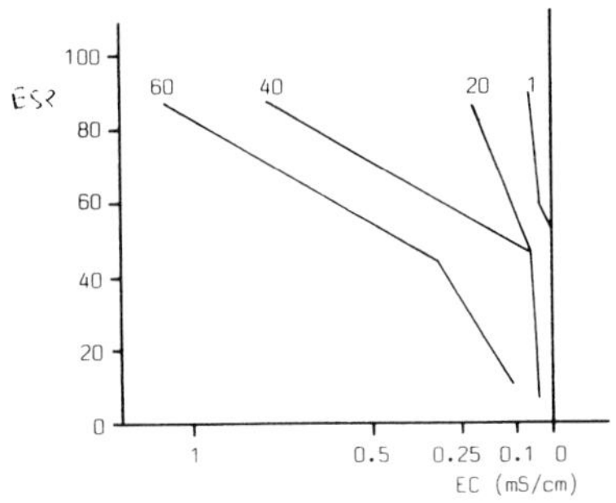

Figure 5c. Iso-perms as function of ESP and EC for Al-treated Winsum soil (MCF=40)

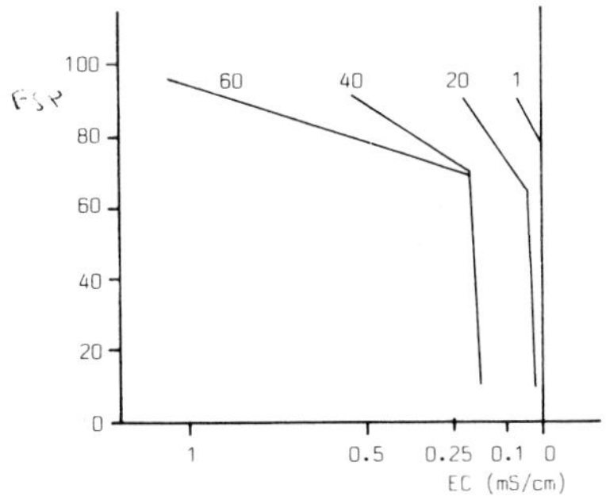

Figure 5d. Iso-perms as function of ESP and EC for Al-treated Winsum soil (MCF=65)

Change of structure and fabrics of clay alluvial soils under agriculture

S.A. Avetjan, B.G. Rozanov,
N.G. Zborishuk
Moscow State University,
U.S.S.R.

Agricultural utilization under irrigation of heavy clay montmorillonitic alluvial soils often leads to a rather sharp decline of their productivity due to loss of the original structure and development of vertic properties. This phenomenon was studied in an area where meadow alluvial soils were brought to cultivation ten years ago. These soils contain 45-60 percent of predominantly montmorillonitic clay, 3-4 percent of humus; they have CEC of about 35-40 meqv per 100g, dominated by calcium with sodium content not exceeding 5% of CEC. Their pH is 7.0-7.5.

Meadow alluvial soils in their virgin state (Profiles 20 and 30) are well structured due to the good development of grass root systems. Shallow ground water and shadowing by plants protect the soil surface from drying out. After ten years of cultivation these soils (Profiles 2, 14, 21) became massive and structureless. When watered up to field capacity and higher, they swell and become sticky, plastic and smeary; after drying, their surface becomes broken by a network of deep (50-70 cm) and wide (3-5 cm) cracks into large polygons with a diameter of about 30, 50 or 70 cm. These polygons are separated by smaller cracks into very hard and compact non-porous aggregates up to 5 cm in diameter. Under strong and rapid dessication the soil surface develops a thin-laminated polyedric crust, while under periodical alternation of drying and moistening it becomes selfmulched. The structural composition of virgin and cultivated soils differs sharply (Table 1).
Destruction of the original structure of these soils as a result of cultivation may be attributed to:

a) decrease of the humus content by 10-20 percent;

b) leaching of the water soluble salts (the virgin soils are slightly saline from the surface);

c) repeated mechanical soil treatment when soils are too wet or too dry;

d) modification of soil aggregates under alternation of swelling and shrinkage in the moistening-drying cycles: re-orientation and re-packing of highly dispersed clay particles.

Table 1. Aggregate composition of the surface horizons of virgin and cultivated soils: results of dry (numerator) and wet (denominator) sieving

Profile	Size of fractions, mm								
Nos	> 10	10-7	7-5	5-3	3-2	2-1	1-0.5	0.5-0.25	<0.25
Virgin soils									
20	10.5	13.3	14.3	24.0	15.7	14.8	4.1	2.5	0.8
	-	-	2.6	8.7	16.7	21.6	24.4	16.2	9.8
30	36.8	16.8	16.3	15.1	6.7	5.9	1.1	0.8	0.4
	-	27.4	8.6	20.8	10.0	17.5	7.2	4.0	4.8
Cultivated soils									
2	71.8	7.1	4.7	2.8	4.7	2.2	4.7	2.8	0.9
	-	29.8	5.4	9.8	11.8	12.6	6.6	3.6	20.4
14	64.6	10.6	12.0	3.2	5.2	2.0	1.2	0.8	0.6
	-	21.0	4.6	9.6	6.2	15.8	8.4	12.0	22.4
21	82.6	6.2	4.3	3.5	1.2	0.9	0.8	0.3	0.2
	-	26.8	8.4	10.0	12.6	2.4	6.8	4.2	28.8

Studies of soil microfabrics and microchemistry with scanning electronic microscope (Figs. 1 and 2) and with electronic micro-analyser revealed that the compactness was of a physical rather than a chemical nature. Re-orientation and re-packing of highly dispersed material and the adhesion-cohesion phenomena are the main factors.

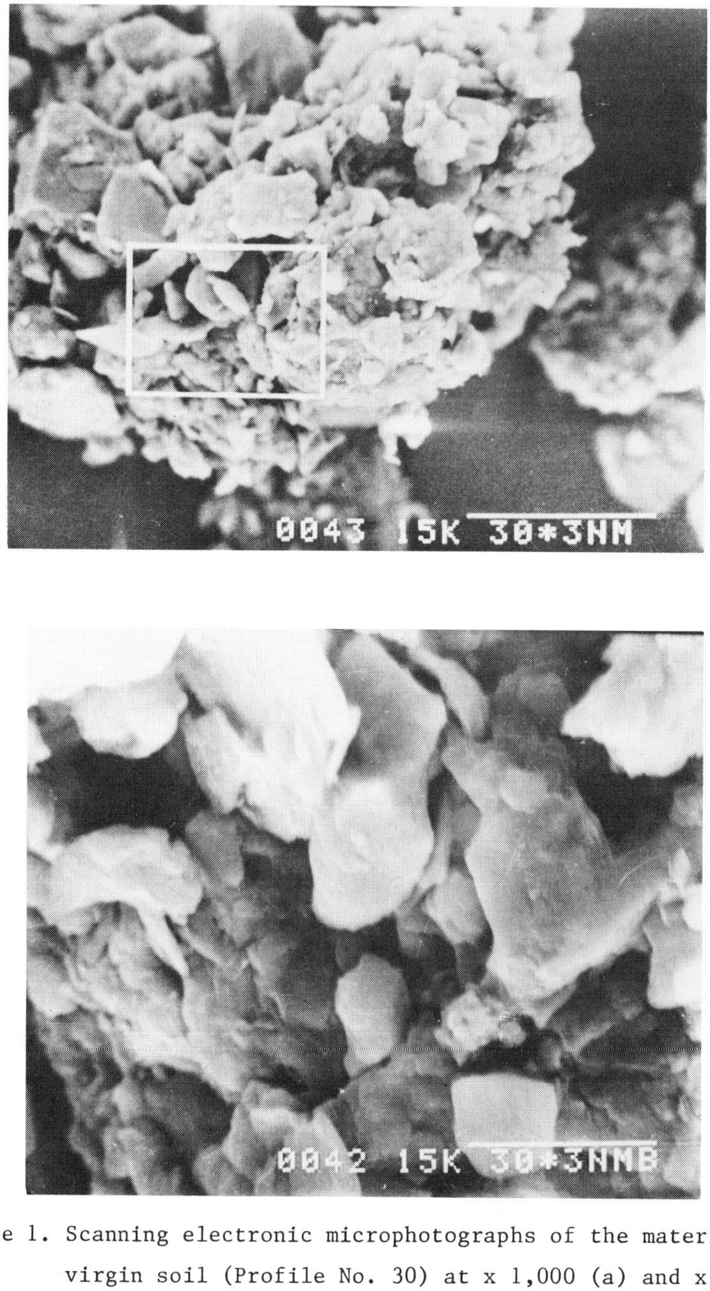

a

b

Figure 1. Scanning electronic microphotographs of the material from virgin soil (Profile No. 30) at x 1,000 (a) and x 10,000 (b)

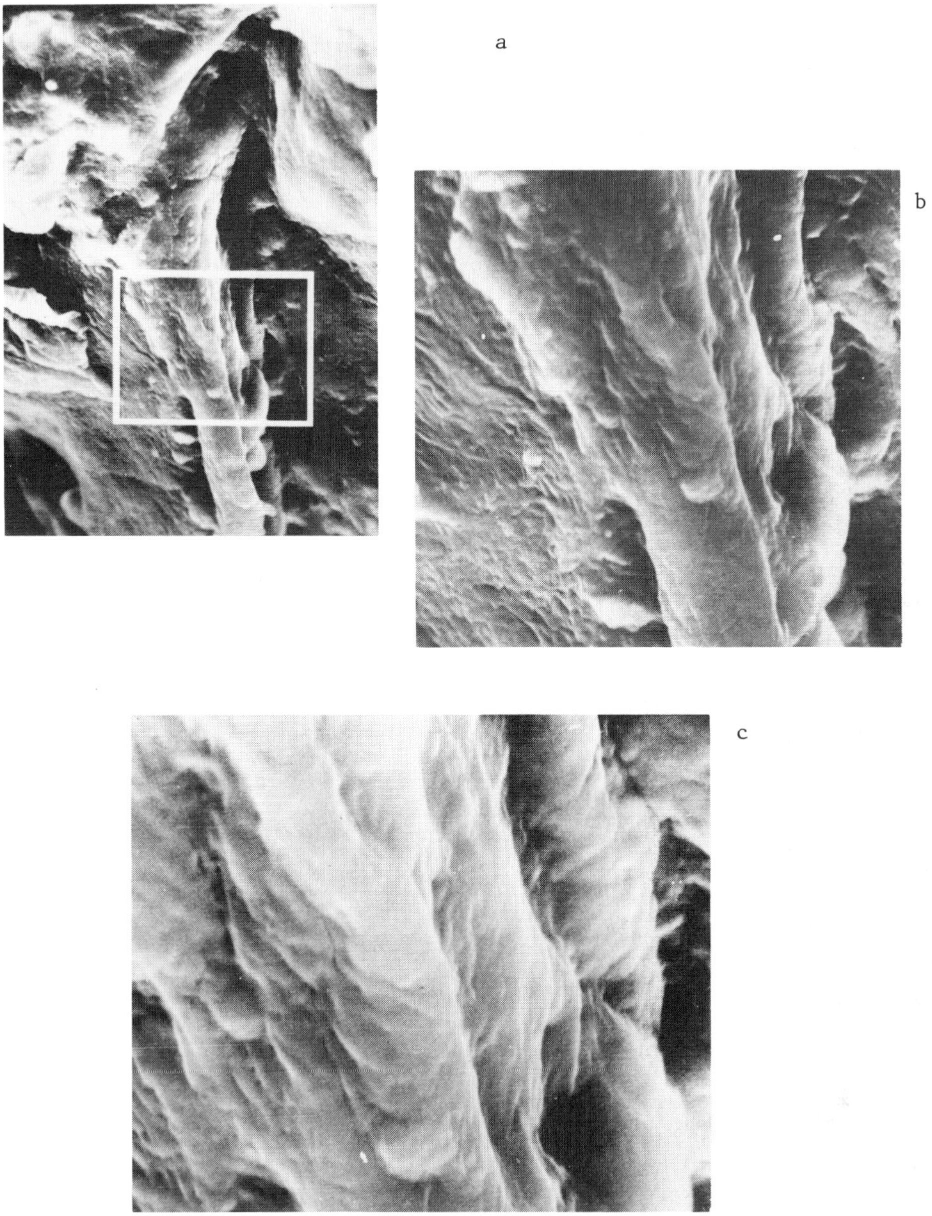

Figure 2. Scanning electronic microphotographs of the material from cultivated soil (Profile no. 14) at x 1,500 (a), x 4,000 (b) and x 7,500 (c)

Relation between the density of heavy clay soil and its moisture content

B.G. Rozanov - N.G. Zborishuk
G.S. Kust - J.L. Meshalkina
Moscow State University USSR

In a study of swelling, alluvial heavy clay soils, we have observed rather striking variability with time of soil density, which was closely related to the dynamics of soil moisture as a function of weather fluctuations. These alluvial soils represent a northern variety of dark hydromorphic vertisols which extensively crack and swell with changing moisture contents. In fact, they are post-alluvial soils, which have been converted, after embanking and drainage, into Pellic Vertisols containing to up 60 percent clay with a dominantly montmorillonitic composition. In this paper, field and laboratory experiments on soil density are compared.

In the first series of field experiments, soil density was determined in undisturbed samples under naturally varying moisture content. The standard method of using a cutting auger with a fixed internal volume was used for measuring soil density and moisture content in the same samples. Ten measurements were made each time for each of the soil horizons being studied. These experiments revealed a decrease of soil density (p) from 1.48 g/cm^3 to 0.92 g/cm^3 with moisture content (W) increasing from 23% (pF3.2) to 55% (w/w) (pF=0). Calculations of range correlation coefficients (R) by the Spirman method show (with a probability of 99.9%) that there is correlation between p and W values. Regression equations describing this correlation for different moisture intervals are presented in Table 1, indicating a wide scatter of the results of individual density measurements which exceed the limits of the normal distribution. This scatter is explained by the low accuracy of the method.

Table 1. Dependence of soil density (p) on soil moisture content (W) (values from 10 replications in each instance, by the field method).

Depth,	W, (%)		p, (g/cm^3)		p = a + b W	Rank Correlation
cm	min	max	max	min		Coefficient (R)
10-15	23.2	26.9	1.47	1.19	p = 2.53-0.05W	0.92
	26.9	40.0	1.19	1.02	p = 1.44-0.01W	0.92
	40.0	52.1	1.02	0.92	p = 2.22-0.03W	0.92
40-50	30.8	42.6	1.48	1.09	p = 3.21-0.05W	0.85
	42.6	53.1	1.09	0.94	p = 1.95-0.02W	0.85

In the second series of laboratory experiments soil density was determined at different moisture contents during artificial drying of the soil. Undisturbed soil aggregates (5 for each soil depth with a volume from 50 to 70 cm^3) were sampled and placed into glass vessels with $V=100\ cm^3$. The soil moisture content was near the maximum field water-holding capacity. The remaining volume of the vessels was filled up with dry quartz sand (fraction 0.25-0.05 mm) of known density ($p = 1.68\ g/cm^3$). The change of soil density under gradual drying was found from the change of the total volume of soil plus sand, which was maintained step by step at a constant level during the drying by adding measured volumes of sand. Determinations were continued up to the end of water loss after drying at 105°C. The accuracy of density determination is significantly higher in this method, as compared with the field method, because the same sample is used during the experiment. Soil density changed from 1.20 to 1.78 g/cm^3 within the interval of soil moisture from 35% to 0%. Results of the regression analysis are shown in Table 2.

Table 2. Dependence of soil density on its moisture content (average values from 5 replications in each instance, by the laboratory method).

Depth, cm	p = a + bW	Correlation Coefficient (2)
10-15	p = 1.88-0.019W	0.95
25-35	p = 1.78-0.015W	0.96
40-50	p = 1.76-0.015W	0.97
60-70	p = 1.75-0.014W	0.99

Substantial differences in numerical parameters of the two sets of regression equations obtained by the two methods are explained by their different accuracies. Particular practical problems occur when measuring soil density with a cutting auger in strongly cracked soil. Having this in mind, it was recommended to the engineers to use the following equation which was based on the result of the laboratory experiments: p = 1.82-0.016W. This equation yields the average soil density value for different soil moisture contents which can be measured experimentally with good accuracy allowing introduction into mathematical models, e.g. for drain spacing calculations. Thus obtained, density values are more accurate and reliable than values obtained with field methods, accurately reflect soil density changes with moisture fluctuations. The above equation is not universal; its parameters a and b should be determined for each soil within a specified range of moisture contents. They depend on soil texture and mineralogical composition.

A distribution function model of channelling flow in soils based on kinematic wave theory

Keith Beven

Institute of Hydrology, Wallingford, Oxon, UK

Peter Germann

Department of Environmental Sciences,

University of Virginia, Charlottesville, VA, USA

Abstract

This paper represents a further development of the kinematic wave approach to modelling flow through macropores or channels in the soil. The macroporosity is assumed to take the form of a number of essentially independent flow paths of unknown geometry. In each pathway a power law relationship between water content and flow velocity is assumed to describe adequately flow conditions in the channels. Variation between the pathways is ascribed to variation in the coefficient of the power law relationship. A probability distribution associated with this coefficient may be derived from experiments on soil columns. Analytical equations for each flow pathway are presented. The model is used to predict the results of experiments on undisturbed soil columns.

1 Introduction

Most attempts to model the occurrence of channelling flow or macropore flow in soils have either been essentially theoretical exercises with little reference to real data (eg Edwards et al, 1979; Beven and Germann, 1981) or else have relied on detailed measurements of flow pathways (eg Hoogmoed and Bouma, 1980). A review of the requirements of a model of channelling flow is given in Beven and Germann (1982). They note the limitations of the two domain (channels/matrix) concepts on which most of the current models are based. The present paper is an attempt to circumvent some of these limitations and yet deal with the complex nature of these dynamic and highly spatially variable flow processes in a relatively simple way.

To do this, we introduce the concept of the channels or macropores as multiple independent flow pathways through the soil. The pathways are assumed to be of complex and unknown geometry, but in each pathway a single valued relationship between water content and flow velocity is assumed to represent the dynamics of flow. Variation between pathways is assumed to be adequately represented by variation in a single parameter, which then takes on the nature of a stochastic variable. A statistical distribution function is ascribed to this parameter.

In this presentation, a simple conceptual model has been assumed to describe losses to the soil matrix from the channels. This allows analytical solutions to be derived to describe flow in the multiple pathways. It also allows a relatively simple procedure for obtaining the parameters of the distribution function model from breakthrough type experiments. The approach taken here is somewhat analogous to the transfer function model for solute transport of Jury (1982) except in that the theory is not restricted to steady state flows.

2 Theory

Consider a block of soil containing continuous but complex channels, initially dry. The application of sufficient water or rainfall to the surface of the block will lead to both infiltration from the surface into the matrix and flow in the channels. Beven and Germann (1981) after considering some simple theory for film flow in distributions of channels suggested that volume flux density, q, and water content w in the channels could be related by the power law function

$$q = b\ w^a \tag{1}$$

Equation (1) is assumed to hold for each flow pathway in the multiple pathway model. Combining (1) with a continuity equation leads to a kinematic wave equation for each pathway

$$\frac{\partial q}{\partial t} + c \frac{\partial q}{\partial z} = c\ s \qquad (2)$$

where t is time, z is distance from the surface, c = dq/dw and s is a term describing losses to the matrix. Germann (1983) has provided solutions to equations (1) and (2) for the case of a single wetting front with s = 0. In this paper a simple loss term function is used to allow analytical solutions to the flow equation to be obtained. One such loss function is

$$s = - rw \qquad (3)$$

Once infiltration into the channel starts, a wetting front develops (mathematically a kinematic shock) that moves with the velocity

$$\frac{dz_{wf}}{dt} = \frac{q}{w_z} = b\ w_z{}^{a-1} \qquad (4)$$

where w_z is the channel water content at the wetting front which, using (3) is given along the characteristics by

$$w_z = w_* e^{-rt} \qquad (5a)$$

$$= w \ \left[1 - r(a-1)z/c\ \right]^{1/(a-1)} \qquad (5b)$$

where w_* is the channel water content at the surface (= $q_*/b)^{(1/a)}$,

$c_* = ab\ w_*^{a-1}$ and q_* is the input rate to the channel. Note that (5b) applies at the shock front but (5a) does not since the shock front moves more slowly than the water content wave speed by a factor a. Substituting (5b) into (4) and integrating assuming that at the surface $w_z = w_*$ at t = 0 gives the depth of the wetting fron z_{wf} at any time as

$$z_{wf} = c_* \left[1 - e^{-(a-1)rt/a}\right]/\left[r(a-1)\right] \qquad (6)$$

Similar analytical equations can also be developed for some other loss functions and for the drying profiles following the cessation of inputs at the surface.

The parameters of this simple model are a, b and r. In the multiple flow path formulation, the parameters a and r are assumed to stay constant for all flow paths. However, it is assumed that the coefficient of the power law relationship, b, varies between flow paths so that each b value is associated with a probability weighting p(b). If we can derive the distribution function for b, we can then characterise the dynamics of the multiple independent flow paths under different flow conditions.

3 Derivation of the model parameters

The aim of this work is to develop a model that reflects the complexities and spatial variability of real channelling flow in soils, but that is sufficiently simple that the parameters can be derived from some simple experimental procedures. In the example calculations presented here, the parameter values are derived from a flow breakthrough or drainage experiment carried out on a column of undisturbed Typic Hapludult soil (10 x 30 x 95 cm). The procedures used were as follows.

i) An estimate of the input flux to the channels (assumed constant for the analytical solution) was made from the known input rate to the column surface and the estimated total surface matrix infiltration (in this case derived from gamma probe measurements of moisture changes in the column). Matrix infiltration from the surface did not appear to penetrate beyond 25 cm in this experiment

(ii) A value of a is assumed, initially 2 which corresponds to the theoretical exponent for saturated flow in vertical cylindrical channels (Germann and Beven, 1981).

(iii) A trial value of r is assumed, and the b values corresponding to the incremental flows up to specified times during the drainage experiments are calculated by inverting equation (6) to give

$$b_i = q_*^{(1-a)} \left[d\ r\ (\frac{a-1}{a})/(1 - \exp(- r(\frac{a-1}{a})t_i)\right]^a \qquad (7)$$

where d is the depth of the column, and t_i is the ith time point on the flow breakthrough curve. Note that it is necessary to assume that the input flux rate is the same for all flow paths.

(iv) The probability weighting $p(b_i)$ is calculated by equating the observed incremental drainage flux associated with the b_i values, with the equivalent calculated drainage flux, which is given by

$$q(b_i) = q_* (1 - r(a-1)d/c_*)^{a/a-1}$$

$$= q_* \exp(- rt) \qquad (8)$$

(v) The value of r is changed until $\Sigma\ p(b_i) = 1$ such that the multiple pathway model exactly reproduces the outflow curve.

Figure 1 shows the measured outflow curve and the derived distribution function for the b_i values, assuming a = 2. The results obtained are dependent on the initial assumptions concerning q_* and a. Figure 2 shows the sensitivity of the derived distribution function to changing a. Note that, from (8), r is independent of a.

4 Model predictions

To use the model to predict the outflows resulting from other input rates it is necessary to make a further assumption concerning inputs to the channel flow paths. At higher input rates it is likely that some

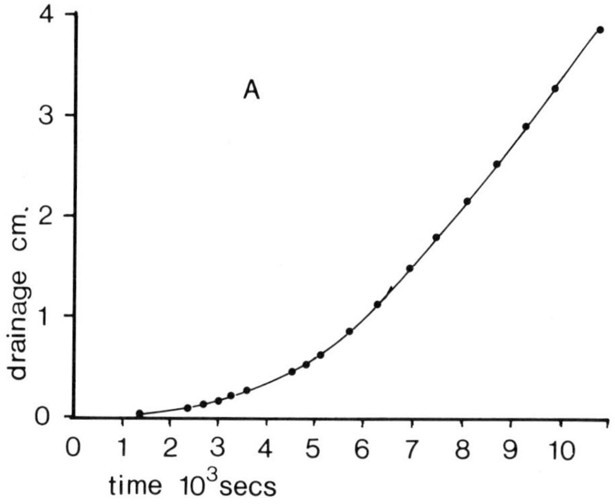

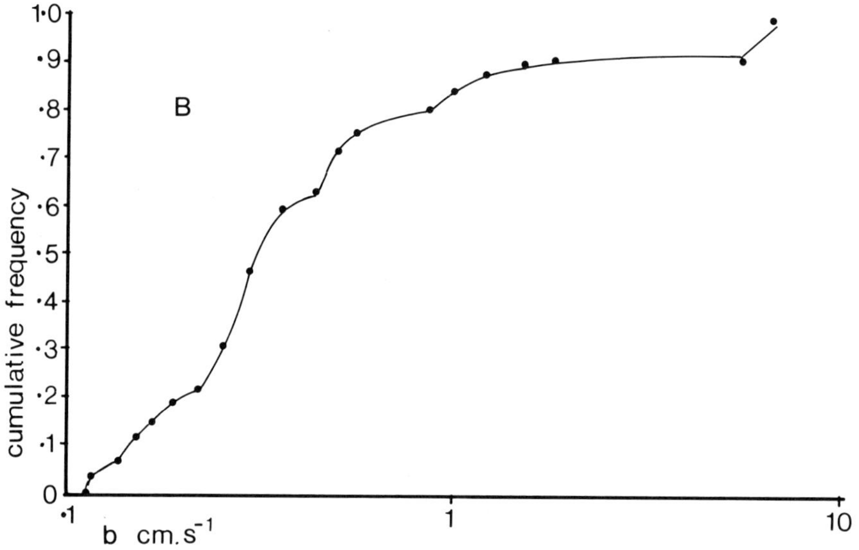

Figure 1 A. Observed outflow curve

B. Derived distribution function for b for $a = 2$, $r = 0.0000555 s^{-1}$.

of the flow paths will saturate and reach flow capacity at the surface. In what follows, it has been assumed that the saturated water content is a constant (w_s) for all flow paths. A first estimate of this channel porosity is provided by the surface water content of the channels with the lowest b values shown in figure 1b; a value of $w_s = 0.089$. At higher input rates therefore these channels will saturate at the surface and there will be some "excess" water available for higher capacity channels. Following a similar procedure to the Germann and Beven (1981) model an operational assumption has been made that this excess water increases the effective input rate to higher capacity channels. Since the distribution function is divided into discrete form for the calculations, this procedure is easily implemented.

Figure 3 shows the predicted breakthrough curves at several input rates using the distribution function of figure 1b. The steepening of the curve arises primarily from the assumption that some of the channel flow paths will reach their flow capacity at higher input rates.

Figure 4 shows the observed and predicted drainage curves for a second experiment at a higher flow rate. The model underpredicts the cumulative drainage unless more water is allowed to take faster flow paths by reducing the effective saturated water content of the channels. The implication is that as input rate increases the faster channel flow paths become more dominant in bypassing flow than predicted by the model.

5 Discussion

The distribution function approach has a number of attractions. The multiple flow path concept allows explicit calculation of different residence times associated with different flow paths, and the way in which these may vary dynamically with flow rate and other conditions. It also provides a framework for easily combining the results from experiments on different samples of the same soil, since if it can be

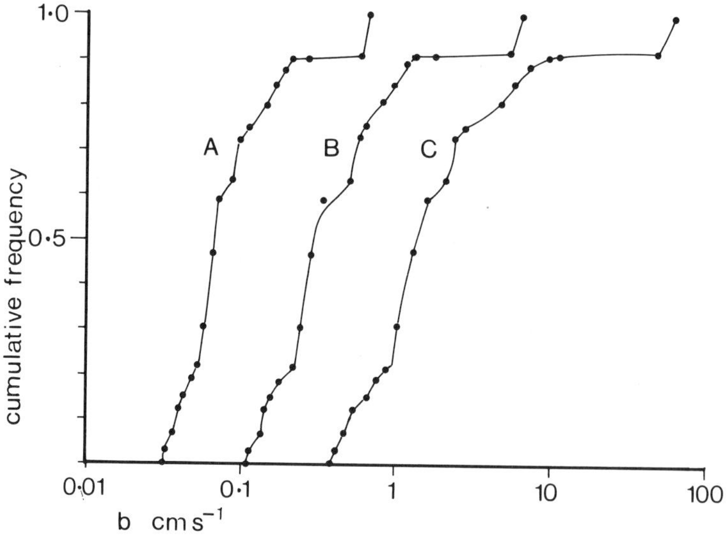

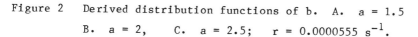

Figure 2 Derived distribution functions of b. A. a = 1.5 B. a = 2, C. a = 2.5; r = 0.0000555 s^{-1}.

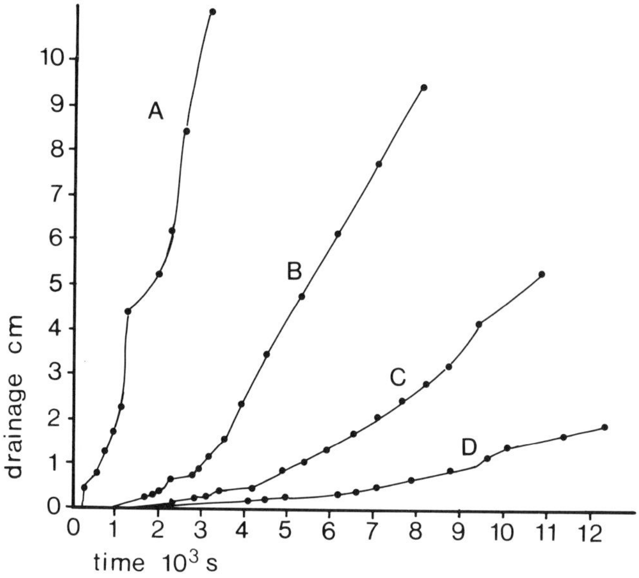

Figure 3 Predicted outflow curves for different application rates with a = 2.0 and r = 0.0000555 s^{-1}

A. q = 0.005 cm s^{-1}; B. q = 0.002 cm s^{-1};

C. q = 0.001 cm s^{-1}; D. q = 0.0005 cm s^{-1}.

assumed that the samples are in some way hydrologically similar, the derived distribution functions are easily combined. Thus, the distribution function can incorporate the effects of spatial variability of flow properties both within and between samples.
The necessity of an analytical solution in deriving the distribution function for the multiple flow path model is clear from the previous section. If an input rate to the channels can be specified, then the model requires only 2 parameters a and r for the distribution function to be derived. Analytical solutions for other flow relationships and loss functions are also possible, and the distribution function model would appear worthy of further study.

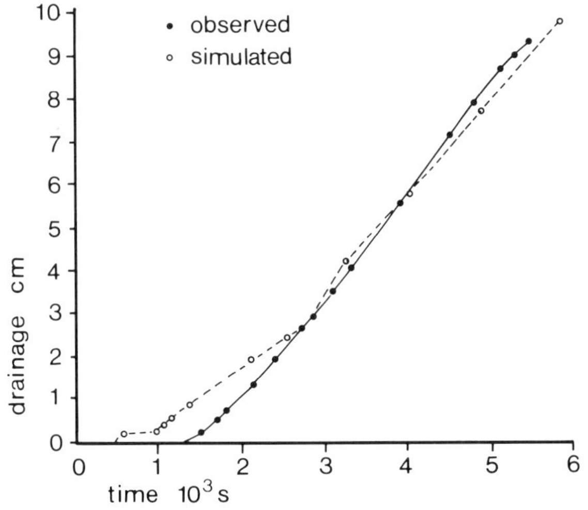

Figure 4. Predicted and observed outflow curves using distribution function of figure 1b, $a = 2$, $r = 0.000055\ s^{-1}$, $w_s = 0.07$.

Acknowledgements

The authors are very grateful to Peter Raats who pointed out an inconsistency in the original derivation before too much damage was done! We would also like to thank Marie Cohen and Eric Hess who helped with the soil column experiments. This work was partially funded by the U.S. Environmental Protection Agency under agreement R810456-01-0.

This information has not been subjected to the Agency required peer review and therefore does not necessarily reflect the views of the Agency and no official endorsement should be inferred.

References

Beven, K J and Germann, P F, 1981, Water flow in soil macropores II A combined flow model. J. Soil Science, 32, 15-29.

Beven, K J and Germann, P F, 1982, Macropores and water flow in soils, Water Resources Research, 18, 1311-1325.

Edwards, W M, R R Van der Ploeg and W Ehlers, 1979. A numerical study of the effects of noncapillary-sized pores upon infiltration. Soil Sci. Soc. Am. J., 43, 851-856.

Germann, P F, 1983, Slug approach to infiltration into soils with macropores. Proceedings of National Conference on Advances in Infiltration, ASAE, Chicago.

Germann, P F and Beven, K J, 1981, Water flow in soil macropores III A statistical approach, J. Soil Science, 32, 31-39.

Hoogmoed, W D and Bouma J, 1980, A simulation model for predicting infiltration into cracked clay soil, Soil Sci Soc. Am J 44, 458-461

Jury, W A, 1982, Simulation of solute transport using a transfer function model Water Resources Research, 18, 363-368.

Discussion

J. Bouma:

The authors are correct in stating that the Hoogmoed and Bouma model relies on detailed measurements of flow pathways. However, S values are used in the simulation, and they represent the total vertical surface area of the peds along which the water flows downwards (m^2 in a plot of 0,5 m^2 per 0.1 m thickness). These S values are a function of the application rate and soil structure. Separate flow pathways are, however, not specifically defined within the plot. We hope to correlate S values with soil structure types as described by pedologists.

Author:

In the same way, we are not attempting to identify our multiple flow paths with actual pores of specific geometry. We are, however, attempting to demonstrate that the dynamics of the reponse of flow pathways of different character may be important in predicting the response under different input rates.

J.J.B. Bronswijk:

Could one use the kinematic wave approach in predicting soil moisture profiles? The loss function s, describing infiltration from cracks to peds, would have to be much more complicated, depending on both the water content of the pores and of the peds, and varying with depth and time. Does this circumstance lead to an intractable distribution function for b?

Author:

It is clearly possible to derive analytical solutions for other loss functions that are functions of total discharge or water content in the macropores, but not for functions that at any depth are functions of time. It is also possible to use numerical simulation to predict the effect of more complex loss functions. However, analytical solutions are initially necessary to derive the distribution function of b.

J.R. Philip:

I take it you envisage the flow in your macropores as film flow, with film thickness small compared with radius of macropore? If it is not small,equation (1) won't be much good. But, if the pore is not full, your column won't drain. Can we reconcile these things?

Author:

We would not wish to pursue too far the theoretical analogy between our power law function and film flow in idealized cylindrical pores. We suspect that in real large pores film flow will break down long before the radius of curvature of the central air/water interface results in a significant drop in potential below atmospheric pressure. However, even if the potential in the water of the film is negative, under steady state flow there is no effect of the inner curvature of the annulus on the potential gradient in the direction of the flow.

With respect to drainage from such a system, if the water close to the exit is at a negative potential, the flow of water will result in a build up of storage until the hydrostatic pressure is equal to atmospheric pressure (even ignoring the kinetic energy of the flow). This requires only sufficient water to result in a very small depth of saturation at the bottom of the channel.

C. Armstrong:

Like you, I have felt that the two-domain concept has its limitations, and that the way forward is to deal with multiple pathways. I wonder if you would like to comment on the posibilities of deriving the distribution functions of these pathways from observational data, e.g. the use of unit hydrographs interpreted as a distribution function of travel times.

Author:

There is a loose link between our work and, say, the transfer function models based on travel-time distributions being used by others. We expect, however, that the transfer function should vary dynamically with flow conditions. We are making a first attempt to include these dynamics in the model.

P.A.C. Raats:

It is perhaps worthwile to point out that as $t \to \infty$, the w-profile becomes steady ,with w remaining 0 for all $z > \{bw_*^{(a-1)}\}/\{r(a-1)\}$.

Author:

This is true. Under the assumptions of the model there is a maximum depth of penetration of the wetting front as $t \to \infty$. This is made clear in a fuller exposition of the model, to be published in Water Resources Research.

Mathematical models of water movement in heavy clay soils

Ya.A. Pachepsky
The Institute of Soil Science and
Photosynthesis
Academy of Sciences of the USSR
Puschino, Moscow Region, USSR
N.G. Zborischuk
The Moscow State University
Soil Science Faculty
Moscow, USSR

Abstract

The contribution presents a set of parameters determining methods for a model describing water transport in heavy clay soils and the way this model can be put to prediction use. The above was applied to clay alluvial meadow soils of an alfalfa-field.

1 Introduction

To enlarge agricultural productivity of heavy soils, by irrigation in particular, it is necessary to predict the moisture regime of the root zone. The peculiar features of heavy clay soils (swelling-shrinking, crack formation, the narrow range of available water, low hydraulic conductivity) manifest themselves depending to a large extent on weather conditions of a certain year (Zaidelman, 1982; Luthin, 1982). That is why observational data on moisture in situ cannot be directly used for predictions which require consideration of different possible combinations of weather and hydrogeological conditions and land reclamation measures. In this situation mathematical modelling appears to be of great help as an addition to other methods of soil research. The moisture regime prediction can be obtained on the basis of an adequate mathematical model. An example of construction of such a model is given below.

2. The subject of investigation

The low flooded part of the Turunchuk island located at the lower part of the Dniester was the territory under study. The island is formed by a thick (26 to 28 m) pack of alluvial deposits which covers a water-bearing sand-gravel horizon. This horizon is hydraulically connected with ground water the level of which normally lies above one-meter depth. On the island there are meadow alluvial and meadow-boggy soils. In argillaceous soil texture clay prevails amounting to 50%. The best part of clay is of montmorillonite origin. With small coverage the drying soil cracks into big blocks. In "wet" periods water remains on the soil surface for a long time. Under this conditions soil moisture regime controlling is of great importance, and this task was considered for perennial grass fields irrigated by sprinkling.

3 Construction of a closed mathematical model of the soil moisture regime

3.1 Selection of space coordinates

A one-dimensional description was used, i.e. soil moisture and soil water potential were assumed to be dependent solely on the vertical coordinate, z. Hoogmoed et al. (1980) considered water transport through macropores to be a two-dimensional phenomenon comprising water movement down the crack walls (the vertical aspect) and water take-up by soil peds from the crack walls (the horizontal aspect); it was proved that the former prevails over the latter. Our own investigations gave the same result. When swelling-shrinking causes vertical shifts of soil surface the vertical coordinate is to be of Lagrange type. In our case no marked shift took place, so z was considered to be an Euler coordinate. The water transport equation has the traditional form:

$$\partial\theta/\partial t = -\partial/\partial z(k\partial\psi/\partial z + k) - R \qquad (1)$$

where

t = time

θ = volumetric moisture content

ψ = suction in unsaturated soil and a value of the hydrostatic pressure component in saturated soil

k = hydraulic conductivity

R = total water uptake rate per unit volume (unit time)

The axis z is directed downwards.

3.2 Model parameters determination

Among model parameters there are hydrophysical characteristics of soil (moisture retention curves, hydraulic conductivity curves and shrinking curves) as well as relationships for calculation of water sources and sinks within soil.

To obtain the moisture retention curve (function $\theta(\psi)$) first there was found a dependence of soil moisture content w (in % of soil dry weight) on pF in two ranges of pF values. With pF less than 2.5 its value was calculated from tensiometer readings in situ. To find the moisture value w samples of soil in immediate contact with the tensiometer ceramic cup were taken with the help of a special sampler. With pF greater than 4.5 the w value was found by the hygroscopic method at a few pF values.

To obtain the volumetric moisture content mentioned in (1) the soil shrinkage curve (function $\theta(w)$) was found. The shrinkage was estimated from data on volume of thoroughly sorted sand necessary to fill up a container with the soil sample being dried. The w-on-θ dependence is shown in Figure 1a. This figure also shows points of the moisture retention curve obtained as a result of weight moisture data transformation into volumetric data. Differences between the moisture retention curves at two depths (15 to 25 and 50 to 60 cm) were insignificant.

Further calculations were based on the fitting moisture retention curve (and not on separate experimental points). The adequate formula was proposed by van Genuchten (1980) and Varallyay et al. (1981):

$$\theta = \theta_r + (\theta_0 - \theta_r) \cdot [1 + (\psi/\psi_*)^m]^{-1} \qquad (2)$$

where

θ_0, m, ψ_* = empirical parameters

θ_r = residual moisture content

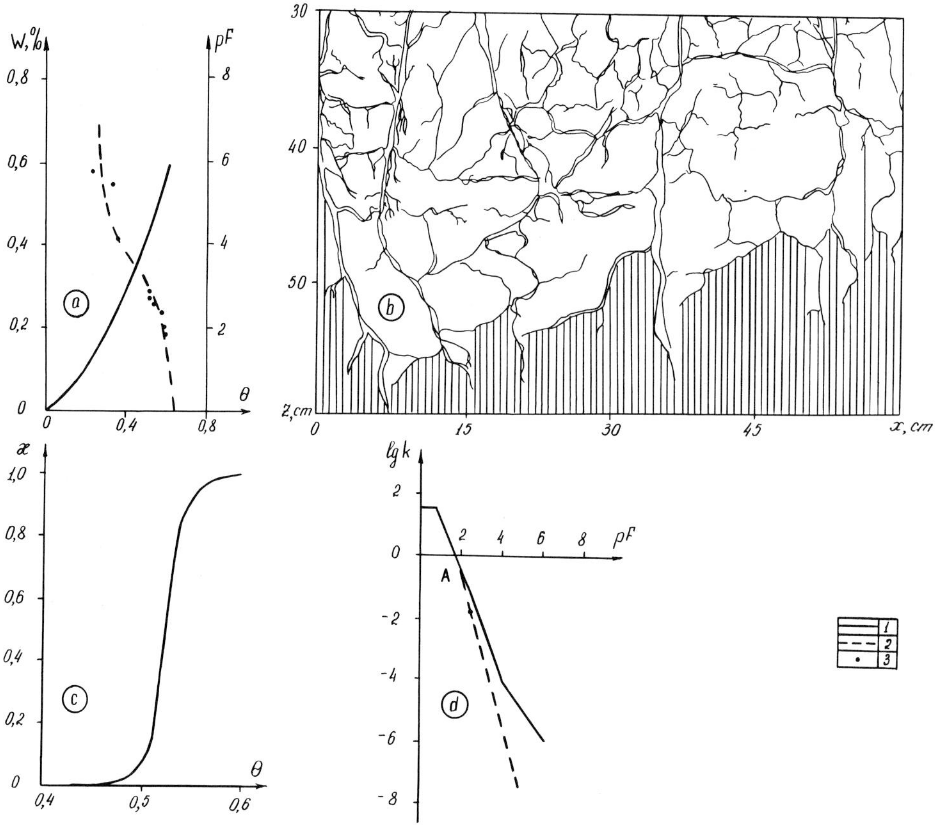

Figure 1. Hydrophysical characteristics of meadow alluvial clay soil: a) shrinkage and moisture retention curves (2 - calculated, 3 - measured); b) sketch of crack formation in a soil block; c) the correction factor κ for crack formation depending on soil moisture; d) soil hydraulic conductivity: 1 - drying curve; 2 - wetting curve; 3 - sound-method

Voronin (1980) showed that with soil moisture retention data found experimentally at low and high pF it is possible to derive an accurate enough prediction of moisture retention at an intermediate pF. When dealing with soils different in structure and texture we found while determining empirical parameters for (2) from moisture retention data in pF ≦ 2 and pF ≧ 4.2 ranges that in this case we could also obtain to a good approximation the moisture retention curve for the intermediate range $2 < pF < 4.2$. This method was used here. Values of θ_0, ψ_* and m were found by the least-squares method, θ_r was assumed to be equal to a value corresponding to the "dead moisture" content. As Figure 1a shows, with $\theta_0 = 0.618$, $\psi_* = 2330$, $m = 0.697$ and $\theta_r = 0.28$ the graph of the function (2) runs rather close to the experimental points of the moisture retention curve.

The wetting curve of θ-on-ψ dependence was not the subject of determination, and hysteresis was neglected. It was assumed that this hysteresis was caused mostly by the swelling-shrinking phenomenon, and is reflected by the $w(\theta)$ function to a greater extent than by the $\theta(\psi)$ function. However, this assumption was not verified experimentally.

The soil hydraulic conductivity was first determined at $pF > 2$. According to Bouma (1981) with these pF an important role is played by hysteresis of vertical hydraulic conductivity which exists due to horizontal crack formation. The wetting hydraulic conductivity curve (function $k_{ads}(\psi)$) was found on the basis of the moisture retention curve. The statistical model of soil porous space was used (Nerpin and Chudnovsky, 1967; Mualem and Dagan, 1978):

$$k_{ads} = M \int_0^{\theta-\theta_r} [u / (\theta_0-u)]^{2/m} \, du \qquad (3)$$

in which

M = the matching factor

ψ_*, θ_0, m and θ_r are the same as in (2)

According to Bouma (1981) the drying hydraulic conductivity curve can be obtained by means of correction for crack formation:

$$k_{des} = k_{ads} \cdot \kappa \qquad (4)$$

Here the correction factor κ is equal to the ratio between the crack-free part of the horizontal soil section and the whole section; it decreases with soil drying.

The κ-on-moisture dependence was found from data of morphometric observations of crack formation. Figure 1b gives a typical sketch of cracks on a profile pit wall between two deep vertical cracks. The cross-hatched region is where ground water can penetrate through capillaries missing, however, horizontal cracks when moving upwards. To find κ-on-θ dependence in figures like 1b line segments were drawn parallel to soil surface between walls of deep vertical cracks with 5 cm depth increment. Then the shaded lengths of each segment were summed up; the square of the ratio of this sum to the total segment length was assumed to be the value of κ for the considered depth. Comparing κ-on-z and θ-on-z dependences we can obtain the sought κ-on- θ dependence. It is shown in Figure 1c.

To find the matching factor M in the k_{ads} and k_{des} formulae there was used the sound method of determination of hydraulic conductivity in situ. The approach by Sudnitsyn et al. (1973) has been modified. The dependence between water flux coming into the tensiometer and time was measured; then by means of repeated numerical solutions of the two-dimensional problem of soil water pouring into a conic nozzle mounted on an impermeable rod we selected M meeting the requirement of best agreement between calculated and experimental fluxes. It was assumed, k_{des} could be found in this way.

According to Pachepsky et al. (1974) at low pF soil hydraulic conductivity cannot be calculated from the moisture retention curve: in this case the main controlling features are the beads-type structure of pore space and the presence of macropores. When moisture is high, k may accurately enough be assumed equal to saturated hydraulic conductivity. The geometric mean of measured values of steady infiltration rate (32 cm daily) was used for calculations. Figure 1d shows the obtained hydraulic conductivity curve in full.

The R value in (1) comprises three parts:

$$R = R_1 + R_2 + R_3 \tag{5}$$

where

R_1 = root water consuming rate

R_2 = rate of water evaporation within soil (from crack surfaces)

R_3 = rate of water absorption when it moves down vertical cracks

To find R_1 we used the model already tested by Averyanov et al (1974) and Feddes et al. (1976) for prediction and epignosis of soil moisture regime:

$$R_1 = T \cdot m_r(z) \cdot f[\psi(z)] \; / \int_0^L m_r(z) \cdot f[\psi(z)]\, dz$$

where

T = transpiration rate

m_r = root specific mass

L = root zone depth

f = factor reflecting decrease of water consumation rate when moisture is too high or too low.

According to Feddes et al. (1976) in case of heavy soil texture function $f(\psi)$ takes the form shown in Figure 2a.

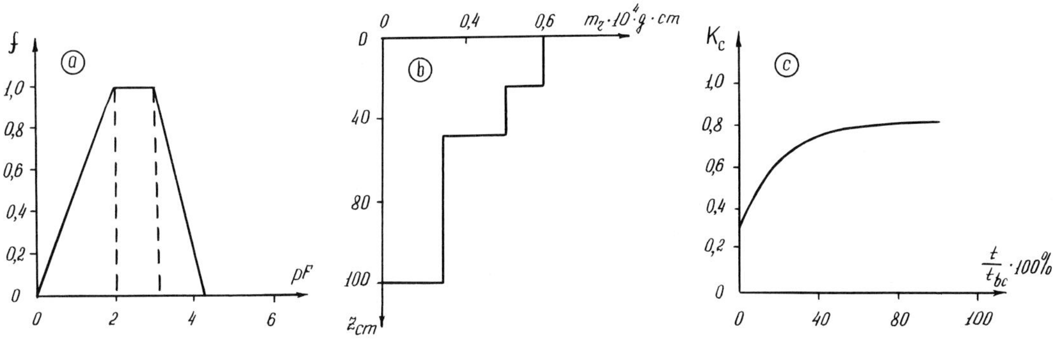

Figure 2. Relationships used to calculate water consumption by third-year alfalfa: a) correction factor reflecting influence of soil moisture potential upon water consumption rate; b) root specific mass (m_r) distribution in depth (z); c) correction factor (κ) reflecting influence of duration of grass growing after a cut (τ) upon transpiration; τ_{bc} is duration between cuts.

Transpiration rate was presented in the form proposed by Hanks and Ashkroft (1980):

$$T = k_c f(\bar{\psi}) E$$

where

E = rate of evaporation from water surface

k_c = factor reflecting biological peculiarities of plants and vegetation phase

$\bar{\psi}$ = average suction in root zone

It follows from data given in the above cited paper that when dealing with third-year alfalfa the k_c-on-time dependence takes the form shown in Figure 2b.

Adams and Hanks (1964) found that surface evaporation inside big vertical cracks is almost the same as from soil surface. Computation in the case of intensive potential evaporation from the surface of soil blocks separated from each other by cracks showed that moisture losses could occur only from thin soil layers close to crack walls. For this reason R_2 term in (5) was neglected.

The R_1 value was taken into account only at the ground level. During the period from t to t+Δt the ground water surface accepts as much water as infiltrated through vertical cracks during Δt starting with the moment $t_1 = t - z_{gw} / v$ (z_{gw} is the ground water depth at the moment t; v is the rate of percolation in vertical cracks).

3.3 Boundary conditions

The quantity of water gone freely through cracks is defined as excess volume beyond the quantity of water infiltrated into soil without puddle formation (no puddles were observed under alfalfa stand). The boundary conditions used in the cases of precipitation and irrigation are based on the above. If evaporation takes place the moisture flux from the surface is equal to

$$Q_0 = -(E - T) = (k_c f(\bar{\psi}) - 1) \cdot E$$

The moisture flux crossing the lower soil layer boundary at z_H depth was calculated under assumption that hydraulic connection between the considered water-bearing horizons is quasistationary:

$$Q_L = -K \cdot (p_1 + z_1 - p_L - z_L) / (z_1 - z_L)$$

Here

z_1 = depth of upper boundary of water-bearing sand-gravel horizon

p_1 = hydrostatic pressure at the lower boundary of the considered soil layer

K = saturated hydraulic conductivity in the ground water horizon

4 Numerical solution of the water transport equation

Scherbakov et al. (1982) compared different algorithms for numerical solution of boundary problems for (1) using the finite-difference method It was shown that the best results could be obtained with local linearization of the moisture retention curve:

$$c_i^{(s-1)} \cdot (\psi_i^{(s)} - \psi_i^{(s-1)})/\Delta t + (\theta_i^{(s-1)} - \theta_i^n)/\Delta t =$$

$$-(\omega/\Delta z^2) \cdot [k_{i+\frac{1}{2}}^{(s-1)} (\psi_{i+1}^{(s)} - \psi_i^{(s)} + \Delta z) - k_{i-\frac{1}{2}}^{(s-1)} (\psi_i^{(s)} - \psi_{i-1}^{(s)} + \Delta z)] -$$

$$[(1-\omega)/\Delta z^2] \cdot [k_{i+\frac{1}{2}}^n (\psi_{i+1}^n - \psi_i^n + \Delta z) - k_{i-\frac{1}{2}}^n (\psi_i^n - \psi_{i-1}^n + \Delta z)] - R_i^n$$

Indices i and n denote values at $z=z_i$, $t=t_n$; Δz and Δt are the space and time increments of the grid; the index (s) denotes a value which is obtained after s-th iteration when finding ψ^{n+1}; $c = \partial\theta/\partial\psi$, $0.5<\omega\leq 1$. The values of $k_{i+\frac{1}{2}}$ at $z = z_i + \Delta z \cdot 0.5$ were calculated as integral means:

$$k_{i+\frac{1}{2}} = \int_{\psi_i}^{\psi_{i+1}} k(\psi) d\psi / (\psi_{i+1} - \psi_i)$$

considering k to be a reverse power function of ψ:

$$k_{i+\frac{1}{2}} = k_i [1-(k_{i+1}\psi_{i+1})/(k_i\psi_i)] \cdot \ln(\psi_{i+1}/\psi_i) \; / \; \{[1-(\psi_{i+1}/\psi_i)] \cdot$$

$$\ln[k_{i+1}\psi_{i+1}/(k_i\psi_i)]\}$$

In particular, when $k \sim \psi^{-2}$ we have

$$k_{i+\frac{1}{2}} = \sqrt{(k_i k_{i+1})}$$

The described finite-difference scheme is the basis for the program MOIST published by Scherbakov et al. (1982). The program was used to compute the moisture regime in question.

5 Verification of the model

The constructed model was verified by comparison of calculated and measured soil moisture distribution data. During 26 days we repeatedly measured the quantity of precipitation, evaporation from water surface, moisture dynamics in the soil profile and the ground water regime. Some results of calculations and measurements are shown in Figure 3. The mean square relative error is 8%, the maximum error value is 19%. It seems the agreement would be better if the layered character of soil and subsoil was paid more attention. The fact that this property has been ignored possibly explains also the discrepancy up to 2o cm between calculated and measured depths of ground water. However, another phenomenon noticed by Bouma et al. (198o) may be of importance here: the level of saturation observed in boreholes in heavy cracked soils lies, as a rule, above the real level of ground water, i.e. above the depth where capillary potential is equal to zero.

6 Application of the model

The constructed model was used for the purpose of polyvariant prediction of meadow alluvial soil moisture regime and alfalfa water supply. The considered variants differed in meteorological conditions of the vegetation period, in the irrigation regime and in conditions of the hydraulic connection between water-bearing horizons. Computations showed that in "wet" years accounting for 2o% the moisture regime of alfalfa-field soils is characterized by excess moisture within the upper one meter thick layer during the whole vegetation period. In "dry" years (accounting also for 2o%) the excess moisture can be expected only in April. Overdrying of the one meter thick upper soil layer in July -

September takes place only in "dry" years at non-irrigated spots where there is no hydraulic connection between ground waters and the lower water-bearing horizon. The large alfalfa coverage diminishes overdrying of the upper soil layer. Irrigation should only compensate the deficit of soil water never exceeding 3o to 35 mm

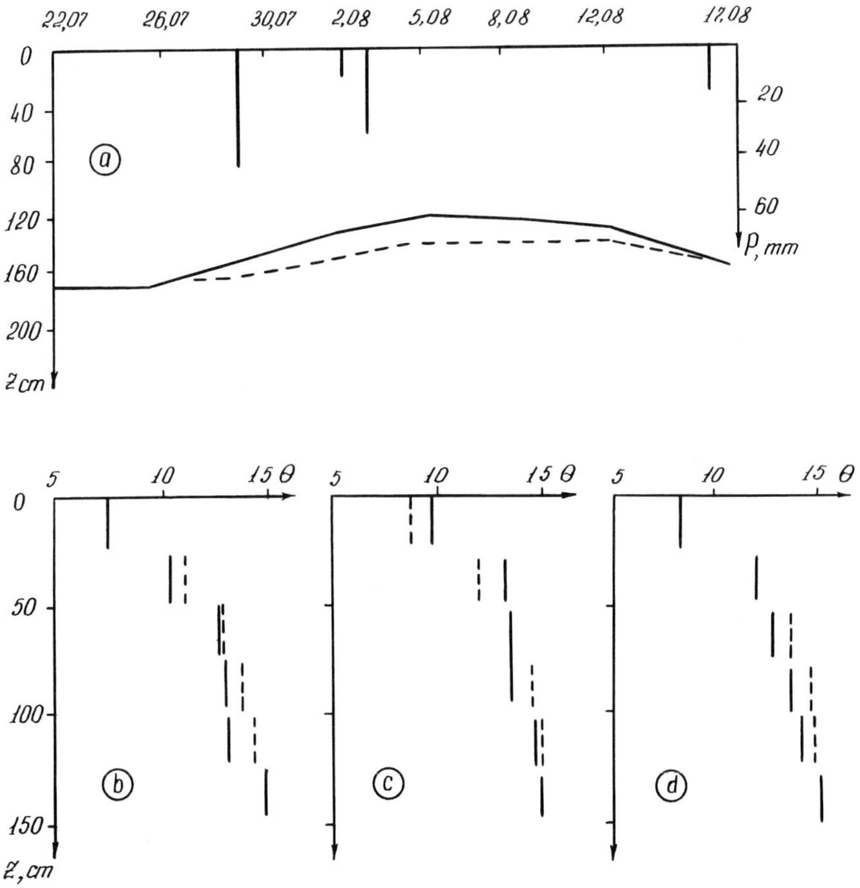

Figure 3. Calculated (1) and measured (2) levels of ground water (a) and moisture profiles (b-d); 3-precipitation and irrigation; (b),(c) and (d) as of July 26, August 5 and August 17, 1982 respectively.

References

Adams, J.E. and Hanks, R.J. 1964. Evaporation from shrinkage cracks. Soil Sci.Soc. Amer. Proc., 28: 281-284.

Averyanov, S.F., Golovanov. A.I. and Nikolsky, Yu.M. 1974. Calculation of moisture regime of reclaimed soils. Gidrotechnika i Melioratsiya, No.3, 34-42.

Bouma, J., Dekker, L.W. and I.C.F.M. Haans 198o. Measurement of depth to water table in a heavy clay soil. Soil Sci., 13o: 264-27o.

Bouma, J. 1981. Soil survey interpretation: estimating use-potentials of a clay soil under various moisture regimes. Geoderma, 26: 165-177.

Feddes, R.A., Kowalik, P., Kalinska-Malinka, K. and Zaradny, H. 1976. Simulation of field water uptake by plants using a soil water dependent root extraction function. J. Hydrol., 3: 13-26.

Hanks, R.J. and Ashcroft, G.L. 198o. Applied soil physics. Springer-Verlag, Berlin, Heidelberg and New York, 152 pp.

Hoogmoed, W.B. and Bouma, J. 198o. A simulation model for predicting infiltration into cracked clay soil. Soil Sci. Soc. Amer. J.,

Luthin, J.W. 1982. Water management of clay soils. Trop. Agric., 59: 1o3-1o9.

Mualem, Y. and Dagan, G. 1978. Hydraulic conductivity of soils. Unified approach to the statistical models. Soil Sci. Soc. Amer. J., 42: 392-395.

Nerpin, S.V. and Chudnovsky, A.F. 1967. Soil Physics. Nauka, Moscow, 583 pp.

Nimah, M.N. and Hanks, R.J. 1973. Model for estimating soil water, plant and atmospheric interrelations: 1. Description and sensitivity. Soil Sci. Soc. Amer. Proc., 37: 525-527.

Pachepsky, Ya.A., Scherbakov, R.A., Varallyay, Gy. and Rajkai, K. 1984. Determination of soil hydraulic conductivity from the moisture retention curve. Pochvovedenie, in press.

Scherbakov, R.A., Pachepsky, Ya.A. and Kuznetsov, M.Ya. 1981. Ion and chemical compounds migration in soil. Water movement. Ecomodel No.7, ONTI, Puschino, 44 pp.

Sudnitsyn, I.P., Muromtsev, N.A. and Shein, E.V. 1973. Sound method used to find soil hydraulic conductivity coefficient. Biologicheskie nauki, No.1, 137-142.

van Genuchten, M.Th. 198o. A closed-form equation for prediction the hydraulic conductivity of unsaturated soil. Soil Sci. Soc. Amer. J., 44: 892-898.

Varallyay, Gy., Mironenko, E.V., Pachepsky Ya.A., Rajkai, K. and Scherbakov, R.A. 1981. Mathematical description of main hydrophysical characteristics of soil. ONTI, Puschino, 27 pp.

Voronin, A.D. 198o. A new approach to determine the dependence between the capillar-sorption water potential and soil moisture. Pochvovedenie, No.1o, 68-79.

Zaidelman, F.R. 1982. Heavy gleyed soils of the USSR humid landscapes. In: Problems of Soil Science. Soviet soil scientists contributions at the 12-th International congress of soil scientists, Nauka, Moscow, pp. 89-96.

Some theoretical and practical aspects of infiltration in clays with D = constant

M. Kutílek
Soil Science Laboratory
Faculty of Civil Engineering
Technical University Prague,
Czechoslovakia

Abstract

In clay soils with highly developed swelling, as e.g. in Na-Vertisols, the water diffusivity approaches a constant value. The physical parameters of a model soil with D=const. are discussed and compared to the observed values. When crusting of the surface due to the rain is considered, the approximate model with D=const. on the surface of the clay soil is developed and the induced shift of the ponding time is examined.

1 Introduction

With increasing alkalinity of clays when the exchangeable sodium percentage ESP rises and the electrical conductivity EC is kept at low values, the relationship between the water diffusivity D and the soil moisture θ starts to be less non-linear /Kutílek, Semotán, 1975, Kutílek, 1983/. In order to demonstrate this phenomenon, the $D(\theta)$ relationship is approximated by the exponential form

$$D = \gamma \exp(\beta \Theta) \qquad /1/$$

where

γ, β = coefficients,

Θ = relative soil moisture, $\Theta = \frac{\vartheta - \vartheta_r}{\vartheta_s - \vartheta_r}$

ϑ_r = residual soil moisture, here in equilibrium with relative vapor pressure $p/p_o = 0.95$,

ϑ_s = saturated soil moisture.

The value of β is related to ESP and to clay content in Table 1. The D (ϑ) data were evaluated from the absorption experiments on confined soil columns. Due to the confinement, the swelling was strongly reduced and restricted to the internal part of the column with slight swelling at the inflow end and with the induced compression of soil at the wetting front, or in the semi-wet portion of the column /Fig.1/.

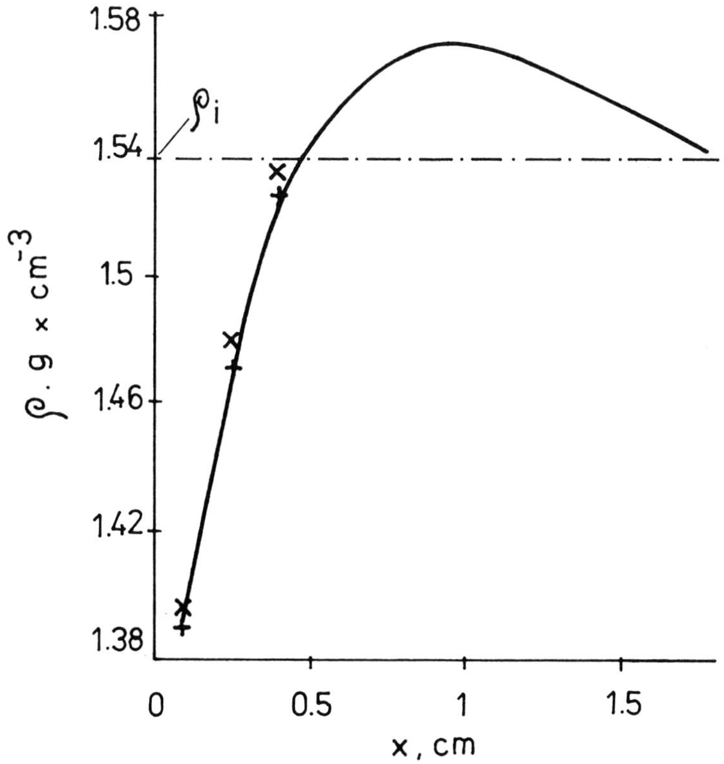

Figure 1. The change of the bulk density of Na-Vertisol /ESP = 27.5%/ at the inflow end of the confined column

Owing to the fact that the free swelling was practically annuled, the results are reported in classical way with θ expressed as volumetric moisture and in Eulerian coordinates.

Table 1. Coefficient β in Eq. /1/ for soils of various degree of alkalinity

ESP,%	EC,mmho/cm	Clay,%	β	
10.5	1.1	33	5.9	
19.5	0.74	35	3.0	
29.0	6.0	43	3.1	
30.0	3.1	32	2.9	
31.0	6.8	44	2.8	
31.0	2.4	29	3.8	
38.0	3.9	42	2.2	
13.0	1.1	59	3.8	
21.5	0.96	62	1.9	/poor relation/
27.5	0.75	68	-1.4	
39.0	0.75	63	-1.5	

The coefficient β decreases with the increase of ESP and when the sodification is combined with high clay content, β reaches even negative values. We can therefore expect existence of a soil with $\beta=0$ and $D=$ const. When the soil with β negative was saturated with Ca^{2+}, the $D(\theta)$ changed substantially, β reaching positive values /Kutílek, Semotán,1975/. We can suppose that the main mechanism changing the $D(\theta)$ from exponential relationship to approx. $D=$ const. is the desaggregation and peptization of clay particles.

Generally, $D=$ const. is a close approximation for the flow of water in confined alkali soils /Fig. 2/ and the discussion of the characteristics of the "linear"soil /Philip,1969/ is appropriate and $D=$const. can be exact for the real soil.

2 Soil with $D=$constant

2.1 Derivation of soil characteristics

Two searched soil characteristics $H(\theta)$ and $k(\theta)$ should comply with the condition

$$D = K\ k(\theta)\ \frac{dH(\theta)}{d\theta} = const. \qquad /2/$$

where

K = saturated hydraulic conductivity,

$k(\theta)$ = relative unsaturated hydraulic conductivity,

H = pressure head,

$H(\theta)$ = moisture retention curve.

Condition /2/ can be rewritten in

$$\frac{d\theta(H)}{dH} = const.\ k(H) \qquad /3/$$

The relationship $k(H)$ should be monotoneous and the same is for $d\theta(H)/dH$ in order to keep $k(H)$ real. This leads to the restriction in the use of the analytical forms of $\theta(H)$. The expression of Brooks and Corey /1964/ meets the condition and will be used in this discussion in the form

$$\Theta = \left(\frac{H_V}{H}\right)^{\lambda} \qquad /4/$$

where

H_V, λ = parameters with physical meaning discussed in detail by Brooks and Corey.

Further on, we follow two types of interpretation of soil physical characteristics. Interpretation I declares $\theta(H)$ as the single basic and sufficient characteristic, from which $k(\theta)$ can be deduced. Interpretation II uses two independent experiments for the determination of $H(\theta)$ and $k(\theta)$. Theoretical relationship of both characteristics is used for the development of approximate expressions.

The $k(\theta)$ relationship deduced from $\theta(H)$ according to physical models of porous media leads after introduction of Eq. /4/ to the general relation

$$k(\Theta) = \Theta^{a/\lambda + b} \qquad /5/$$

where

a = 2, b = 3 in Burdin´s /1953/ method,
a = 2, b = 2 in Childs and Collis-George´s /1950/ method,
a = 2, b = 2.5 in Mualem´s /1976/ method.

Substitution of Equations /4/ and /5/ in Eq. /2/ gives

$$D(\Theta) = -\frac{H_V}{\lambda} \frac{K}{\theta_s - \theta_r} \Theta^{(a-1)/\lambda + (b-1)} \qquad /6/$$

The condition D=const. will be satisfied if
/1/ either $\lambda = -(a-1)/(b-1)$
/2/ or $a = b = 1$.
The first condition is in accordance withthe Interpretation I, and e.g. for the Burdin´s method we get $H = H_V \Theta^2$ and $k = \Theta^{-1}$. However, both characteristics are not physically real. We get similar physically not real results for other methods of $k(\Theta)$ evaluation from $H(\Theta)$.
In the second condition, a, b are very different from the theoretical values of models developed for soils with strong non-linear $D(\theta)$ relationship. It means that the soil with D=const. differs from other soils not by parameter λ with a specific $H(\Theta)$ relationship, but that the difference is in a special method of prediction of $k(\Theta)$.This development is in accordance with the Interpretation II. Since this second condition is physically acceptable, we get the following characteristics of a soil with D=const.

$$H = H_V \Theta^{-1/\lambda} \quad /7/ \qquad k = \Theta^{1/\lambda + 1} \quad /8/ \qquad D = -\frac{H_V}{\lambda} \frac{K}{\theta_s - \theta_r} \quad /9/$$

If λ=1, the conductivity function is quadratic and such a soil is identical with the soil described by Burger´s equation /Clothier et al., 1981/. It means that the Burger´s equation describes a specific case within the family of soils with D=const.

2.2 Numerical and analytical solutions of a model soil

The above derived characteristics in Equations /7/,/8/,/9/ of a soil with D=const. were used for the analytical and numerical solution of the horizontal infiltration /absorption/. In the analytical solution of the diffusivity equation only Eq. /9/ was applied while in the numerical solution of the capacity equation both Eq. /7/ and /8/ were used. The results obtained by two different procedures serve as a proof of correctness of the derived characteristics and the comparison of results indicate the range of error due to the numerical procedure, too.

In the numerical method, the implicit scheme of final differences was used for the solution of the capacity equation

$$\frac{d\theta}{dH}\frac{\partial H}{\partial t} = \frac{\partial}{\partial x}\left[k(H)\frac{\partial H}{\partial x}\right] \qquad /10/$$

In the analytical solution of

$$\frac{\partial \theta}{\partial t} = D\,\frac{\partial^2 \theta}{\partial x^2} \qquad /11/$$

with $\theta = \theta_i \qquad x > 0 \qquad t = 0 \qquad /12/$

$\theta = \theta_s \qquad x = 0 \qquad t \geq 0 \qquad /13/$

we obtain according to Crank /1956/

$$\theta = \theta_i + (\theta_s - \theta_i)\,\mathrm{erfc}\,\frac{x}{2\sqrt{Dt}} \qquad /14/$$

Further on, sorptivity S was calculated according to Philip /1969/

$$S = 2(\vartheta_s - \vartheta_i)\sqrt{\frac{D}{\pi}} \qquad /15/$$

The model soil was characterized by $D = 0.04167\ cm^2 min^{-1}$, $H_V = -1$ cm, $\lambda = 1$, $\vartheta_s = 0.45$, $\vartheta_r = 0.05$, $\vartheta_i = 0.10$, $K = 0.0167$ cm min^{-1}, $k = \Theta^2$. Following results obtained by the two methods were compared:

/1/ The moisture profile $\vartheta(x)$ at the sequence of time intervals were compared and selected data at t = 61 min and t = = 900 min are in Table 2.

Table 2. Moisture obtained by analytical, ϑ_A and numerical, ϑ_N procedure for soil with D = const.

x, cm	t 61 min		t 900 min	
	ϑ_A	ϑ_N	ϑ_A	ϑ_N
1	0.33009	0.33799	0.41783	0.41816
3	0.16416	0.17980	0.35516	0.35618
6	0.10273	0.10598	0.27095	0.27280
9	0.10002	0.10016	0.20454	0.20678
18	-	-	0.11318	0.11417
27	-	-	0.10064	0.10075

Better agreement between the exact and numerical solutions was found for greater time as it can be expected.

/2/ Cumulative infiltration I was determined in both cases by numerical integration of the moisture profiles using the Simpson´s rule and the results are compared in Table 3 with exact I obtained from S which was computed according to Equation /15/. The error due to the numerical integration is negligible when compared to the error induced by the procedure of final differences, but, generally, good agreement exists between the numerical and analytical procedures and the agreement increases again with time.

Table 3. Cumulative infiltration I, cm.

t, min	I from $\int \theta_N \, dx$	$\int \theta_A \, dx$	S = 0.080615 exact $I = S\, t^{1/2}$
61	0.69214	0.62949	0.62963
900	2.45085	2.41843	2.41846

/3/ When the time dependency of the cumulative infiltration I(t) and of the infiltration rate v(t) as obtained by final differences was evaluated, the values of S and α in the equation $I = S\, t^{\alpha}$ were not in agreement with the general theory /α=0.5/ nor with the exact data obtained analytically. Both S and α were time dependent, gradually approaching the theoretical data after great time, as we can again expect from the nature of the numerical method.

The comparison of the results gained by two different procedures of solutions shows that the final differences offer data only slightly different from the exact solutions and the reason for it is in the approximative nature of the numerical procedure. The correctness of the derivation of the soil characteristics in Eq. /7/, /8/ and /9/ can be taken as confirmed.

2.3. Application to the real soil

Na-Vertisol with ESP = 27.5%, EC = 0.75 mmho cm^{-1} is used in this study. Its $D(\theta)$ was obtained by horizontal infiltration and is plotted in Figure 2. The Crank´s /1956/ expression for calculation of mean $\bar{D}$

$$\bar{D} = \frac{5}{3(\theta_s - \theta_i)^{5/3}} \int_{\theta_i}^{\theta_s} (\theta - \theta_i)^{2/3} D(\theta)\, d\theta \qquad /16/$$

offers $\bar{D} = 2.944 \times 10^{-4}$ $cm^2\ min^{-1}$ when θ_s = 0.49 and θ_i = 0.12.

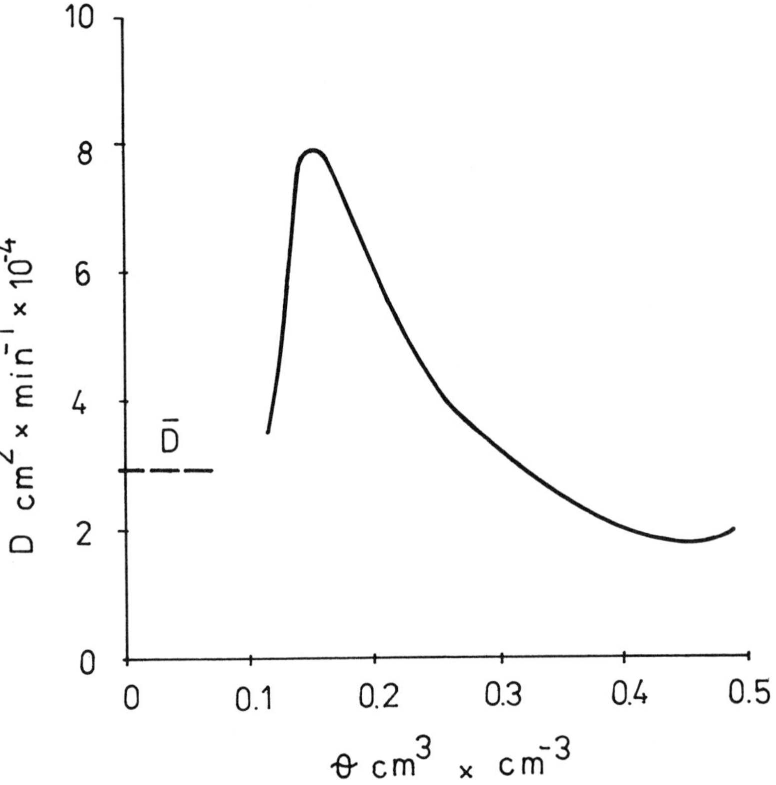

Figure 2. The relationship between water diffusivity D and the volumetric moisture in Na-Vertisol /ESP 27.5%/. $\bar{D}$ indicates the weighted-mean diffusivity

Sorptivity is according to Eq. /15/ $S=7.164 \times 10^{-3}$ cm $min^{-1/2}$. The values of the consumption of water during the experiment give $S=8.504 \times 10^{-3}$ cm $min^{-1/2}$. However, the systematic error due to the evaporation through the walls of the tubing etc. cannot be excluded since the experiment ran slowly and lasted for 30 days. Therefore the term consumption instead of cumulative infiltration is used. From the experimentally determined moisture profile after t=43320 min we get by integartion I=1.44 cm and $S=6.919 \times 10^{-3}$ cm $min^{-1/2}$, while the exact value obtained by Philip´s procedure /1955/ gives $S=7.318 \times 10^{-3}$ cm $min^{-1/2}$. When the approximative expressions for sorptivity were tested and plotted as $S(\theta_i)$, we have found that the Philip´s/1969/ solution, Eq./15/based on $\bar{D}$ was the closest one to the exact values. The equation of Parlange /1971/ offered S lower by approx. 15%, the expression of Parlange /1975/ gave the results higher approx. by 15%, and the theory of Philip and Knight /1974/ resulted in data higher approx. by 30%, when the exact values of S obtained by Philip´s /1955/ procedure were taken as the basis for comparison. Here again, we can see the difference from the relations derived for soils with $D(\vartheta)$ strongly non-linear /Elrick, Robin, 1981/. We can conclude that $\bar{D}$ computed from Eq. /16/ offers reliable results.

Greater differences between the experimental and computed ϑ data occur when the moisture profiles $\vartheta(x)$ are compared /Table 4/.

Table 4. Comparison of the experimental and computed $\vartheta(x)$ for t=43320 min, ϑ_i=0.12, ϑ_s=0.49

x, cm	ϑ exper.	ϑ comp. for D=const.
0.75	0.420	0.446
3.0	0.278	0.324
6.2	0.199	0.201
11	0.145	0.131

In spite of lower accuracy in computing the moisture profiles, we can see that the general shape of the computed moisture profile is identical with the experimental one.

We are allowed to conclude that the approximation of D=const. is appropriate. When the physical nature of the alteration of the physical properties is considered, we can assume that the approximation with D=const. is suitable in all instances when the desaggregation and peptization of clay particles occur in heavy soils.

3. Crusting of soils due to the rain infiltration

The main mechanism of crust formation due to the rain is the desaggregation of the surface, the washing in of fine particles and compaction of the immediate surface by raindrop impact /McIntyre, 1958/. Morin et al. /1976/ suppose that the sealing efficiency of the crust is increased by the action of a very high hydraulic gradient across the crust, and they conclude from their experiments that the crust thickness ranges from 10^{-2} to 10^{-1} cm. The first number is in agreement with finding of McIntyre. Further on, the authors state that the conductivity of the crust on loess and on sandy loam is lower by 3 or 4 orders of magnitude, i.e. for loess, the crust conductivity should be in ranges 7×10^{-5} to 7×10^{-6} cm min^{-1}. These data lead to the assumption that the crust can be defined as a soil desaggregated to such an extent that the crust hydrodynamic characteristics are close to the characteristics of desaggregated Na-Vertisol. Accepting this premise, we can discuss the influence of the crust upon practical behavior of the soil during rainstorm infiltration using the theoretical development on soils with D= const. The most instructive is the influence of the soil crust upon the shift of the ponding time.

Let us assume in accordance with the results of McIntyre and Morin et al. that the crust hydraulic conductivity K_c decreases exponentially with the cumulative rain. If the rain intensity is taken constant, v_r, the crust conductivity K_c can be expressed by

$$K_c = K_o \exp(-c_1 v_r t) \qquad /17/$$

where

K_o = conductivity of the undisturbed soil at $t = 0$ at $z = 0$.

The final value of the conductivity of the crust at the ponding time t_p is K_{cf}. Analogically to Eq. /17/, the decrease of the water diffusivity in the crust is expressed by

$$D_c = D_o \exp(-c_2 v_r t) \qquad /18/$$

where

D_o is computed acc to Eq. /16/ as the mean-weight diffusivity of the undisturbed soil at $t = 0$ at $z = 0$.

Since $K_{cf}/K_o \approx 10^{-3}$ to 10^{-4} and H_V/λ is expected in accordance with Na-Vertisol to be 10^{-1} while in microaggregated clays it is 10^0, we suppose that $D_{cf}/D_o \approx 10^{-2}$ to 10^{-3} and $c_1 > c_2$.

Provided that the Philip´s /1969/ algebraic equation $v = St^{-1/2} + A$ is applicable with v = infiltration rate, the ponding time t_p is calculated from /Kutílek, 1980/

$$t_p = \frac{S^2}{A^2}\,\frac{(2b-1)}{4b(b-1)^2} \qquad /19/$$

where

$b = v_r/A$

When mean diffusivity and conductivity of the crust are used in time interval $\langle 0, t_p \rangle$, we get with Eq. /15/ and taking $A = (1/3)K$

$$t_p \approx \frac{(\theta_s - \theta_i)^2 D^* (6v_r - K^*)}{v_r (3v_r - K^*)^2} \qquad /20/$$

where $D^* \approx \dfrac{D_o}{t_p c_2 v_r}$ $\qquad K^* \approx \dfrac{K_o}{t_p c_1 v_r}$

If the crust formation process was not considered as developing in time and the final values of D_{cf} and K_{cf} were uncorrectly considered, the t_p could be shortened up to approx. by one order of magnitude. The moisture profile below the crust is computed from the integrated equation of continuity inserting for v_r the reduced flux below the surface film. The water storage in the crust is neglected.

Acknowledgements

For the analytical data on ESP and EC and for the selection of samples, I am indebted to E.L. Strmecki, FAO. For the numerical analysis by final differences, the program of J. Mls from our Laboratory was used. I acknowledge with gratitude the cooperation of T. Vogel on Chapter 2.

References

Brooks, R.H. and A.T. Corey 1964. Hydraulic Properties of Porous Media. Hydrol. Paper 3, Fort Collins Colorado State Univ.

Burdine, N.T. 1953. Relative permeability calculations from pore-size distribution data. Petr. Trans. Am. Inst. Min. Metall. Eng. 198:71-77.

Childs, E.C. and N. Collis-George 1950. The permeability of porous materials. Proc. Roy. Soc. 201A:392-405.

Clothier, B.E., J.H. Knight and I. White 1981. Burgers' equation: Application to field constant-flux infiltration. Soil Sci. 132:255-261.

Crank, J. 1956. The Mathematics of Diffusion. Oxford Univ. Press, London.

Elrick, D.E. and M.J. Robin 1981. Estimating the sorptivity of soils. Soil Sci. 132:127-133.

Kutílek, M. 1980. Constant rainfall infiltration. J.Hydrol. 45:289-303.

Kutílek, M. 1983. Soil physical properties of saline and alkali soils. In: Int. Symp. on Isotope and Radiation Techniques in Soil Physics and Irrigation Studies, IAEA-FAO, in press.

Kutílek, M. and J. Semotán 1975. Soil water properties of Gezira soils. In: Int. Symp. New Development in the Field of Salt-affected Soils. A.R.E. Cairo, pp. 299-308.

McIntyre, D.S. 1958. Soil splash and the formation of surface crusts by raindrop impact. Soil Sci. 185:261-266.

Mls, J. 1982. Formulation and solution of fundamental problems of vertical infiltration /in Czech/. Vodohosp. Čas. 30:304-311.

Morin, J. Y. Benyamini and A. Michaeli 1976. The dynamics of soil surface crusting and water movement in the profile as affected by raindrop impact. Prelim. Report, Izrael Min. Agric. Hakirya, Tel-Aviv.

Mualem, Y. 1976. A new model for predicting the hydraulic conductivity of unsaturated porous media. Water Res. Res. 12:513-522.

Parlange, J.-Y. 1971. Theory of water movement in soils: 1. One dimensional absorption. Soil Sci. 111:134-137.

Parlange, J.-Y. 1975. On solving the flow equation in unsaturated soils by optimization: Horizontal infiltration. Soil Sci. Soc. Am. Proc. 39:415-418.

Philip, J.R. 1955. Numerical solution of equations of the diffusion type with diffusivity concentration dependent. Trans. Faraday Soc. 51:885-892.

Philip, J.R. 1969. Theory of infiltration. In: Adv. Hydrosci. 5:215-305.

Philip, J.R. and J.H. Knight 1974. On solving the unsaturated flow equation. 3. New quasi-analytical technique. Soil Sci. 117:1-13.

Discussion

P.A.C. Raats:

Is the curve in Figure 1 based on only the pairs of data points shown? Is the Figure intended to show how the measured expansion in the interval 0 - 0.5 cm might be compensated for by compression in the interval 0.5 - 2 cm?

Author:

The data in the domain of compression were measured, but they are scattered, just indicating the compression phenomenon. The line is drawn on the basis of equality of areas.

P.A.C. Raats:

The linkage of the Brooks/Corey representation of the physical properties to the Burgers' equation is interesting. From a theoretical point of view, Burgers' equation deserves attention since 1) it contains a physically realistic, minimally nonlinear, gravitational term, and 2) thanks to the Cole/Hopf transformation, analytical solutions can be obtained for a wide variety of conditions. My question is: for the same soil and the same initial and boundary conditions, would retaining gravity have a significant influence upon the results of your calculations?

Author:

The influence of gravity is supposed to be less significant than in strongly nonlinear soil. However, the computation has not yet been done.

Hydraulic conductivity and structure of three Australian irrigated clays

J.M. Cooper

CSIRO, Soils Division, Canberra, A.C.T.

Swelling clay soils are widespread in the eastern Australian hinterland,occupying about 10^8ha of alluvial plains, gently sloping fans, and undulating to gently rolling uplands. The three sites to which reference is made are on irrigated areas of alluvial plains. In the management of irrigated swelling soil, hydraulic conductivity in the swollen ('saturated') state is significant for profile drainage, salt leaching, and the re-establishment of aerobic root zone conditions following irrigation or heavy rainfall. The aim was to examine the relationship beween hydraulic conductivity, structure and macroporosity.

At Kerang irrigated cereal cropping (not rice) and pasture production, which began some 80 years ago, soon led to shallow watertables and salinization of the soil. By judicious management and drainage, partial reclamation has been achieved and allows some crop and pasture production. At the site sampled (a Chromustert) the upper 0.2-0.3m was dark coloured subangular blocky (10-30mm) clay, with many grass roots, some with rusty staining. This graded into a brown, friable, angular and subangular blocky (10mm breaking to 3mm) zone, with many fine (<1mm) and some larger (3mm) cylindrical voids, often with dark soil infills. These voids continued beyond 0.8m. Below 0.5m some slickensided surfaces were apparent, many aggregates were sub-rounded or tubular in form and some soft and nodular carbonate occurred.

At Benerembah the area has been irrigated for some 25 years, mainly for rice, other cereals and pasture, and the watertable is at 10-15m and rising. At the sample site (a shallow surfaced Natrustalf, with vertic characteristics when cultivated,cf. Chromustert), the brown, weakly structured surface layer passed sharply to a dense, massive red-brown clay with few widely spaced vertical planar voids and with few roots and few visible voids. Below about 0.3m there were occasional oblique planar surfaces with slickensides, increasing in frequency with depth, and also a few small carbonate nodules.

At Narrabri irrigation began about 20 years ago for cotton production and since then some soil structural degradation has occurred as a result of tillage and other cropping operations. Watertables do not occur. The sample site (a Pellustert) had 0.15m of very dark grey, crumbly, self-mulching clay, overlying columnar (115 x 70mm) clay, breaking to subangular blocky. Strong parallelepipedal structure with many slickensided surfaces occurred below 0.25m. Only occasional fine cylindrical voids were visible. Scattered small carbonate nodules occurred throughout the profile.

Standard analytical data (Table 1) were obtained for profiles sampled at the three sites. 'Saturated' hydraulic conductivity (K) was measured by two techniques (Table 2). Dye solutions (first methylene blue, then Rhodamine WT) were allowed to soak into the undisturbed blocks which, subsequently, were horizontally sectioned at 0.05m intervals and examined by incident light microscopy.

Table 1. Analytical data for Kerang (Kg), Benerembah (B),and Narrabri(N)

Depth	Clay content (%)			pH[1]			EC[1] (dS/m)			ESP[2]		
(m)	Kg	B	N	Kg	B	N	Kg	B	N	Kg	B	N
0-0.1	51	38	62	7.0	7.0	7.9	0.56	0.10	0.09	11	5	1.7
0.1-0.2	-	59	-	7.4	7.4	7.8	1.71	0.24	0.09	-	11	1.8
0.2-0.4	71	60	63	7.6	8.2	8.3	3.59	0.49	0.09	20	16	3.5
0.5-0.7	65	47	-	7.9	8.1	8.6	4.37	1.3	0.16	19	20	7

[1] pH and electrical conductivity of 1:5 soil:water suspension

[2] Exchangeable sodium percentage

The dye investigations showed that the flow in all 3 soils occurred mainly in cylindrical voids which often contained roots and/or faecal pellets. Dyed voids were far more frequent in the Kerang soil than the others. When dyed roots intersected planes, the dye sometimes spread out in dendritic patterns over part of the planar surfaces but, in general, these surfaces appeared not to provide continuous pathways for flow.

Table 2. 'Saturated' hydraulic conductivity (mm/day) by two methods

Depth	Undisturbed blocks[1]			Well permeameter[2]		
(m)	Kg	B	N	Kg	B	N
0.2-0.45[3]	70	3.5	1.6	150	0.3	3.5
	370	4.9	2.1	250	6.7	4.5
	420		33	250	6.7	18
				350		420
0.5-0.75[4]	23	6	<1	50	0	0
	210	7	<1	65	3.3	0
	280	12	6	80		0
				90		0
				90		

[1]Method of Bouma & Dekker (1981) except that the excavated block was first coated with rapid setting cement before encasing it in gypsum.
[2]A constant head method due to W.D. Reynolds & D.E. Elrick (priv.comm.).
[3]Irrigation waters used for Kg, B and N of EC 150, 40,and 200µS/cm and SAR 2.7, <1 and 1.4 respectively.
[4]'Drainage' waters used for Kg, B and N of EC 50000, 950 and 1150µS/cm and SAR 51, 5 and <1 respectively.

For the undisturbed blocks and the well permeameters (Table 2), the Kerang soil, with many cylindrical voids, had the highest K values, despite high clay content and ESP (Table 1). The Narrabri soil showed greater variation in K values, at least some of which was probably associated with previous, widely spaced ripping. Zero values for the well permeameter method at 0.5-0.75m may have been caused by smearing of the walls of the permeameter holes. Though the Narrabri soil had very frequent planar voids and low ESP, its K values were generally low. Cylindrical voids were rare. The low K values for the Benerembah soil also were associated with rare cylindrical voids and, in addition, few planar voids and high ESP.

The identification of cylindrical rather than planar voids as the significant pathways for saturated waterflow in the Kerang soil, provides an explanation for the rapid development of watertables and consequent salinity following the advent of irrigation. Conversely the presence of the cylindrical voids should allow rapid drainage and desalinization. Destruction of their continuity by tillage or other means should be avoided, otherwise very low profile permeability is likely to develop as a result of the high sodicity levels. Thus tillage should be restricted both in depth and frequency, and should be accompanied by an appropriate ameliorant, e.g. gypsum, to maintain the permeability of the tilled layer.

The Benerembah soil, on the other hand, presents a more difficult management problem. Porosity created by deep ripping or deep ploughing in an attempt to improve water entry into, and drainage from the profile is unlikely to persist for long because of high subsoil ESP values. Massive doses of gypsum would be required to lower the profile sodicity levels appreciably.

The Narrabri soil with low ESP values, might be expected to benefit from deep tillage because the porosity created is likely to be persistent and provide the needed rootzone drainage for re-establishing aerobic conditions after irrigation and rain. In this soil, unlike the others, mole drainage could perhaps be a proposition.

Acknowledgements

A visit by Dr. J. Bouma (Netherlands Soil Survey Institute) to Australia stimulated this work.

References

Bouma, J. and L.W. Dekker. 1981. A method for measuring the vertical and horizontal K_{sat} of clay soils with macropores. Soil Scil Soc. Am. J., 45: 662-663.

Rapid changes in soil water suction in a clayey subsoil due to large macropores

H.H.Becher
Lehrstuhl für Bodenkunde, TU München,
8050 Freising-Weihenstephan, F.R.G.
W.Vogl
Lehrstuhl für Bodenkunde, TU München,
8050 Freising-Weihenstephan, F.R.G.

A fertilization study on a clayey soil in southern Bavaria showed no definite effects of K- and P-fertilizer levels on crop yields during a four-year period since its start in 1974 (Vogl, 1982). This finding initiated in 1978 a study for evaluating the soil water regime and its effect on crop yields. Monitoring soil water tensions during the growing seasons 1978-1980 showed unexpectedly rapid changes in soil water tension (Vogl, 1982), the reason of which was at first glance not detectable.

The soil of the study site in the so-called Tertiary Hill Country in southern Bavaria is an Aquic Chromudert derived from Tertiary calcareous (up to 57% carbonates) clay (37-62% clay, 22-41% silt, 4-22% sand) overlain by a thin loess layer (28% clay, 39% silt, 33% sand). Smectites dominate in the Tertiary clay. During 1978-1980 the crops on the site were winter wheat, winter barley followed by rape as intercrop, and corn, respectively.

Mercury tensiometers were installed at 8 different depths between 25cm and 160cm (in 3 replicates down to 65cm and 2 replicates at the other depths). Self-made tensiometers constructed from PVC-pipes (26mm outer diameter) fitted tightly in auger holes (25mm diameter), thus obviating precipitation running off into the soil along the tensiometer shaft. They were arranged such that any interaction between two neighbouring depths was eliminated. Soil water characteristics and saturated hydraulic conductivities were determined on undisturbed soil core samples in the laboratory, and unsaturated hydraulic conductivities were calculated using the equation from Campbell (1974). Tensions were read and precipitations recorded daily during the growing seasons. The resulting soil water tension/time curves for each depth were smoothed by fitting a third order polynom using the procedure of Savitzky and Golay (1964).

The smoothed soil water tension/time curves for depths >85cm showed a typical feature in the seasons 1978 and 1979. Soil water tensions slowly increased by about 200cmWC within 3-6 weeks and rapidly decreased by the same amount within 7-10 days. This saw-tooth like feature was more pronounced in 1978 than in 1979 and was only faintly developed in 1980 (Figure 1). For depths <85cm the saw-tooth like feature could only be observed in 1978 before harvesting, and it vanished more and more with decreasing distance to the soil surface. After harvesting, soil water tensions in ≦85cm remained between 0 and 50cmWC, if no intercrop had developed. In 1980, due to much precipitation during the first growing stages of corn (which is well developed late in the season), soil water tensions were near or below 0cmWC in all depths until corn was well developed. Periods with high precipitaion coincided in 1978 and 1979, with a large decrease of soil water tension at all depths (Figure 1).

Initially, poor tightening between soil and tensiometer shafts was considered to be responsible for the saw-tooth like behaviour of the tension/time curves. Two reasons, however, were contradictionary to this. Firstly, a tensiometer installation check showed no possibility for water to flow deep into the soil along the tensiometer shaft. Secondly - and this is very important - if there would be a poor tightening then the tensions should change more rapidly and drastically in the upper horizons than in the deeper horizons, particularly after harvest when no transpiration occurs. Water flow through macropores having a high continuity, which may develop as cracks in this clayey soil during a dry period, seems to be a more plausible cause for the saw-tooth like feature of the tension/time curves. This is confirmed by the soil water characteristics and unsaturated hydraulic conductivities of the horizons, field observations on cracking, and water tensions in the upper horizons.

The low unsaturated hydraulic conductivities of the upper horizons even at low soil water tensions cause most of the water of a higher precipitation following a dry period to flow downward, along the walls of the cracks developed during the dry period, and not into the aggregates of the upper horizons. The very large changes of soil water

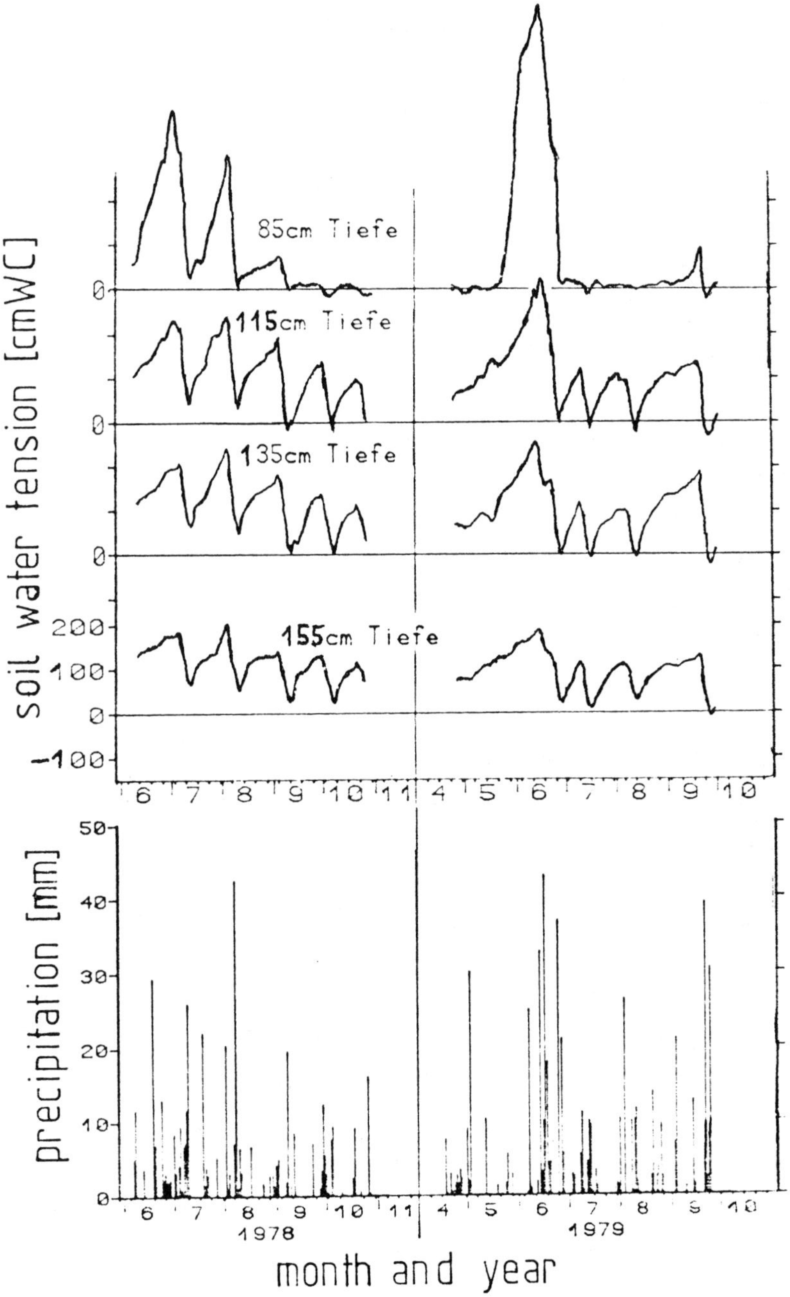

Figure 1: soil water tension/time curves for the 85, 115, 135, and 155cm depths in the seasons of 1978 and 1979, and corresponding precipitation

tension (e.g. 260cmWC in the range between 100 and 400cmWC) due to a small change in water content (2%b.v. H_2O) of the horizons >85cm depth, also cause a rapid decrease in the soil water tension in the subsoil horizons. This phenomenon was also observed by Germann and Beven (1981) and Kutilek (1980; personal communication) and explained in the same way. The relatively dry soil aggregates do not swell quickly enough on water supply to hinder surface water to flow down. Thus, after a longer rainfall period, water-logged soil conditions may occur in some parts of the subsoil, even though the soil horizons above and below of this part are relatively dry. Drainage of the water takes place only to a small degree, because the unsaturated hydraulic conductivities of the horizons >150cm are low even near saturation. No cracks develop below this depth.

After harvest when evapotranspiration is strongly reduced, soil water tensions within rooting depth decreased to about 30-50cmWC for the rest of the year. Although the aggregates of the upper horizons swelled to some extent due to wetting up of the soil, the cracks are more or less intact in the surface soil as well as in the subsoil. The low unsaturated conductivities of the surface soil even at low water tensions, and the high continuities of the cracks cause heavy precipitation to cause the saw-tooth like feature of the water tension/time curves in the subsoil horizons.

After a wet spring and planting a crop that is well developed late in the season (e.g. corn), no cracks will develop and, thus, as in 1980, no saw-tooth like feature of the water tension/time curve will be observed.

References

Campbell,G.S. 1974. A simple method for determining unsaturated conductivities from moisture retention data. Soil Sci. **117**:311-314

Germann,P. and K.Beven 1981. Water flow in soil macropores. I.-III. J. of Soil Sci. **32**:1-39

Savitzky,A. and M.J.E.Golay 1964. Smoothing and differentiation of data by simplified least squares procedures. Analyt. Chem. **36**:1627-1640

Vogl,W. 1982. Stoffflüsse in repräsentativen Böden des Unterbayerischen Hügellandes (Hallertau). Ph.D. Thesis, Techn. Univ. München

Rainfall infiltration into swelling soils

J.V. Giráldez
Instituto Nacional de Investigaciones
Agrarias, Apdo. 240, Córdoba, Spain
G. Sposito
Dept. of Soil and Environmental
Sciences, University of California,
Riverside, CA 92521, USA

Water movement through swelling soils exhibits features different from those observed in rigid soils, mainly due to the presence of the envelope-pressure potential. The governing equation for soil water movement is the generalized Richards equation,

$$\frac{\rho_{bo}}{\rho_\omega}\frac{\partial\theta}{\partial t} = \frac{\partial}{\partial z}\left[D\,\frac{\partial\theta}{\partial z}\right] + \frac{\partial H}{\partial z} \qquad (1)$$

where ρ_{bo} is the dry bulk density in some reference state, chosen to be the initial one, ρ_ω is the mass density of soil water, θ is the gravimetric water content, D is the soil water diffusivity, z is a material coordinate taken positive upwards, and H is a gravity-envelope-pressure parameter given by:

$$H = K(1 - \rho_{b\omega}\bar{V}_\omega) \qquad (2)$$

K representing the hydraulic conductivity, $\rho_{b\omega}$ the wet bulk density, and $\bar{V}_\omega$ the slope of the shrinkage curve. For a rigid soil H is equal to K, but in deformable soils H is lower than K, becoming negative at high water contents.

Rainfall infiltration can be described by integration of (1) in a soil profile, whose initial water content profile is known, subjected to an external flux condition at the soil surface, r being the rate of rainfall,

$$-(D \frac{\partial\theta}{\partial z} + H) = r \qquad (3)$$

This upper boundary condition holds up to the time when the surface water content reaches its maximum value, i.e. the ponding time. After that moment the surface water content remains constant. The integration can be carried out numerically through a Crank-Nicolson finite difference scheme. Alternatively the approximate analytical method proposed by Smith and Parlange (1978) in their study of rainfall infiltration in nondeformable soils, may be used. The integration of (1) by both methods, which agree closely, allows a comparison of the infiltration into swelling and nonswelling analogous soils, i.e. having the same properties, except that $\bar{V}_\omega = 0$. The most relevant results are the ponding time, and the post-ponding infiltration rates.

In general, ponding occurs earlier in swelling than in nonswelling analogous soils. This property can be seen in the approximate equation for ponding time, t_p,

$$\int_o^{t_p} (r + H_0)dt = \frac{\rho_{bo}}{\rho_\omega} \int_{\theta_0}^{\theta_m} \frac{D(\theta - \theta_0)}{r_p + H} d\theta \qquad (4)$$

where the subscript o stands for the initial and m for the maximum values. r_p is the rainfall rate at the ponding time. This equation is a generalization of an equation of Smith and Parlange, valid for both deformable and nondeformable soils. Due to the negative value of r, the fact that H is smaller than K in swelling soils explains the shorter t_p for them.

The post-ponding infiltration rates are lower in swelling than in the nonswelling analogous soils. This can be seen easily from a generalization of the Smith and Parlange equation. For the infiltration rate f corresponding to time t,

$$t - t_p = \frac{\rho_{bo}}{\rho_\omega} \int_{\theta_0}^{\theta_m} \frac{D(\theta - \theta_0)}{(H - H_0)^2} [Ln \frac{(r_p + H_0)(f + H)}{(f + H_0)(r_p + H)} +$$

$$\frac{(H - H_0)(f - r_p)}{(r_p + H)(f + H)}] d\theta \qquad (5)$$

In rigid soils the ultimate rate of infiltration is equal to the hydraulic conductivity at saturation. In swelling soils there is ultimately a balance between the gradient of the matric potential and the force associated with gravity and the envelope pressure. As a result, ultimately the rate of infiltration approaches zero and the water content profile approaches the equilibrium profile with free water at the surface.
The case of instantaneous ponding is obtained by taking the limit in (5) when $r_p \downarrow -\infty$ and $t_p \downarrow 0$, that is

$$t = \frac{\rho_{bo}}{\rho_\omega} \int_{\theta_0}^{\theta_m} \frac{D(\theta - \theta_0)}{(H - H_0)^2} \left[\mathrm{Ln} \frac{f + H}{f + H_0} - \frac{H - H_0}{f + H} \right] d\theta \tag{6}$$

which, again, is a generalization of a previous formula of Parlange (1971). Simpler results may be obtained for constant rainfall rate by using the interpolation formula of Parlange et al. (1982) for the hydraulic conductivity

$$\frac{dK}{d\theta} = \frac{2\alpha}{S^2} (K_m - K_0)(\rho_{bo}/\rho_\omega)^2(\theta - \theta_0)D \tag{7}$$

where S is the sorptivity approximated by Talsma and Parlange (1972)

$$S^2 = 2(\rho_{bo}/\rho_\omega)^2 \int_{\theta_0}^{\theta_m} D(\theta - \theta_0)d\theta \tag{8}$$

and α is a parameter. When $\alpha \simeq 0$, K will increase slowly as θ approaches θ_m; when $\alpha \simeq 1$, K will increase rapidly near the maximum water content. In this case the dimensionless ponding time, T_p,

$$T_p \equiv \frac{2(\rho_{bo}/\rho_\omega)}{S^2} H_m^2 t_p \tag{9}$$

is expressed as a function of $H_p \equiv H_m/r_p$, neglecting H_0, by the relation

$$T_p = \begin{cases} H_p^2/(1 + H_p) & \text{"} \quad \alpha \simeq 0 \\ H_p \,\mathrm{Ln}\,(1 + H_p) & \text{"} \quad \alpha \simeq 1 \end{cases} \tag{10}$$

Likewise the post-ponding infiltration rate, $f_p \equiv f/r_p$, becomes implicitly,

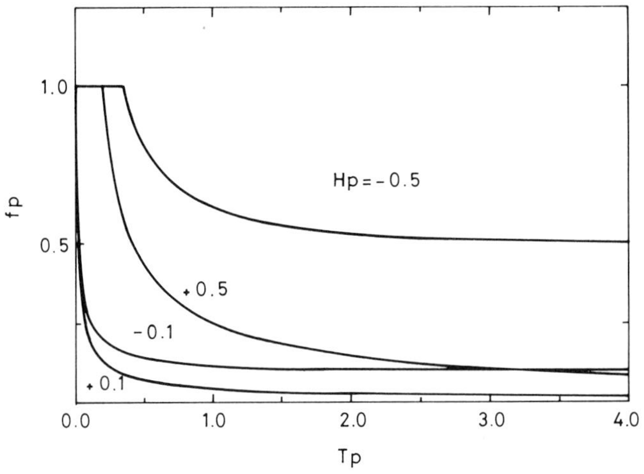

Figure 1. Normalized post-ponding infiltration rate, f_p, versus normalized time, T, for two normalized H_m values, H_p, according to (10) and (11) for swelling (positive H_p values) and nonswelling (negative H_p values) soils.

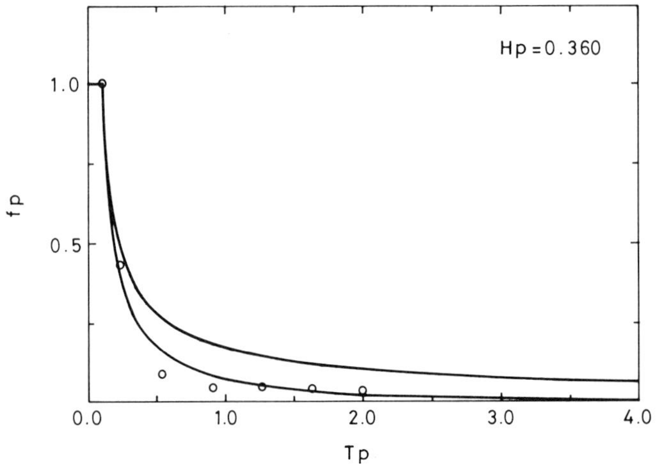

Figure 2. Normalized post-ponding infiltration rate, f_p, versus normalized time, T, for Putnam clay (Neal, 1938) compared with the predictions of (11), with $\alpha \simeq 0$, lower curve, and $\alpha \simeq 1$, upper curve, respectively.

$$T - T_p = \begin{cases} \mathrm{Ln}\,[(f_p + H_p)/f_p/(1 + H_p)] + H_p(f_p - 1)/(1 + H_p)/(f_p + H_p) & , \ \alpha \simeq 0 \\ (1 - f_p)H_p/f_p - \mathrm{Ln}\,[(f_p + H_p)/f_p/(1 + H_p)] & , \ \alpha \simeq 1 \end{cases} \qquad (11)$$

Figure 1 shows the different infiltration rates according to (10) and (11) for swelling, positive values of H_p, and nonswelling soils, negative H_p values, for the case of $\alpha \simeq 1$. Figure 2 exhibits the close approach of (11) to the experimental data of Neal (1938).

References

Neal, J. 1938. The effect of the degree of slope and rainfall characteristics on runoff and soil erosion. Missouri Agric. Exp. Sta. Res. Bull. 280.

Parlange, J.-Y. 1971. Theory of water infiltration in soils: 2. One-dimensional infiltration. Soil Sci., 111: 170-174.

Parlange, J.-Y., I. Lisle, R.D. Braddock, and R.E. Smith. 1982. The three-parameter infiltration equation. Soil Sci., 133: 337-341.

Smith, R.E. and J.-Y. Parlange. 1978. A parameter-efficient hydrologic infiltration model. Water Resour. Res., 14: 533-538.

Talsma, T. and J.-Y. Parlange. 1972. One-dimensional vertical infiltration. Aust. J. Soil Res., 10: 143-150.

Field evidence for a two-phase soil water regime in clay soils

Adrian C. Armstrong & Robert Arrowsmith
Field Drainage Experimental Unit,
Ministry of Agriculture, Fisheries and Food, Anstey Hall, Trumpington,
Cambridge, England

1 Introduction

In clay soils, an important component of soil water movement is the network of paths for rapid movement termed 'macropores'. The division of water into two phases, of high and low mobility, has proved useful for modelling soil water regimes (Armstrong 1983); subdividing the soil into a system of high conductivity and low drainable porosity representing the macropores, and a second system with very low conductivity but high drainable porosity representing the soil matrix. This paper considers the degree to which the behaviour of the soil water system predicted by the model is reflected in observation.

2 Methods

In the past, measurement of soil water regimes has assumed a single phase model, recording a single water table level. However increasing experimental evidence has indicated that in clay soils, such techniques give conflicting results. If it is accepted that water levels measured in open auger holes record only the level of water in the macropores, the apparent conflict between the low conductivity of these soils and the rapid response of these observations is resolved. Corresponding measurements in the soil matrix can be obtained using the neutron probe, an instrument designed specifically to measure the total water content of the profile. Since the bulk of the soil water is in the matrix, neutron probes will reflect in the main the water content of the micropores.

3 Site details

The Napton Holt moling study provides simultaneous observations of the levels in auger holes and neutron probe measurements of soil water content. The soil of the site is a Denchworth series clay, (an Aeric Haplaquept or Eutric Gleysol), and the site was in grass for the duration of this study. Neutron probes were monitored from December 1979 to November 1981, on an approximately fortnightly basis, whereas rainfall and the water levels in the auger holes were read weekly. Access tubes were installed to a depth of 1m, and read at 10 cm increments to a depth of 80 cm. The auger holes were installed with a 80 mm auger to a depth of 850 mm and lined with perforated pipe. Neutron probe readings were calibrated against a sand standard, and no allowance made for soil differences between access holes.

The pattern of weekly/fortnightly readings (which was dictated by the objectives of the moling experiment), was sufficient to offer a first test of the conceptual two-phase model. However, the computer implementation of the two-phase model requires rainfall data on a daily basis, and further soil information that was not available for this site. It was thus not possible to simulate the behaviour of this site directly.

4 Results

Results are presented for one neutron probe access point and its associated auger hole only, chosen because of its distance from the moling operations on the site. Figure 1 shows the weekly rainfall, the total profile water content (mm of water in the top 850 mm of soil), and the water level in the auger hole. The pattern of water movement within the profile is given by figure 2, which graphs the percentage water content for each of the successive levels. These two graphs are typical of the whole data set, and can be compared with the type of soil water regime predicted by the two-phase model as shown in figure 3 (from Armstrong 1983).

The water levels in the auger holes fluctuate rapidly in response to rainfall events. In the absence of mole drainage near the auger hole, these fluctuations are relatively shallow, and the water level drops below 50 cm only after the profile starts to dry out in the spring. The rapid rise and fall of water levels in response to individual rainfalls cannot, however, be resolved by the weekly observations.

In contrast, the total profile water content, as revealed by the neutron probes, shows only slow variation, and changes only

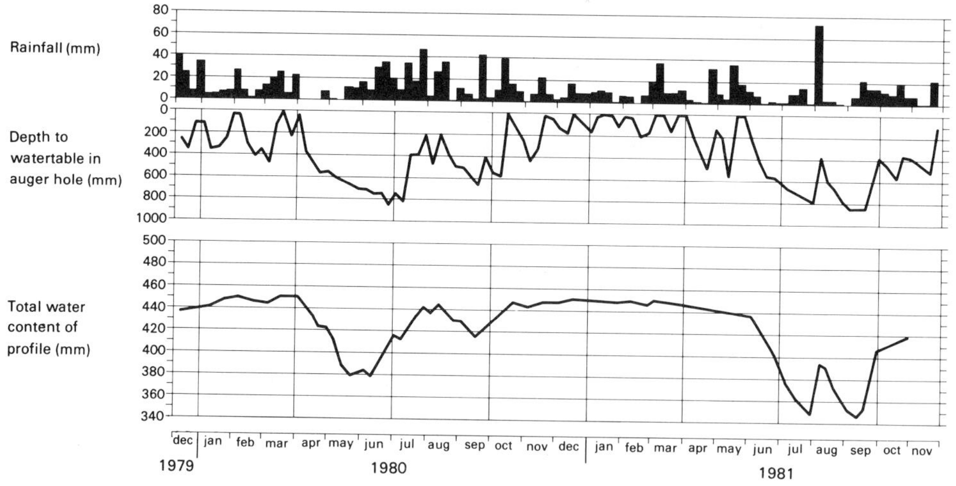

Figure 1. Weekly rainfall, depth to water table, and total water content of profile.

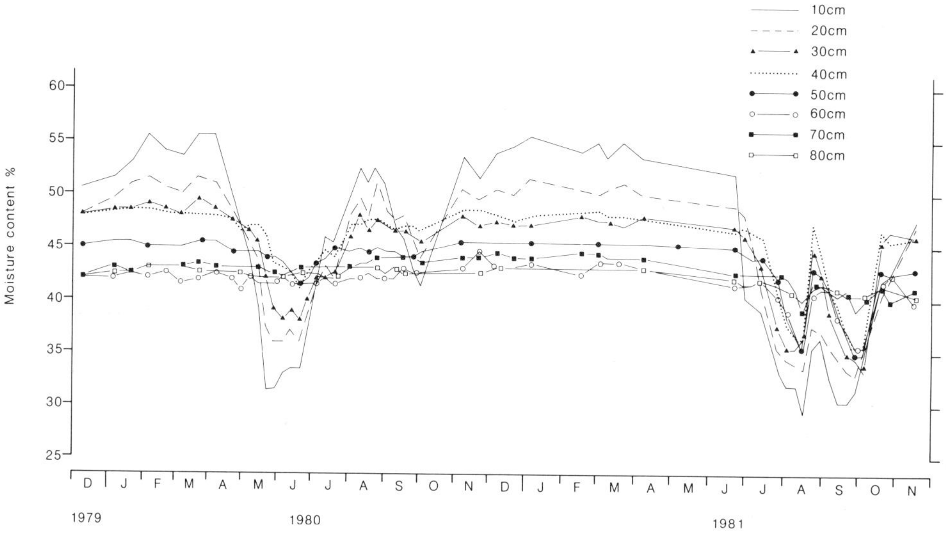

Figure 2. Pattern of water movement within the profile.

slightly during the winter period. Changes in the total water content of the profile are recorded only in the summer period, but even then they are restricted to the upper layers. There is only very small variation in water content at depths greater than 50 cm, and despite considerable wetting and drying of the profile, the soil below 50 cm took little part

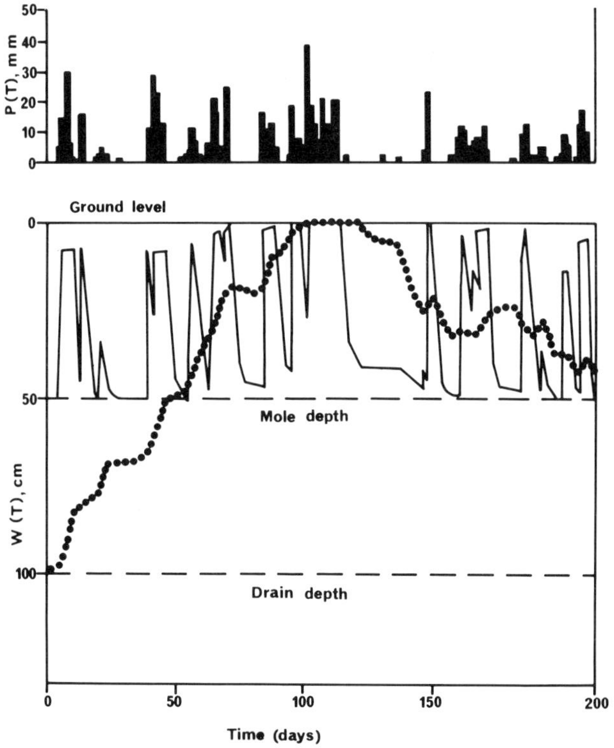

Fig 3. Results from the model. Daily rainfall values,P(T), and watertable levels,W(T),plotted for micropores(•••••)and macropores(———)

in the soil water system. One major exception is the very rapid fluctuation in water contents in August 1981, in response to 70 mm of rainfall after a dry period, when rapid variation extends down to the deeper levels, suggesting that the influence of the previous drying was to open up the macropores to depth, so that 'short-circuited' water can move down the profile and wet up the subsoil.

Comparing these results with the two-phase model in figure 3, it is clear that the division of soil water into rapidly varying and slowly moving phases is supported by the observational data. The rapid fluctuation in the simulated macropore phase is well matched by the auger hole levels. The comparison between the predicted micropore water levels and the neutron probe values is less promising. However, this reflects the use in the simulation of a water table model as a heuristic device to explore the two-phase model. It is suggested that the observational data confirm the conceptual model, but indicates an inadequacy in the particular mathematical representation of the micropore phase in this realisation. A more realistic representation must then be sought.

5 Conclusions

These data show that in a clay soil there is a distinct difference in mode of behaviour between water levels recorded in an open auger hole and the profile water content measured by the neutron probe. This overall pattern is very similar to that predicted by the two-phase model. Whereas this evidence cannot be taken as direct confirmation of the model, since no attempt has been made to match model predictions against observations, it does give confidence that the conceptual model has some validity.

6 Reference

Armstrong, A.C. 1983. A heuristic model of soil water regime in clay soils in the presence of mole drainage. Agricultural Water Management, 6, 191-201.

Discussion

L.P. Wilding:

1. How was the neutron probe calibrated? Our experience is that it is very difficult to calibrate neutron probes in clayey shrinking-swelling soil systems. The problem is shrinkage cracks about the neutron probe and simultaneous varying in moisture-content volume relationships.
2. We have determined the K sat. of the soil ped matrix (micropores) versus macropores by installing into the same horizon solid wall piezometer tubes with and without cavities extracted below the tubes. Those tubes with cavities intersect macrovoids, while those tubes without cavities have a high probability of terminating within the soil ped matrix, if the peds are large. Then, K sat. is measured during periods of saturation by rate of recharge, once the water in the piezometers has been removed.

3. Comment on continuity or interaction of the 'Two-Porosity' system. While at very high levels of resolution we are dealing with a completely interconnected porous system, in a physical context the cutans along major macrovoids essentially separate the porosity into two network systems. Hence there is microscopic thin-section evidence to support the limited interaction of the two pore systems.

Author:

1. The neutron probe was initially calibrated using a sand standard, i.e. 8 m^3 of oven-dry sand, which defined the 300 point, and a smaller sample of completely saturated sand. The probe was assumed to be linear between these two points. This calibration was retrospectively corrected by reference to volumetric samples taken adjacent to those of the access tubes at the end of the experimental period. Any inadequacies of this procedure, however, do not invalidate the observations which are concerned with relative changes, not absolute values of water content. Cracking around the access tube can be minimized by careful field installation. Examination of the profile by excavation at the end of the experiment suggested this was not a problem.
2. No comment.
3. I agree that interaction between the two components is difficult to model. The model as implemented uses the Donnan equation, assuming the macropores act as fully penetrating ditches. Insertion of a physically reasonable estimate of the frequency of macropores results in an interaction that is too strong, and the two systems behave as one (Armstrong, 1984). In order to achieve realistic results a wider spacing must be used, indicating that the interaction is surprisingly weak. This reflects your microscopic observations.

Reference: Armstrong, A.C.1984. Reply to V. Chour's note on a heuristic model of soil water regimes in the presence of mole drainage. Agricultural water Management, 9: 158-164.

Infiltration of water into cracked soil

V. Novák, A. Šoltész
Institute of Hydrology and
Hydraulics of Slovak Academy of
Sciences, Bratislava,
Czechoslovakia

An empirical method is presented for determining the time course of the rate of infiltration into heavy, swelling soils. The initial water content and precipitation intensity are taken into account.

It is assumed that:

- water infiltrates into heavy soils primarily through the cracks, infiltration through the soil matrix is relatively small, (Bouma et al. 1977, Bouma and Dekker 1978, Bouma et al. 1978);
- the geometry of cracks, expressed by the volume fraction of cracks P_c, is a function of the water content, (Doležal and Kutílek 1972, Novák 1976).

The relationship between the rate of infiltration v_i and the time t is given by

$$v_i = v_z \qquad \text{for } t \leq t_p \tag{1}$$

$$v_i = v_v + (v_z - v_v) \exp(-\beta(t - t_p)) \qquad \text{for } t \geq t_p \tag{2}$$

where

t_p = ponding time
v_z = rate of precipitation
v_v = steady infiltration rate which can be considered to be equal to the hydraulic conductivity K of the soil at saturation

β = empirical parameter

The measurements have shown that the rate of infiltration v_i depends on the volume fraction of cracks P_c, this being a function of the initial water content w_i and the rate of precipitation v_z (Mein and Larson 1973, Peschke and Kutílek 1976). Equation (2) implies that the rate of infiltration v_i is influenced by the initial water content w_i and the rate of precipitation v_z by means of the ponding time t_p and the parameter β. Identification of the characteristics t_p and β consists in finding the relationships between them, crack porosity P_c and v_z. The characteristics of the infiltration process were determined by infiltration of simulated precipitation into heavy soil at the locality of Zemplínske Hradište. Relations $P_c = f(w_i)$ - thirteen curves - were unambiguous and therefore the relationship $t_p = f(P_c, v_z)$ was replaced by $t_p = f(w_i, v_z)$. For constant rate of precipitation it was found that:

$$t_p = t_{p,max}\{(w_s - w_i)/(w_s - w_h)\}^n \qquad (3)$$

where

w_h = water content of soil for $P_c = P_{c,max}$ (w_h = 6,06)
w_s = water content at saturation (w_s = 45%)
n = empirical coefficient (n = 6)
$t_{p,max}$ = period of crack closing during water excess for given locality ($t_{p,max}$ = 48 h)

Values $t_{p,max}$ and $t_{p,min}$ were estimated from the relationships:

$v_z = K$ when $t_p = t_{p,max}$ and $v_z = v_{z,max}$ when $t_p = t_{p,min}$

In our case $K = 1{,}05$ mm.h^{-1}, $t_{p,min} = 0{,}01$ h, $v_{z,max} = 60$ mm.h^{-1}. For arbitrary w_i the relation $t_p = f(v_z)$ - Figure 1 - was calculated from the values presented above:

$$t_p = t_{p,max} - (v_z - K/v_{z,max} - v_{t,min})^m \, (t_{p,max} - t_{p,min}) \qquad (4)$$

where

m = an empirical coefficient characterizing the impact of w_i (Table 1) and all other values are known from measurements.

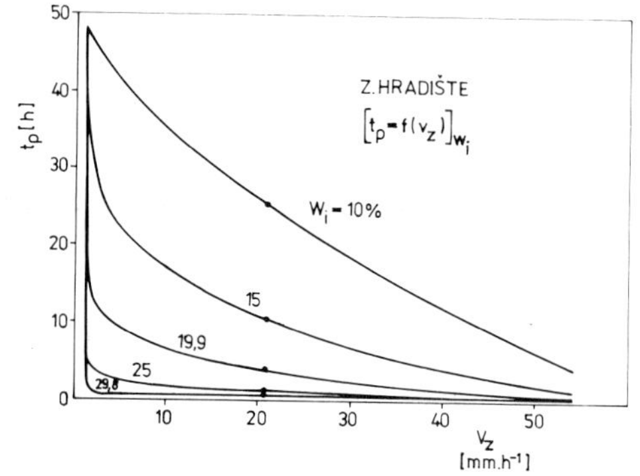

Figure 1. Ponding time (t_p) against precipitation intensity (v_z) for different initial water contents (w_i) - equation (3).

The parameter β, as a function of the water content, was determined from the results of measurements on the assumption that $\beta = \beta_0$ for $P_c = 0$; $w_i = 34,9$; $\beta_0 = 0,175$. Figure 2a presents the relation $\beta = f(w)$. Relation $v_i = f(t)$ for $v_z = 20,8$ mm.h^{-1} is presented in Figure 2b, and its parameters are given in Table 1.

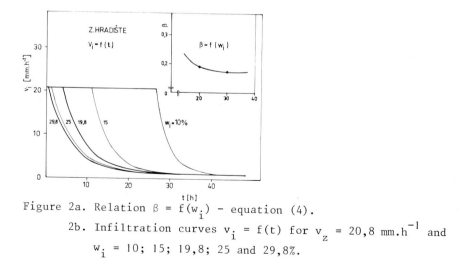

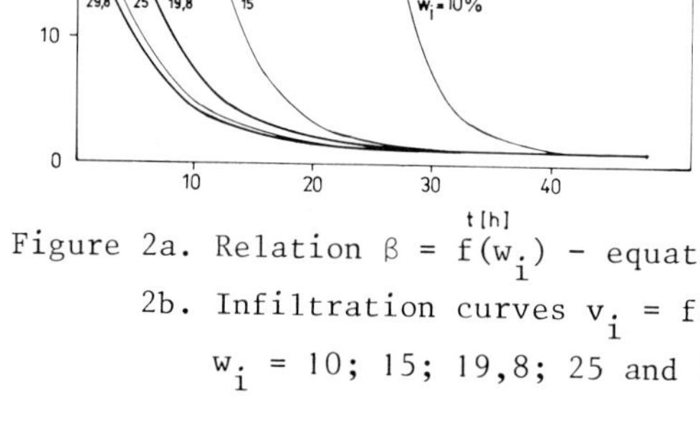

Figure 2a. Relation $\beta = f(w_i)$ - equation (4).

2b. Infiltration curves $v_i = f(t)$ for $v_z = 20,8$ mm.h^{-1} and w_i = 10; 15; 19,8; 25 and 29,8%.

Table 1. Characteristics of the calculated infiltration curves

	$v_z = 20,8\ (mm.h^{-1})$				
w_i	10	15	19,8	25	29,8
t_p (h)	26,06	10,76	3,9	1,09	0,22
m	0,716	0,232	0,077	0,021	0,004
	0,305	0,229	0,189	0,180	0,176
P_c	0,162	0,155	0,135	0,114	0,075

The suggested method enables one to calculate the relation $v_i = f(t)$ for heavy cracked soils when the soil characteristics K, $P_c = f(w_i)$ and the results v_z, t_p, w_i from at least one infiltration test are known.

References

Bouma, J. et al. 1977. The function of different types of macropores during saturated flow through four swelling soil horizons. Soil Sci. Soc. Amer. J., 41, pp. 945-950.

Bouma, J. and L.W. Dekker 1978. A case study on infiltration into dry clay soil, I. Morphological observations, Geoderma, 20, pp, 27-40.

Bouma, J. et al. 1978. A case study on infiltration into dry clay soil II. Physical measurements, Geoderma, 20, pp. 41-51.

Doležal, F. and M. Kutílek 1972. Flow of water in swelling soil. In: Proc. second Symp. Fundamentals of transport phenomena in porous media. Univ. of Guelph, Ontario, Canada, v.1., pp. 292-305.

Kutílek, M. and V. Novák 1976. The influence of soil cracks upon infiltration and ponding time. In: Proc. of the Symp. Water in heavy soils, Kutílek, M. and Šútor, J. (ed.), Bratislava, v.1., pp. 126-134.

Mein, R.G. and G.L. Larson 1973. Modeling infiltration during a steady rain. Water Resources Res. 9, 2, pp. 384-394.

Peschke, G. and M. Kutílek 1976. The role of preferential ways in infiltration. In: Proc. of the Symp. Water in heavy soils, Kutílek, M. and Šútor, J. (ed.), Bratislava, pp. 70-85.

The role of earthworm channels in water flow on a drained clay soil

J. Urbánek[1], and F. Doležal[2]

The geometry of tubular channels made by earthworms and their influence upon the drainability of a clay soil was investigated at Praha 4 - Opatov, on a meadow, with a drainage system installed 40 years ago. Soil type (according approx. to FAO classification) was Gleyic Luvisol, clayey, with approx. 50 % of particles below 10 microns in topsoil and 70 % in subsoil, wet, developed in quarternary deluvial sediments over ordovic shales. The lateral drains were laid manually into a trench that was 0,95 m deep, and 0,22 m wide at the bottom. Drain tiles of internal diameter 50 mm and of length 0,33 m were covered by humous soil material to a depth of about 0,15 - 0,20 m above the bottom, and the trench was further filled with mixed soil backfill material. The distance between lateral drains was 10 m.

Investigations were made down to a depth of about 1,1 m. Location A was above a drain, and location B was in the middle between lateral drains. On a vertical plane section, perpendicular to the drain, the old backfill soil was studied, and undisturbed cylindrical soil core samples with a volume of 100 cm^3 were taken from the most interesting points. It is clear from the results in Fig. 1 that some sort of eluviated zone had developed in the backfill, at the walls of the trench. The soil in this zone shows higher permeability, more coarse pores and less shrinkage than both the soil in the middle of the backfill and the undisturbed subsoil.

---

[1] State Bureau for Land Reclamation, Prague, Czechoslovakia

[2] Design and Construction Institute of the Czechoslovak Ceramic Industries, Prague, Czechoslovakia

On square plots with an infiltration area of 1 m^2 at depth 0,35 m, infiltration rates were measured at locations A and B. Infiltration rates after 8 hours were 0,13 mm/min above the drain, and 0,003 mm/min between the drains.

As no cracks were observed, we concentrated on the earthworm channels, which had, for depths greater than 0,35 m and except for the vicinity of drains, approximately vertical directions. The numbers and sizes of the earthworm channels were registered in horizontal soil sections just below infiltration plots. Only the channels with diameters above 2 mm were taken into consideration. The maximum diameter observed was 10 mm. In Fig. 2, vertical profiles are given of the frequency of the channels, of their areal fraction (macroporosity) and of their laminar hydraulic conductance:

$$\Omega_T = \frac{3}{2}\pi \cdot \sum_i n_i R_i^4$$

where R_i ... mean radius of channels of the i-th class,

n_i ... number of channels of the i-th class in unit cross-sectional area.

The maximum depth to which the earthworms borrowed their channels between drains was 1,15 m, where a hard layer occured. Most of the channels had their lower ends at depths from 0,4 to 1,1 m, in the impermeable subsoil. We found here spherical voids, where living earthworms were sometimes present, wound into a little ball. It was therefore evident that earthworm channels between drains, without mutual hydraulic connection and without any internal drainage, did not contribute substantially to water movement in the soil. The contrary was valid, however, for location A, where we removed the soil material above the drain to a depth of 0,7 m below surface. From here we followed carefully the individual earthworm channels, in order to ascertain, whether thay had a hydraulic connection with the drain. Investigations were made for a length of about 4 m, examining 11 joints between drain tiles. The methods of investigation depended mainly on the detailed structure of macropores in the vicinity of individual joints. Visual observation prevailed, with the help of photography. We measured sometimes the flow of water by pouring it into big channels by means of a funnel. We used also gypsum to obtain casts of the macropores. From these 11 joints, some marks of hydraulic connection to the interior of the drain were present in 6 cases, and in one of them (No. 10), the connection was clearly present:

Joint No.	Commentary
0	A channel, visually observed, led to the joint, apparently permeable, containing small roots.
2	Two channels, visually observed, led to the joint. The joint itself seemed to be rather clogged. Small-scale infiltration test in the backfill, 5 cm above the drain: 0,84 cm^3/s on the area of about 200 cm^2.
4	A channel ended near the joint in a spherical void with a big earthworm in it. Observed visually, hydraulic connection not sure.
7	Hydraulic connection indicated by breakthrough of gypsum into the drain. After removing the drain tile, channels filled with gypsum were visible below the drain.
8	The gap between the tiles was wider than in other joints. Not only water, but also soil and earthworms could pass through it. Not typical.
10	Both gypsum suspension and water flowed rapidly into a channel of diameter 9 mm (at the rate of 0,6 cm^3/s). A cast of the channel continued nicely into the joint and into the drain.

From the measurements of hydraulic conductivity on large soil cores (diameter 80 mm, height about 70 mm) with channels of earthworms in them, the hydraulic conductance of individual channels (with diameters about 7 mm) appeared to be approx. 8 cm^3/s. Apparently, the natural flow path of water into the drain was much more tortuous. Unsufficient permeability of the joints themselves, partially clogged with soil, was the main reason for the reduction of flow rate. Nevertheless, the positive effect of earthworms upon the hydraulic properties of the backfill is evident. In the vicinity of the drain, the channels of earthworms led also parallel to the drain. Similar observations were reported by Taylor and Goins (1967).

References:

(1) URBÁNEK, J. 1972: Physical properties of a drained heavy soil, Ph. D. thesis, Prague, 111 p. (in Czech)

(2) URBÁNEK, J. 1976: The description of spatial pattern of soil macropores and its exploitation for the assessment of hydraulic characteristiques of the soil, In: Colloquium on Stereology, Vysoké Tatry, Czechoslovakia, 25 - 29 October, 1976, Vol. 1, pp. 251 - 258, The House of Technology, Košice.(in Czech)

(3) TAYLOR, G. S.,and GOINS, T. 1967: Field evaluation of tile drain filters in a humid region soil, Ohio Agric. Res. and Dev. Centre, Wooder, Ohio, Research Circular 154.

Seasonal changes in soil-water redistribution processes affecting drain flow

I. Reid and R.J. Parkinson*

Dept. of Geography, Birkbeck College, University of London WC1E 7HX, U.K.

(*now at Seale-Hayne College, Newton Abbot, TQ12 6NQ, U.K.)

Abstract

Annual changes in soil-water content in a cracking clay are shown to control the character of tile drain outfall through control of the mechanisms of soil-water redistribution. Hydrographs generated by tile drain networks and simple tile laterals over a period of four drainage seasons are compared for a) winter versus non-winter storm differences and b) simple versus complex storm differences. The observed variations are large and related to pre-rainfall soil-water contents and soil conditions.

Introduction

Agricultural underdrainage modifies the soil-water regime, resulting in larger water content variations at drained than at undrained sites (Robinson and Bevan, 1983). These water content fluctuations are of particular importance in soils that contain significant quantities of expandable clay minerals, since there is considerable enhancement of shrinkage cracks in these soils as water content decreases. As the majority of drainflow generating water movement in clay soils occurs in macropores (Bouma, 1981; Leeds-Harrison et al, 1982), seasonal patterns of drainflow response should result from the changing status of these water transmission pores. The results presented here do indeed demonstrate that seasonal variations in drainflow characteristics occur due to changes in soil-water content. In addition, short term variations in drainflow hydrograph response following closely spaced rainfall also relate to pre-storm soil-water contents.

Tile drain outfall was recorded over four drainage seasons, from 1977 to 1981, on a mixed arable/grass farm 18 km north of central London U.K. Ground slopes vary from 0 to 5°. The soil is a surface water gley (haplaqualf), and typically comprises 55% clay, 40% silt and 5% sand and gravel. Saturated hydraulic conductivity is variable, values of $10^{-3}m.s^{-1}$ for the topsoil and $10^{-6}m.s^{-1}$ for the subsoil being representative. Drainflow was continuously recorded from two herring-bone type networks (26.5 and 6.0 ha) and six laterals, ranging in drainage area between 0.21 and 0.55 ha.

Drainflow from the laterals was recorded using V notch weirboxes or tilting buckets linked to event recorders. The results of one network and two tile laterals are presented here to illustrate changes in water handling by the soil. Rainfall was measured continuously using an autographic gauge. From the chart trace, the rainstorm centroid was identified (P_{50} in Figure 1).

Drainflow hydrographs were analysed to extract time of start, t_s, and time of concentration, t_c, which were related to the rainstorm centroid. Peak drain discharge, Qp, was also determined. Over the four year period there were 215 storms. Two distinct phases of the annual water balance were identified using soil-water contents from neutron probe records: the winter period, when soil-water contents range narrowly around a winter mean value, and a non-winter period when soil-water contents are not fully recharged but drains also ran (Reid and Parkinson, 1981).

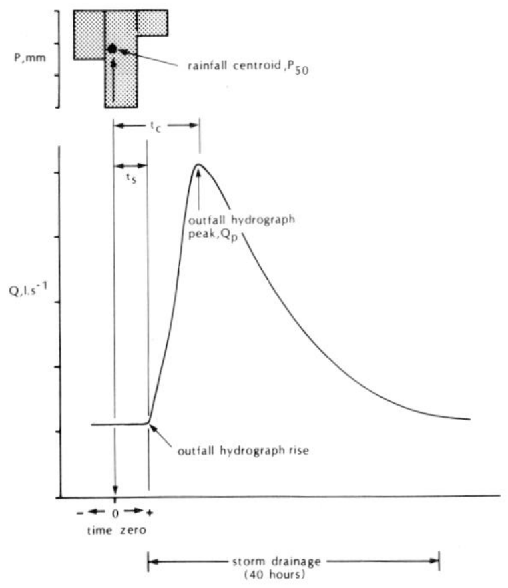

Figure 1. Drainflow hydrograph parameters.

Variations in drainflow characteristics

Drains in this clay soil typically run from late October until early June. The winter period, identified from soil-water records, usually occurs from early January until late May/early June. Therefore the majority of non-winter drainflow occurs in the autumn wetting-up period. Table 1 illustrates that median values of drain outfall hydrograph parameters are controlled by soil-water contents.

	Simple winter storms			Simple non-winter storms		
	t_s (hrs)	t_c	Qp ($l.s^{-1}$)	t_s (hrs)	t_c	Qp ($l.s^{-1}$)
Network	-0.16	3.16	2.37	0.25	4.08	1.60
Lateral A	0.58	3.67	0.24	0.37	5.26	0.09
Lateral B	0.68	3.16	0.09	0.58	4.29	0.08

Table 1. Median values of drain outfall hydrograph parameters for simple winter and simple non-winter storms

In the autumn, first drainflow appears less than one month after the lowest water contents in the year, and some two months before the soil is fully wetted (Reid and Parkinson, 1984). This drainflow by-passes the soil matrix, flowing down desiccation cracks before soil moisture deficits have returned to zero. As a result t_s values for laterals A and B are smaller in non-winter than winter. In the larger network, which features a bowl-shaped hollow, water contents are maintained at higher levels in the non-winter. Soil cracks are less well developed and the hydrograph time of start is delayed by comparison. All systems show a delayed t_c in non-winter compared with the winter, due to the fact that greater matrix absorption in non-winter slows the arrival of the pulse of rainwater at the tile drain. The effect of this absorption is further illustrated by the much smaller median peak discharges of the non-winter period.

In addition to changes in drain hydrograph characteristics over a drainage season, short term variations in drainflow occur during the winter period. Winter storms have been divided into <u>simple</u>, which includes the first hydrograph rises of complex storms, and secondary rises in complex storms. Despite the fact that the soil is fully recharged - water contents are at the winter mean level, $\theta = 0.454 m^3.m^{-3}$ - Table 2 demonstrates that the temporary soil water storage that exists at the beginning of a rainstorm is fully occupied by the time a secondary rainstorm commences. As a result, secondary storms show a quicker initial response, t_s, and

	Simple winter storms			Secondary winter storms		
	t_s (hrs)	t_c	Qp ($l.s^{-1}$)	t_s (hrs)	t_c	Qp ($l.s^{-1}$)
Network	- 0.16	3.16	2.37	-0.31	2.04	5.26
Lateral A	0.58	3.67	0.24	0.33	2.46	0.80
Lateral B	0.68	3.16	0.09	0.56	2.09	0.42

Table 2. Median values of drain outfall hydrograph parameters for simple rises and secondary rises in complex winter storms

reach a peak discharge more rapidly (t_c). The same behaviour is exhibited by both the simple laterals and the network. The clearest illustration of the difference between simple and secondary storms is shown by the median peak discharges, which for secondary storms are at least two times greater than for simple storms. But the simple storm response shows that clay soils, even at field capacity, can and do store water temporarily and thus reduce or delay drainflow.

References

Bouma, J. 1981. Soil morphology and preferential flow along macropores. Agric. Wat. Manag., 3: 235-250.

Leeds-Harrison, P., Spoor, G. and Godwin, R.J. 1982. Water flow to mole drains. J. Agric. Eng. Res., 27: 81-91.

Reid, I. and Parkinson, R.J. 1981. Too wet, too dry: clay soil probems. Soil and Water 9: 7-9.

Reid, I. and Parkinson, R.J. (1984). The nature of the tile drain outfall hydrograph in heavy clay soils. J. Hydrol. (in press).

Robinson, M. and Bevan, K.J. 1983. The effect of mole drainage on the hydrological response of a swelling clay soil. J. Hydrol. 64: 205-223.

Impact of water relations of vertisols on irrigation in Sudan (Field studies on Gezira clays)

Osman A. Fadl
University of Gezira, Wad Medani,
Sudan

The water relations of the vertisols in central Sudan have influenced every aspect of development of crop production and gravity irrigation. Of those fairly uniform soils the Gezira Scheme covers 900 000 hectares which are irrigated from the Blue Nile by extensive canals, paradoxically, with no drainage system. Fadl and Farbrother (1973) and Ali and Fadl (1977) concluded that irrigation in the Gezira with Blue Nile water for 55 years moved the salts downwards, reduced exchangeable sodium and increased exchangeable calcium.

In the desiccating summer the top 3 cm breaks up into loose granular mulch which overlies deeply cracked soil with a weak subangular structure. Aggregates in a freshly prepared seedbed slakes down immediately after irrigation or rains and the structure becomes massive. Nevertheless production records of the rotations on those unique soils have shown bumper yields of cotton, wheat, sorghum, groundnuts, forages and horticultural crops. Early research recommended planting of summer crops on ridges to maintain their susceptible seedlings above the level of standing water during the rainy months of July to September. The decline of yield in the late 1970's was attributed to economic factors and to the inadequacy of the old infrastructure to keep pace with new policies.

In the commercial area of the Gezira moisture profiles showed that, at cotton planting, as much as 5160 m^3/ha of total water was retained in the 150 cm depth (Figure 1). By September, however, irrigation and rains raised that quantity to 7040 m^3/ha.

In an experimental site at the Gezira Research Station flooding of dry plots in April gave net increases of between 1236 and 1562 m^3/ha

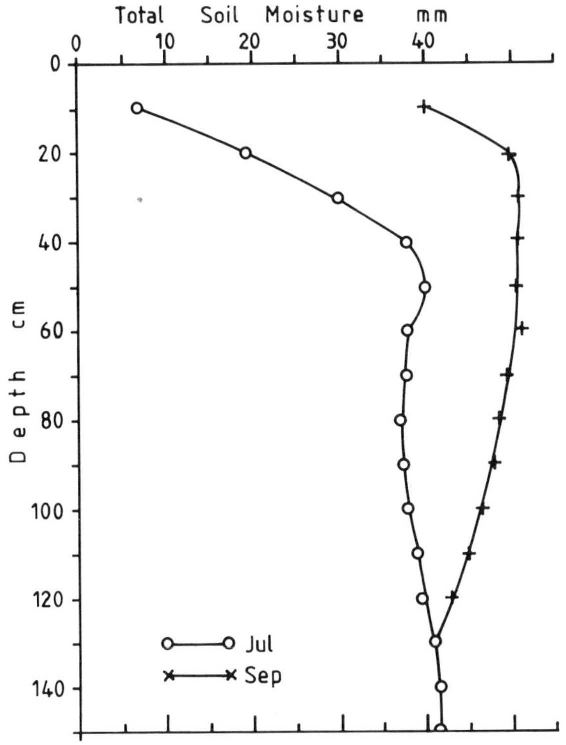

Figure 1. Typical moisture profiles under a Gezira tenancy in which mid July data represented cotton planting date and September the crop establishment (measurements made with a neutron probe)

of water stored to 110 cm depth. A second flooding of the same basins in May gave an average increase of 458 m^3/ha while the third in July added 833 m^3/ha. By the end of July there was 1700 m^3/ha of available water to 100 cm depth. Field measurements showed that subsequent irrigations varied between 700 and 1500 m^3/ha. The rates of application depended on the crop, the season and the preceding period of drying. Thus mean water intake of those soils was primarily influenced by the moisture content at the time of water application. The agricultural significance of that can be illustrated by typical field management situations.

The first agronomists in the Gezira introduced descriptive terms, e.g. "heavy" and "light", to describe watering rates in which approximately 100 mm and 40 mm respectively of free water stood on the surface. From the previously given water intake data let us analyse cases where different rates were applied to commercial or experimental plots. Thus a "heavy" watering of an already wet profile would amount to 1458 m^3/ha of which free water equals 1000 m^3/ha and soil uptake 458 m^3/ha. However, a "light" watering of a dry extensively cracked site would be 1799 m^3/ha of which free water constituted 400 m^3/ha and soil uptake 1399 m^3/ha. Such quantitative approach is now employed by agriculturalists and engineers in the Gezira to predict well in advance the water requirements of the cropped area over the season. Because of the overlapping of crops estimates of planting water depended on the drying period between the last rains or pre-sowing water and the sowing date. This procedure gained authenticity since it was utilised to evaluate the validity of the plans of cropped areas and to schedule irrigation which would have to satisfy the requirements of those crops. The supplies could under no circumstance exceed the known capacity of the distributory system.

Farbrother (1972) did not find free water to 3 m under a canal which conveyed water continuously for more than 30 years. Fadl (1974) followed moisture movements to 250 cm depth under paddy rice and found that flooding for 32 days gave increments of 1574 m^3/ha, for 60 days 2446 m^3/ha and 92 days 2472 m^3/ha. Such long flooding had a mean intake rate of 0.02 cm/hr. On the other hand a detailed study by Fadl and Ali (1977) on a saline-sodic site, which was flooded for the first time, indicated that the wetting front did not give defined plane boundaries and the layers started to transmit water before they were saturated. Thus one hour of flooding raised the moisture content in the 10 cm layer from 7 mm to 36 mm and in the 20 cm from 12 mm to 20 mm. After 4 hours the 10 cm reached saturation at 49 mm while the moisture at 20 cm was 33 mm and an increase of 6 mm were detected at 30 cm depth.

The effect of the wetness of a Gezira profile on infiltration rates is shown in Figure 2a and b. The data was collected with a neutron probe from access tubes in a plot which had been prepared for rice

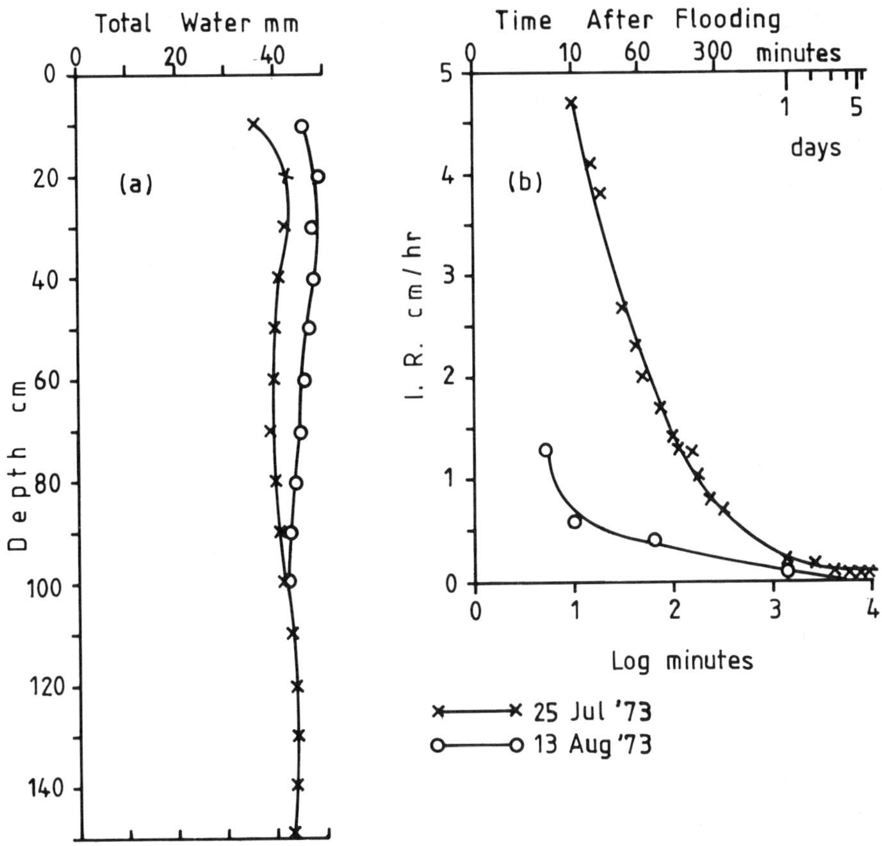

Figure 2. (a) Two moisture profiles at one site prior to flooding and (b) the corresponding infiltration rate (I.R.) in the Gezira

growing. In Figure 2a the wetness on 25 July was caused by an early heavy irrigation because the total rainfall to that date was 22 mm. The curve representing 13 August illustrated water distribution immediately before a prolonged flooding which continued to early November. The corresponding infiltration rates under the same profiles of 25 July and 13 August were given in Figure 2b. The nuclear technique measures the vertical changes in moisture irrespective of the magnitude of divergent flow.

The initially high infiltration rate in July was in response to both the comparatively low antecedent moisture and the numerous cracks. Those measurements were taken under a depth of 100 mm free water which dropped to 20 mm in four days. Rainfall during this period was 17 mm and some water was drained off the surface. After three days of flooding the mean rate of infiltration was 0.09 cm/hr and by the sixth day it was 0.05 cm/hr. Figure 2b indicates yet again that the infiltration rate was faster in July than in August because the former had a drier profile than the latter. Thus infiltration rate in August dropped to 0.06 cm/hr after one day. It can be safely concluded, therefore, that those swelling clays take up water to a certain maximum "wetting" while the excess remained on the surface to evaporate. The lack of deep drainage has worked towards water conservation to the extent that a tenant farmer cannot "over-water" his land though he could water-log it.

The low permeability in those soils and the restricted lateral movement of water have made it possible for engineers to adopt low cost methods of canal construction namely, building of banks from borrowed spoil which is the clay dug out from a (would be) canal bed. More than 10,500 km of unlined canals of different capacities now command the irrigated Gezira, some of them since it was commissioned in 1925, without seepage on the adjacent roads.

Figure 3 illustrates the moisture depletion as a percentage of total loss. The daily requirements of cotton were 9 mm and of groundnuts 7 mm. After irrigation the water stored in the upper 30 cm was adequate to satisfy 95 per cent of the water used by the shallow rooted groundnuts but only 70 per cent of the requirements of cotton. The roots of cotton supplemented the difference from the deep water reserves to 80 cm. However, the pattern of losses under bare land in mid-summer was identical to that of groundnuts.

The effect of moisture changes on dry density was measured by Fadl and Ali (1977) with neutron and gammaprobes in a field which was irrigated for the first time. Prior to flooding the volumetric moisture cotent at 10 cm depth was 5.3 per cent and the dry density 1.33 g/cm^3. The moisture at 60 cm was 18 per cent and the corresponding density 1.51 g/cm^3. However, after flooding and swelling the volumetric moisture at 10 cm increased to 44.9 per cent and the dry density lowered to 1.17 g/cm^3. Similarly, at 60 cm the moisture

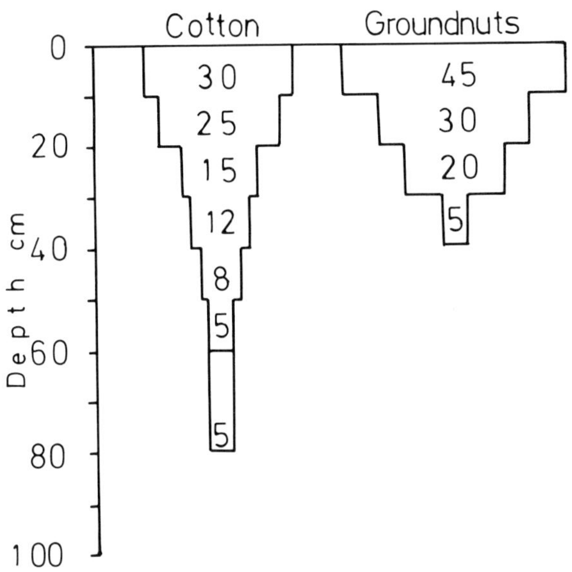

Figure 3. Moisture depletion expressed as per cent of total loss from the profile in mid season under vigorously growing crops adequately supplied with water

reached 46.0 per cent and the density 1.33 g/cm^3. Abdine and Farbrother (1967-68) published a table of dry density with the aim of using it to convert gravimetric data to volumetric. The changes of the density of a layer depended on whether the swelling was free or confined and on the size of the overburden. Such changes which were caused by wetting and drying cycles on those soils resulted in the extensive cracks characteristic of the cultivated areas.

References

Abdine, A.Z. and H.G. Farbrother 1967-68. Cotton Research Reports, Republic of the Sudan. Cotton Research Corp.

Ali, M.A. and O.A. Fadl 1977. Irrigation of a saline-sodic site in the Sudan Gezira II-salt movement. Trop. Agric. (Trinidad), 54: 279-283.

Fadl, O.A. 1975-76. Gezira soil: wetting under paddy rice and drying of bare land. Gezira Research Station, Ann. Rept.

Fadl, O.A. and M.A. Ali 1977. Irrigation of a saline-sodic site in the Sudan Gezira I-water movement. Trop. Agric. (Trinidad), 54: 157-165.

Fadl, O.A. and H.G. Farbrother 1973. Water management in the Sudan. In: Research needs for on-farm water management, D.F. Peterson (ed.), Proc. Internl. Symp., Utah State University, pp. 53-60.

Farbrother, H.G. 1972. Field behaviour of Gezira clay under irrigation. Cott. Gr. Rev. 49: 1-27.

Seasonal changes of hydric and structural behaviour in clay soils with saline watertables on the coast of Languedoc, France

Ramifications for mechanical properties and translocation phenomena.

J.C. Favrot - R. Bouzigues - Ph. Lagacherie
INRA - Soil Science - Montpellier.

INTRODUCTION

In Languedoc, to the west of the "Petit-Rhone" River, halomorphous clay soils take up nearly 25 000 hectares. They are located in a flat area and often below sea level between the coastal salt pans and the Mediterranean Sea. They are partly cultivated after important hydro-agricultural works were carried out around 1960 (building of dikes, draining, pumping stations, irrigation).
The regional climate, which is of the temperate Mediterranean type ($T° \approx 13°8$, $P \approx 700$ mm; 2 750 hours sunshine, winds) causes wide seasonal fluctuation in soil water content and depth of water table. Swelling-shrinking movements result in pronounced cracking phenomena. The incidence of those structural and poral dynamics on the changes in consistency and on the translocation of clay, iron and salts has been studied on drainage experimental fields.

MAIN CHARACTERISTICS OF SOILS

Formed from fluvio-marine material deposited under brackish conditions, the Languedocian heavy soils (sodic solonchaks) have a rather constant morphology: an Ap horizon (0-30 cm), clayey and containing 4 to 6% organic matter, followed by a (B) Go horizon, very clayey, greyish and with a distinct prismatic structure, which overlies at about 70-100 cm a II A, peaty horizon 10-20 cm thick. The underlying III CG horizons have a clayey, sandy-clayey, or sandy-texture with shell debris. They are saline (EC: 10 to 50 mS/cm) and sodic (100 Na/T 15 to 50%) before reclamation. Their soil water properties are characterized by the following data: water holding capacity at pF 3 varying between 30 and 40%, wilting point at pF 4.2 between 20 and 30%. With regard to soilmechanics, Proctor tests indicate a slight sensitivity to compaction ($D^{c} \approx$ 1.35 for $W^{c} \approx$ 22% and $D^{M} \approx$ 1.45 for $W^{M} \approx$ 28%). The consistency thresholds are typical for clay. The liquid limit (LL) is 55%, the lower plastic limit (LP) about 32%, the shrinking limit 18%, and the dry bulk density of paste D = 1.79 g/cm^3. In the upper part of the profile (B)Go horizon: LL $\approx$ 70%, LP $\approx$ 30%, LR $\approx$ 16.5%, D $\approx$ 1.84 g/cm^3.

VARIATION IN WATER CONTENT AND EVOLUTION OF SENSITIVITY TO COMPACTION AND OF CONSISTENCY

Taking into account the nature and order of the soil horizons, the monitoring of the water content with a neutron probe is only possible between 20 and 70 cm. The calibration of the apparatus gives rather similar calibration curves (i.e. straight lines) when carried out by different methods: (1) gravimetry + clod density or gamma densimetric probe; (2) absorption-diffusion measurements of samples + clod density or (3): gamma densimetric probe.

During the first half year of 1983, which had little rain, the overall evolution of sensitivity to compaction of the Ap horizon resulted in a rather long period of sensitivity besides the 0-1 cm layer, which always remained below the value of W^{c} (and of the wilting point). The evolution of soil consistency during the same period showed that the Am horizon never reached the plasticity limit, whereas the (B) horizon remained in a state of plasticity for about 100 days, an unfavorable condition for operating a trenchless draining machine. The latter situation seems to be linked to the presence of coatings on the faces of prisms, thus acting as a screen against the outward diffusion of water contained in the micropore system of peds. No significant differences were observed between drained and undrained plots.

WATER CONTENT AND THE EVOLUTION OF POROSITY

Periodic measurements of bulk density were made both on clods (paraffin-coated prisms from 1 500 to 2 500 cm^3) and with a gamma densimetric probe. Taking into account the cracks and the volume of soil used for measurements, the total porosity inside the prisms, calculated by using 2.56 as the value of real density, decreased in the whole profile, when the soil dries out, from 54% at the beginning of February to 45% at the end of June for the Ap horizon; the reverse was observed with the gamma densimetric probe. In the surface horizons the porosity increased from 48% at the beginning of February to 57% at the end of June for a water content decreasing from 36% to 17% by volume. The position of the probe in a crack explains this phenomenon.

It was observed that drain trenches were systematically and clearly more porous than the undisturbed soil less than one year after the trenches were made (a difference of 8 to 10% in absolute value).

CRACKING AND SOLID PHASE TRANSLOCATIONS

Rather thick clay deposits from 5 to 25 mm were observed in drain pipes laid in these clay soils.

The presence of calcareous clayfilms on the walls of cracks and of voids in the drain trench (and in soil in contact with slightly or non-calcareous horizons) on the one hand, and the lack of sorting of deposits inside the pipes (observed in thin section) on the other hand, indicate clayey aggregates from the soil surface or aggregates dislodged from the crack walls after drying are transported downwards as sludge by rain or irrigation water. There was hardly any dispersion of sodic clay (very thin oriented clay beddings).

These phenomena are due to the abundance and size of the cracks, which are 5 to 7 cm in width and 70 to 100 cm in depth (down to the peat horizon). Even during the winter under saturated conditions, some cracks are not completely closed.

Iron clogging was often associated with clay clogging. In particular, it occurred when the drain pipes were in contact with peat horizons just after lowering of the water table. The presence of bacterial coatings in the ferric glas (Gallionnella, Lepthothrix) observed with a scanning electron microscope confirmed the bio-chemical nature of the process.

DRAINAGE AND DESALINIZATION

There was a large decrease in salt (NaCl) from the upper horizons, which lost their saline and sodic characteristics down to a depth of 50-60 cm, due to the lowering of the saline water table by the drainage system and the leaching by irrigation water. Within a few years the electrical conductivity of the saturated paste decreased from 25 mS/cm to 4 mS/cm.
It should be noted that owing to the percolation of water through the cracks, irrigation always caused a slight rise of the saline water table.

CONCLUSION

The clear evolution of structural behavior caused by rainfall and water tabel fluctuation from one season to another was accompanied by solid phase and solute transport as well as bio-chemical processes and affected directly the efficiency of the irrigation and drainage system.

REFERENCES

1. Al Addan F. (1983): Comportement hydrique et propriétés mécaniques d'un sol argileux à nappe salée du littoral languedocien - Mémoire DAA - ENSA Montpellier, 45 p.
2. Favrot J.C. - Bouziques R. - Vaquie P.F. (1984): Efficacité et pérennité des réseaux de drainage en sols argileux halomorphes du Languedoc. 12ème Congrès Irrig. Drainage. Fort Collins USA (à paraître).
3. Lagacherie Ph. (1982): Colmatages argileux et ferriques des réseaux de drainage. Cas des solontchaks sodiques argileux du littoral languedocien. Mémoire ENSA Montpellier, 37 p.
4. Lagacherie Ph. - Bouziques R. - Favrot J.C. (1984): Caractérisation et conditions de formation des colmatages argileux et ferriques en sols argileux halomorphes du littoral languedocien. 12ème congrès Irrig. Drainage. Fort Collins USA (à paraître).
5. Vaquie P.F. (1982): Le drainage en Bas Languedoc. Mémoire ENITRT Strasbourg, 142 p.

Water regime of a mole drainage experiment on a heavy clay soil of the Sologne area (France)

B. LESAFFRE*, M. NORMAND* and G. VALENCIA**

(*) CEMAGREF, Divisions Drainage et Assainissement Agricoles, et Hydrologie-Hydraulique, Parc de Tourvoie, 92160 ANTONY, France.

(**) SRAE, Centre, Cité Administrative Coligny, 131 rue du Faubourg Bannier, 45042 ORLEANS CEDEX, France.

1 Experimental layout

The procedure of "reference areas" (FAVROT, 1984) to find out the most suitable drainage methods has shown that the classical drainage methods are unsuitable for the heavy clay soil of the Sologne area located 150 km south of Paris (HOREMANS and LESAFFRE, 1984). Therefore SRAE (Regional Division for Water Management) and CEMAGREF (National Center for Research on Water and Forest Management and Agricultural Engineering) decided to set up an experiment with mole drainage.

Mole drainage seems appropriate for this soil on account of the low permeability, whereas the high clay content, the structural stability and the high plasticity should allow a good durability of the moles (CEMAGREF and SRAE, 1981).

Table 1 presents some physical properties of the soil and table 2 gives a short description of the experimental field. The monitoring system includes a rainfall gauge, a discharge recorder, a moisture measuring location for a gamma-neutronic probe and a piezometer. The last two instruments were doubled in December 1982. During heavy rainfall the outlet of the drainage system was submerged. So the highest discharges could not be accurately recorded from 1980 to 1982. The system was improved in summer 1982.

Table 1. Physical and chemical properties of the clayey pseudogley of Viglain

	Depth	Soil texture			Structure Stability	Atterberg Limits		
		Clay 0-2 μ	Silt 2-50 μ	Sand 50-2000 μ	(log_{10} Is) Hémin	WP Plastic limit	WL Liquid limit	IP Plasticity index
	0-20	19.9	39.8	40.1		-	-	-
Western	20-40	20.8	32.7	46.4		-	-	-
soil sample	40-50	39.9	30.7	29.4		26.3	61.8	35.5
	50-75	46.9	32.5	20.7		29.2	81.2	52.0
Eastern	20-40	23.8	37.6	38.6	1.35			
soil sample	40-50	53.1	25.3	21.6	0.89			
	> 50	58.2	23.0	18.7	0.89			

Table 2. Some characteristics of the experiment location in Viglain.

Available measuring periods: 2 (1980-1981 and 1981-1982)

- A period begins in September and finishes in August.

Climate: altered oceanic.

- Average yearly rainfall: 710 mm (Marcilly-en-V.; Villemurlin, 1951-1970),
- Three days' rainfall of return period one year: 43 mm (average 14,3 mm per day

Parent material: Sologne clayey and sandy soils (plus alluvial sediments from the beginning of the quaternary area).

Soil: Clayey pseudogley with a vertic tendency (typic Haplaquent).
Texture: clay (below 40 to 50 cm).

Drainage lay-out:

* A 1.32-ha-large plot, drained with a 30-m pipe-spacing, gravel trench backfills, P.V.C. drain pipes at a depth of 0.8 to 1 m, installed by a trench machine in June 1980. Moling was set up in September 1980 with a 2-m spacing at about 0.6 m depth, and again in September 1982 in the eastern part of the field.

- Design drainage rate: 1.1 l/s/ha.
- Average ground slope: 2.5%.

* A reference plot drained with P.V.C. pipes at a spacing of 10 m, installed by a trench machine.

2 Analysis of the first experimental results

2.1 Moisture content

Figure 1 presents the observations of the moisture content to a depth of 1.60 m. The variation in moisture content is rather small, and is limited to the upper 50 cm of the soil profile. If no rainfall occurs

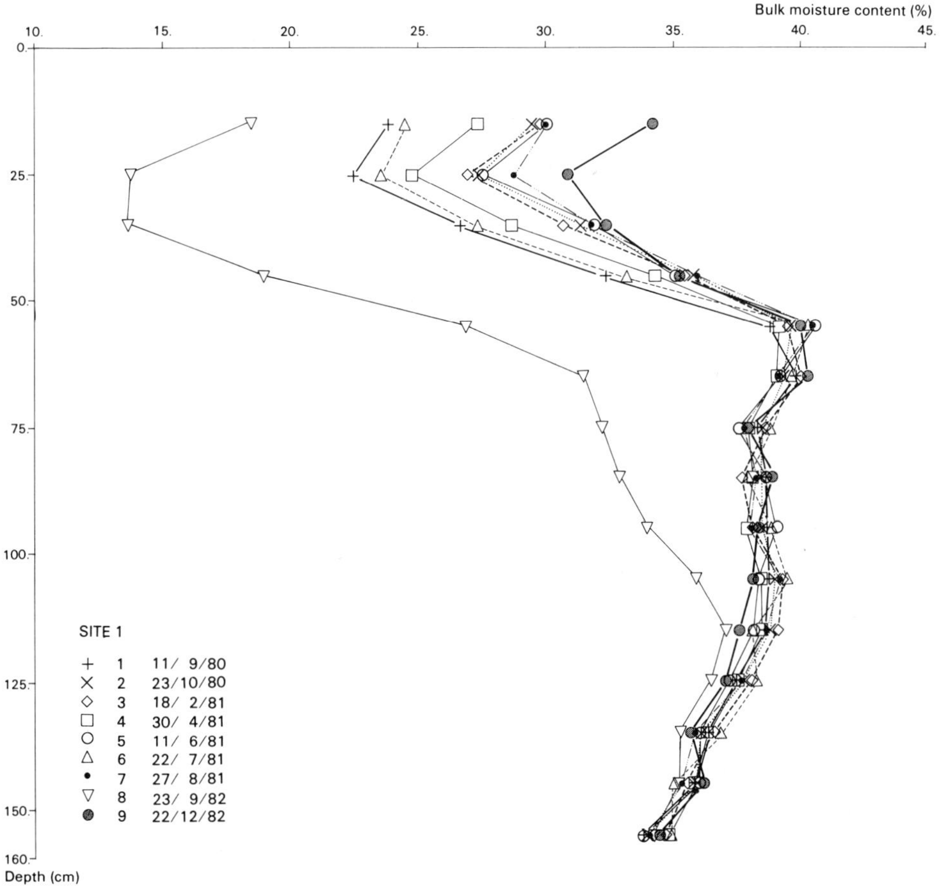

Fig. 1 VIGLAIN field experiment: moisture profiles

in summer over a long time and a deeply rooted crop is grown, the moisture content can decrease to a depth of 1.25 m, yielding a moisture deficit of 140 mm. Drainage starts when the total moisture content above drain level (0.85 m) attains about 300 mm.

2.2 Discharge

It is difficult to compare the discharge recession curve and the watertable drawdown curve, probably because the piezometer recording shows a delay owing to the low permeability and the high percentage of small pores.

The analysis of the tail recession curves according to the method for drains located on or in an impervious layer (Guyon, 1963, 1966, 1981; van Schilfgaarde, 1963, 1965) permits the hydrographs to be fit to theoretical curves with an initial discharge q_0 below 0.15 l/s/ha (Table 3).

Table 3. Tail recession curves fitting to the equation $q_0 = q\ /(1 + \beta_t)$

TAIL RECESSION CURVE						
Beginning		End		q_0 (l/s-ha)	β (h -1)	$\beta / \sqrt{q_\cdot}$
16-12-80,	3 h	17-12-80,	4 h	0.12	0.021	$6.0\ 10^{-2}$
12-01-81,	11 h	13-01-81,	4 h	0.055	$7.9\ 10^{-3}$	$3.4\ 10^{-2}$
16-01-81,	1 h	16-01-81,	10 h	0.15	0.014	$3.6\ 10^{-2}$
3-03-81,	4 h	4-01-81,	3 h	0.125	0.016	$4.4\ 10^{-2}$
15-04-81,	10 h	16-04-81,	0 h	0.08	0.05	0.18
8-05-81,	2 h	8-05-81,	13 h	0.09	0.047	0.16
18-07-81,	20 h	19-07-81,	1 h	0.14	0.25	0.66

Since $\beta / \sqrt{q_0}$ is proportional to $\sqrt{K/\mu}$, where K is the saturated hydraulic conductivity and μ the effective porosity (Guyon, 1972, 1981), some conclusions that are related to other experimental results

(Guyon et al.,1984) can be drawn from the analysis of the tail recession curves.

The low value of the initial discharge q_0 shows that only a minor part of the water flows through the undisturbed part of the soil below the tilled layer. Most of the water runs over the surface and through the tilled layer towards soil cracks, the moling gear coulter track, and the drain trenches.

The ratio $\sqrt{K/\mu}$ remains constant during the winter when the soils are saturated and increases in spring, indicating that the hydraulic conductivity increases faster than the effective porosity. This phenomenon is probably caused by cracks developing during the dry season. Guyon (1972) noticed the same phenomenon in silty soils, but the differences were less pronounced.

2.3 Outflow coefficients and peak flows

The analysis of the outflow coefficients, or restitution rates, (Table 4 and of the peak flows (Table 5) shows two important points:

- Except during winter, when the outflow coefficient reaches almost 100% in February, it remains rather low because part of the rain serves to replenish the moisture deficit (Herve et al.,1984).
- The highest peak flow rates occur during rainfall of more than 23 mm per day. If the rainfall exceeds this value, surface runoff starts, causing at the same time a decrease of the outflow coefficient.

Table 4. Outflow coefficients (ratio rainfall/measured drain discharge).

Dates year 1980/81	10/11 to 11/06	11/06 to 12/03	12/03 to 12/29	12/30 to 01/30	01/30 to 02/26	02/26 to 03/31	03/31 to 04/27	05/05 to 05/19	06/05 to 06/29	Total
Outflow coefficient %	17.4	10.5	54.0	65.7	(100)	53.4	16.2	37.7	10.9	42.8

Dates year 1981/82	09/22 to 10/08	11/27 to 02/11	02/11 to 02/28	03/01 to 03/29	03/30 to 04/09	05/01 to 05/25	06/01 to 06/30	Total
Outflow coefficient %	6.6	53.7	12.5	18.7	30.2	0.8	0.8	31

The high value of the peak flows, far superior to the initial tail recession discharge q_0, confirms the above-mentioned analysis of the water movement in this heavy clay soil.

Table 5. Peak flow rates exceeding 2.2 l/s/ha

Dates year 1980/81	12/18	01/04	01/15	01/20	02/04	03/02	03/08	05/11	08/02	08/08
Peak flow l/s/ha	2.5	(3.9)	(2.7)	(3.5)	(4.2)	2.8	2.4	(2.8)	2.2	(3.8)

Dates year 1981/82	12/24	12/29	12/30	01/08.09	01/10.11	
Peak flow l/s/ha	(3.57)	2.23	2.64	(3.67)	2.90	(rough values)

3. Conclusion

The first experimental results show the relation between the water movement in the heavy clay soil of Sologne and the operation of mole drainage. The present experimental program aims at:

- Continuing the measurements and improving the measuring methods;
- Comparing the agronomical results of successive years;
- Studying the evolution of the structural behaviour of the moles.

References

CEMAGREF et SRAE Centre, 1981. Cestre T., Hervé J.J., Normand M., Baillon J.M., Valencia G. 'Le drainage-taupe: exemple d'une expérimentation en Sologne'. Etudes du CEMAGREF n° 485-486, mars-avril 1982, 61 pages.

Devillers J.L., Favrot J.C., Poubelle W., 1977. 'Approche originale d'une opération d'assainissement-drainage à l'échelle d'une petite région naturelle'. Communication pour le Xème Congrès des Irrigations et du Drainage. ICID, Athènes, 1978.

Favrot J.C. 1984. Acquisition des données nécessaires au drainage des sols difficiles par la méthode des secteurs de référence. XIIème Congrès International des Irrigations et du drainage. Fort Collins.

FEODOROFF A. 1970. Aptitude des terrains à recevoir un drainage-taupe. Bulletin Technique de Génie Rural n° 102.

GUYON G. 1963, 1966. Considérations sur l'hydraulique du drainage des nappes. Bulletins Techniques de Génie Rural n° 65 et 79.

GUYON G. 1972. Expérimentations sur le drainage entreprises par le CEMAGREF. Bulletin Technique d'Information du Ministère de l'Agriculture n° 273-274, septembre 1972, pp 921-946.

GUYON G. 1981. Le drainage agricole. Essai de synthèse en 1980. Bulletin Technique de Génie Rural n° 126. CEMAGREF.

GUYON G., LESAFFRE B., BOUYE J.M., DUMITRIU A., MAMECIER A. 1984. Courbes de tarissement du drainage en sols limoneux lessivés hydromorphes battants peu perméables, XIIème Congrès International des Irrigations et du drainage, Fort Collins.

HERVE J.J., LESAFFRE B., ALDANONDO J.C., LAURENT F. 1984. Restitution et débits de pointe d'un réseau de drainage en sols limoneux lessivés hydromorphes battants peu perméables, XIIème Congrès International des Irrigations et du drainage, Fort Collins.

HOREMANS P., LESAFFRE B. 1984. Le drainage des sols de Sologne : application de la méthode des secteurs de référence à une région aux sols difficiles. XIIème Congrès International des Irrigations et du drainage, Fort Collins.

SRAE Centre avec la collaboration du GVA de Sully-sur-Loire et du CEMAGREF 1981. Suivi de réseaux expérimentaux de drainage à Viglain et Isdes, SRAE-EPR Centre, 84 pages.

VAN SCHILFGAARDE J. 1963. Design of tile drainage for falling uster table. Journal of the Irrigation and Drainage Division. Proceedings of the ASCE 89 (IR. 2), pp 1-12.

VAN SCHILFGAARDE J. 1965. Transient design of drainage systems. Journal of the Irrigaiton and Drainage Division. Proceedings of the ASCE 44 58 (IR. 3), pp 9-22.

Soil variability and hydraulic restrictions in the marshland of the west central atlantic region of France

P. Collas
CEMAGREF Groupement de Bordeaux
Division Hydraulique Agricole,
Gazinet, 33610 CESTAS Principal
L. Damour, Y. Pons,
INRA-SAD INA-PG, Saint-Laurent-de-la-Prée, 17450 FOURAS

1 Introduction

The marshlands of the French west central atlantic region cover an area of nearly 250 000 hectares between the Vilaine in the north and the Gironde in the south.
The marsh soils are known as "Bri" and are usually clay soils (40 to 60 % clay), often saline and sodic. The marshes are not everywhere drained to the same extent and even where the collective management of main drainage system is efficacious, it is not systematically extended to field drainage, although this is an essential requirement for any step towards more intensified agriculture.
Defective water management, deficiencies in electrical energy availability, the unsuitability of farming structure and the difficulties of soil tillage, explain the prevalence of natural grassland and its extensive use (0,6 UGB / hectare), which leads to poor economic results.

2 Soil variability

The different ways of approaching the possibilities and requirements for intensive agricultural cultivation, and, in particular, the experience of the Reference Farms, have made it possible to emphasize an interesting global potentiality but above all, a large difference in situation and according to the type of soil (collectif 1980, L.

Damour 1981).
The suitability of a soil for more intensive land use depends not only on the usual fertility factors ; in the case of these marine clay soils the structure of the soil, and its permeability are of prime importance. The permeability and its evolution are conditioned by the structural stability of the soil which in turn depends on the lime content, the sodicity and the organic matter level ; the salinity indicates especialy the hydrological history of the area.

3 Hydraulic characteristics

The soils of the plots used for characterizing a different hydraulic performance are representative for a stable situation (A), dispersion (C) and an intermediate situation (B). The plots have subsurface drains (spacing 15 to 20 m) and show similar characteristics with regard to topography, soil profile (only slightly differenciated), particle size distribution (50 to 60 % clay) and clay minerals (illite, kaolinite, smectite) which give a C.E.C. of nearly 30 meq/100 g.
The soils are, however, different from a hydraulic point of view - Collas (1983) - as shown by the mean piezometric level during the winter period (1st January - 31 st March), the average time needed for lowering the water table by 20 cm after rainfall, and by the hydraulic conductivity measured in situ - Guyon method (1981).
These differences are compared in table 1 with differences in three chemical parameters (Na_e $CaCO_3$, salinity expressed by EC). It can be noted that the most calcareous, least saline and sodic plot (A) has the best hydraulic performance.

4 Water flow in the soil. Study of a poorly permeable soil

4.1 Material and method

The different hydraulic performance as well as the study of drain discharge, leads us to suggest a model of water movement in the soil during a period of saturation after rainfall - Concaret et al (1976), Damour and Pons (1979) - . Two types of flow can be distin-

Table 1. Chemical characteristics and in situ hydraulic performance of the plots

Plot	Horizons (cm)	Chemical characteristics: Salinity EC (1/5) µS/cm	CaCO$_3$ (%)	Na$_e$/T (%)	Hydraulic in-situ caracteristics in saturated period (January to March 1982): Mean Piezometric level (cm)	Time for fall of water table by 20 cm (day)	Hydraulic conductivity (m/day)
Stable	0- 20	230	0	1			
(A)	40- 50	210	7	1	110	0,5	0,8
	70- 80	210	11	3			
	110-120	230	15	7			
Inter-	0- 20	230	5	2			
mediate	40- 50	360	5	14	45	7	0,3
(B)	70- 80	700	8	37			
	110-120	1070	4	39			
Disper-	0- 10	1170	0	3			
sion	40- 50	1150	0	28	30	14	0,1
(C)	70- 80	1390	0	31			
	110-120	3780	0	31			

guished : surface flow (S) drained off by the trenches and flow through the soil mass (M) defined as "water table flow" by Lesaffre and Laurent (1983). In type (A) which is well permeable, flow M is predominant, whereas in soil type (C), surface flow S predominates. To show the relative importance of the two types of flow, we take as an example the most impermeable plot (C) and compare the results obtained by two independent methods.

The first method involves the use of a natural tracer (water salinity) - Collas and Pons (1980). In fact, plot (C) is characterized by low surface salinity and high salinity in depth. It can then be assumed that the water flow through the soil mass (M) is more saline than surface flow through the tilled layer and the drainage ditch (S).

The second method compares the flow of plot (C) with an adjoining plot which has surface drainage (open ditches 40 cm deep, at a spacing of 15 m), in order to directly estimate the surface circulations (S).

4.2 Results

Fig. 1 shows, for the plot with subsurface drainage a relation of the type log E.C. = a + b log Q, the parameters a and b changing sharply at a characteristic flow of 0,05 l/s/ha (Qc). When the discharge rate is lower than Qc, there is little variation in the salinity of the water (E.C. $\simeq$ 11 000 μS/cm).

Fig. 2 shows that the open ditch flow is of the same order as that of the subsurface drains when the latter exceeds 0,05 l/s/ha. When the drain flow is lower than Qc, the open ditch flow becomes zero.

4.3 Discussion

When the discharge rate is less than Qc, there is no flow from the open ditches, the E.C. of the drain water is close to that of the water table and the origin of the drainage water is solely M ; the quantity of water then evacuated by the drain is very low. When the flow exceeds Qc, the E.C. of the drain water strongly decreases with increasing flow, the origin of the water being mixed S and M which means that M water is diluted by not very saline water S (E.C. maximum open ditch water $\simeq$ 4 000 μS). In this particular case, the flow of water from the soil mass quickly becomes negligible compared to surface flow. These results illustrate the general water flow model outlined above, strongly suggesting that in this poorly permeable soil, the drain behaves as a covered open ditch.

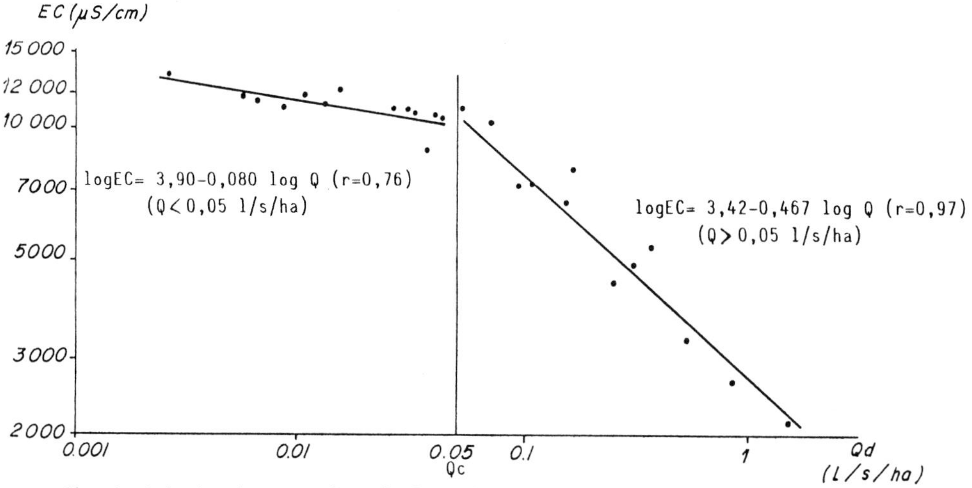

Fig. 1 Relation between electrical conductivity of drainage water and discharge rate

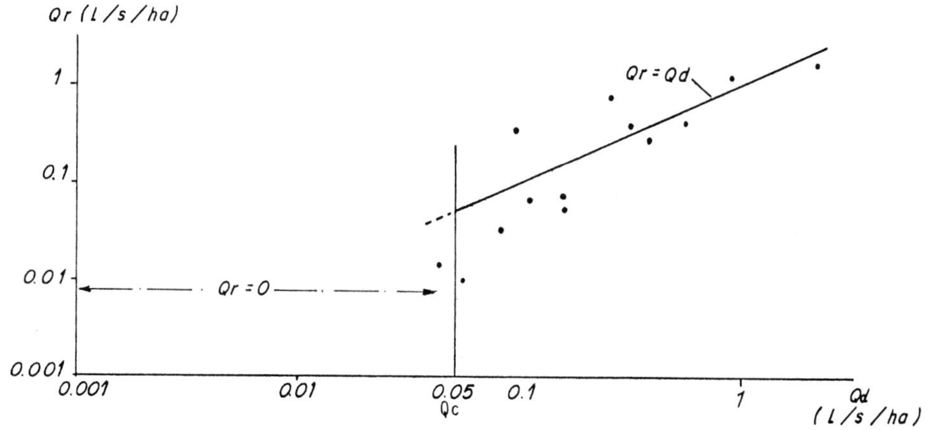

Fig. 2 Open ditch flow (Qr) versus flow of subsurface drain (Qd)
(Instantaneous weekly sampling 27/12/82 to 04/07/83)

5 Conclusion

The observations on the drained plots also revealed the extreme interdependence between soil type, drainage system, soil tillage and crop sequence. The different hydraulic performance of the three plots therefore lead to a whole series of consequences of which the most important are presented in table 2. Knowledge about the performance and their effect becomes of prime importance since it enables one to adjust the means and systems of production to the possibilities of the soil.

Table 2 Conséquences of hydraulic performence of soils on management and development

Structure performance / Consequences for :	Stable (A)	Intermediate (B)		Dispersion (C)
Main drainage system	Draining action	Lowering of water level without draining action		
Field subsurface drainage	possible	possible		not advised
- spacing	large (30 m)	average to narrow	{10m / 20m	(adapted open ditches)
- slope	slight < 2°/oo	steep > 3°/oo		
- collectors	subsurface	open		
- construction	by trenchless machine	by trenching machine		
Soil evolution after drainage	rapid and deep	slow and shallow in relation with the dominant surface flow		
Tillage	as soon as possible after drying-up	only in the dry season shallow ploughing or to		only in the dry season be avoided
Range of suitable crops	spring and automn	autumn dominant		autumn exclusive perennial fodder advised
Environment	minimum risks	risks of solid elements being carried away in the drainage water (dominant S flow) filling of hydraulic structures and harmful effects to shell-fish breeding		

References

Collas, P; and Y. Pons 1980. Conductivités électriques des eaux de drainage en sol de marais. INRA-DRSAD - Domaine de Saint-Laurent-de-la-Prée ; 25 pp.

Collas P. 1983. Contribution à l'étude du drainage des sols des Marais de l'Ouest. Etude du CEMAGREF n° 504 ; 80 pp.

Collectif 1980. Les Marais de l'Ouest. Etude des conditions et conséquences de l'adoption des techniques d'assainissement agricole dans un réseau d'Exploitations de Références. Rapport DGRST, INRA-SAD - Unité INA-PG ; 172 pp.

Concaret, J., J. Guyot, and C. Perrey 1976. Circulation dans les sols de l'eau excédentaire ; conséquences sur la technologie du drainage, INRA Science du Sol, Dijon ; 82 pp.

Damour, L. and Y. Pons 1979. Les Marais de l'Ouest et les aspects agronomiques de l'après-drainage. Revue Drainage n° 19-20 : p 31/34.

Damour, L. 1981. L'application du drainage dans les Marais du Centre-Ouest Atlantique. In : Drainage agricole : théorie et pratique, Chambre Régionale d'Agriculture de Bourgogne, pp 457-470.

Guyon, G. 1981. Hydraulique des nappes des sols drainés. Bulletin Technique du Génie Rural n° 127, CEMAGREF ; 81 pp.

Lesaffre, B. and F. Laurent 1983. Valeur et durée des débits de pointe du drainage agricole en sols à pseudo-gley. Compte-rendu Académie d'Agriculture n° 16, 69, pp 1371-1380.

Discussion

P. Bullock:
Do you have any evidence that, following drainage of saline soils, their stability increases? For example, do the soils move from the unstable to the moderately stable or stable classes?

Author:
The structure of the top layer of sodic soils improves as a result of drainage and tillage for arable crops. However, in the case of sodic, non-calcareous soils, the improvement is only possible by applying gypsum to the top layer. The underlying layers do not show an improvement of their structure. Without gypsum application there is no change from the unstable to the moderately stable class.

Simulation of the hydraulic behaviour of a plot of drained and tilled marshland

J. Duprat *

1. INTRODUCTION

In the course of a study carried out by the C.E.M.A.G.R.E.F.in Bordeaux, a model of the hydraulic functioning of a drained marshy soil was set up and put into operation. This study used results of measurements obtained on an experimental plot of the Domaine de l'Institut National de la Recherche Agronomique (Département Structures Agraires et Développement) at Saint-Laurent-de-la-Prée in Charente-Maritime. The plot has a surface area of 1 ha and buried drains positioned at a mean depth of 0.80 m with a spacing of 20 m. The soil is a slightly saline vertisol with a compact structure containing about 60% clay.

The mathematical model used in the simulation was constructed by accurately following a pedological model which diagrammatically represents the soil profile and its variations in behaviour owing to varying water contents throughout the year. The soil profile was split up into three layers referred to as A, E and U respectively. Layer A, which ranges from 0 to 50 cm in depth, is limited at the base by a compacted floor. Layer E (from 50 to 90 cm) is considered to be constant in volume so that the total porosity is fixed and the microporosity develops, on swelling, with loss of macroporosity. U, below 90 cm, with its lower limit fixed

* Retired engineer in Agronomy. Paper presented by J.C. Chossat, engineer in Agronomy in the agricultural engineering branch of the Centre National du Machinisme Agricole, du Gente Rural des Eaux et des Forets in Bordeaux (France).

at 210 cm, is considered to be an ordinary reservoir in which no distinction is made between free and immobilized water. Input data are the daily rain and P.E.T. (Potential Evapotranspiration) to which have been added temperatures during spells of frost.

The operation of the hydraulic model requires a certain number of parameters, such as the minimum and maximum volumes of the microporosity and the macroporosity for A and E, the porosity of the reservoir U, as well as parametric relationships. These were derived from measurements carried out for this purpose in the field and/or from model adjustment.

2. MEASUREMENTS

Climatology	: rain and evapotranspiration
Hydraulics	: Piezographs and flow recordings of continuous drains
State of the soil:	Measurements of moisture content using neutron probe and of apparent density in the profile and on clods at various degrees of humidity or shrinkage
Bearing capacity of soil	: observations made in relation to soil moisture content and penetrometer values

3. MAIN PARAMETRIC RELATIONSHIPS (fig. 1)

The rain falling on the surface may be added to the free water already present in horizon A, which remains from the previous day. If there is no free water in A, only rain will be taken into account. A part of this free water will become immobilized in horizon A. Some free water may reach horizon E. Hence the first series of parameters which relate to the daily possibility of water immobilization in the microporosity of horizon A or E. For each of these two horizons, daily immobilization will be compared with the amount of free water present in the soil, in order to determine the actual immobilized water.

The water which is not immobilized in A will drain into deeper horizons. The compacted zone, at the base of horizon A, checks this flow. It is therefore necessary to know the daily filtration possibility through the compacted zone. This aspect, like the former, is a function of the wetness of the zone concerned. In both cases the higher the water content, the lower the downward flux.

The clay soil develops deep cracks during the dry season. Part of the rain can then pour into the fissures and pass straight through horizon A or E. Thus the need to determine parameters which define the percentage of rain passing via this rapid circuit. They are preferential flow by fissures in A or E. These parameters depend on the amount of rain and the moisture content of the soil. Another form of preferential flow can exist, this time mainly in wet weather, when there is a surface sheet of water. Then, preferential flow towards the drain may occur via the drainage ditches. This flow, which in some cases can constitute the largest part of drain discharge, depends on the amount of free water present in the surface layers. It is also necessary to ascertain how much water is being transpirated by plants to the atmosphere, by noting the relative proportions of P.E.T. in A and in E (or E and U). This fairly complex calcultion bases its value on the P.E.T., the nature of the crop, its stage of cultivation and the conditions of each horizon.

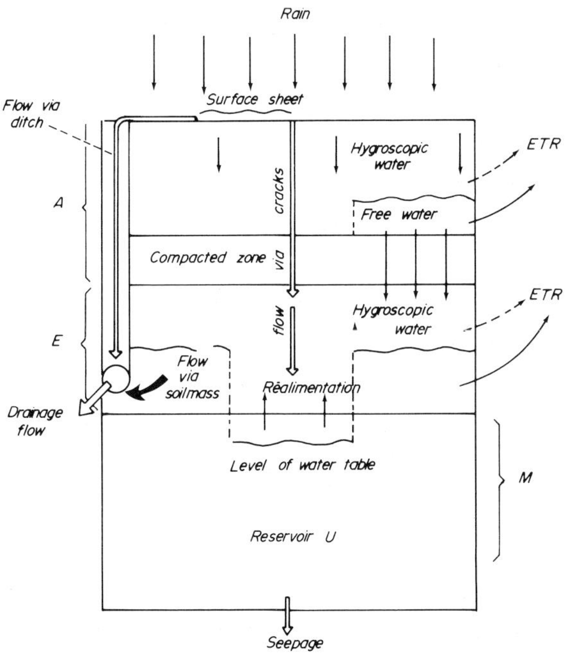

Fig. 1 This diagram is theoretical; all these phenomena could not occur simultaneously

The simulation of the drain discharge takes into account the flow via the drainage ditches and in every case the flow via the soil mass. The calculation, based on standard laws, involves a hydraulic conductivity which depends both on the moisture content of the soil and on the time of year. Finally, two other flows are taken into account: (1) Wetting of layer E by capillarity from the water table in U (referred to as reallimentation in figure 1). This depends on the water immobilized in E, on the volume of water contained in U and on the depth of the water table. (2) Seepage. This consists of flow through the mass of the subsoil towards the sea, which has a mean level 2 m below that of the soil of the plot under study.

4. OPERATION OF THE PROGRAMME

The programme is run on a daily basis. Taking into account conditions prevailing the day before, it follows the flow and determines whether or not the water is immobilized. The value of the fixed parameters and the general parametric relationships were initially approximated on the basis of knowledge of the hydropedological characteristics of the soil being studied. Then parameters and relationships were adjusted by operating the programme over the six years for which we possessed very frequent control measurements.

5. STUDY OF SURFACE SOIL CONDITIONS

One of the most important objectives of the study was to define mechanical stability of surface soil. For this purpose the moisture content of the soil between 0 and 20 cm was calculated. Calculations involve initial moisture contents, the amount of P.E.T. transfered to layer A, the water immobilized in the entire horizon and the free water in transit.

6. EVALUATING THE POSSIBILITIES FOR TILLAGE

From observations in the field it was possible to define 5 classes of moisture conditions corresponding to 5 classes of farming operations:

Class 5: No tillage possible

Class 4: Use of tractor and surface working authorized

Class 3: Possible use of cultivator and rotary harrow

Class 2: Ploughing possible

Class 1: Subtilling possible.

By simulating for a period of 30 consecutive years results can be presented in the form of a table which gives the probabilities of being in one class or the other throughout one whole year. The table of probabilities depends on the choice of crop. When there is a succession of different crops, the first part of the probability calender must be calculated by including the water requirements of the previous crop. In the simulation the temperature on days when it is freezing are included, which can lead to the definition of additional days when it is possible to carry out farming operations. This is of interest when drawing up the calendar for a specific year, but does not lead to a considerable increase in the available days when one is considering a period of 30 years.

7. PRACTICAL USE OF THE SIMULATION OVER 30 YEARS

By comparing the calendar of tillage possibilities and the farming calendar, one can draw conclusions about the feasibility of the different farming operations. The critical periods and the most favourable times become apparent. The table shown in fig. 2 is an example of such a comparison between the two calendars. The calendar of farming operations is shown in the top part of the table. The lower part is a probability graph of tillage possibilities for a winter cereal sunflower sequence.

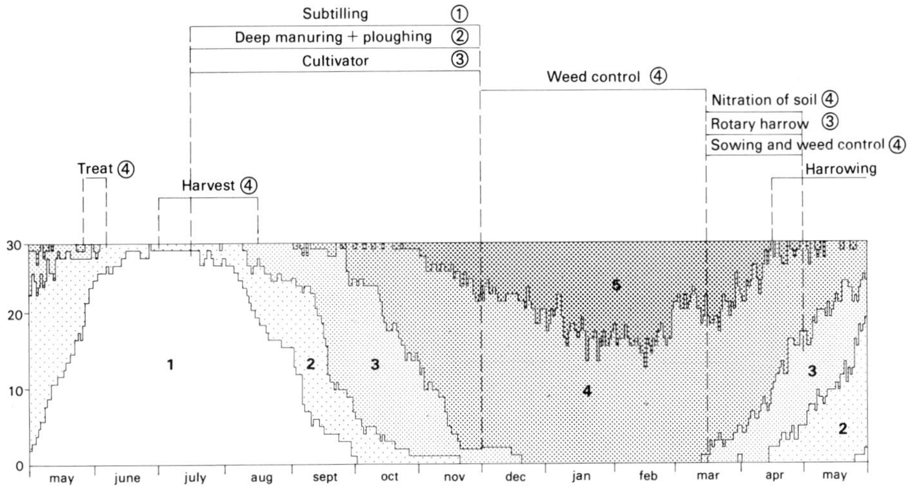

Fig. 2 Daily probabilities of the five classes of intervention probabilities (winter cereal sunflower sequence)

For each day is shown the number of years (over a total of 30) for which soil conditions correspond with each of the five classes. Thus it can be seen that for the month of January the soil was exclusively in classes 4 and 5. Weed control operations only require class 4, and can therefore be carried out without too much difficulty. But the use of the rotary harrow (class 3) in March-April offers serious restraints. In the same period weed control (requiring class 4) is limited in time, since it must be completed before the other tillage operations.

8. EXTENSION OF THE SIMULATION

The procedure was carried out for local, defined conditions. The extension of the simulation to other types of marshy soil requires a fresh adjustement of the parameters and parametric relationships. Basically, though, the approach can be similar.

Hydraulic and hydrological operating of a field experiment in Lorraine heavy clay soil over a period of eight years

by : B. LESAFFRE, R. MOREL (*), A. KINJO, L. FLORENTIN, F. JACQUIN (**)

(*) CEMAGREF, Division Drainage et Assainissement Agricoles, Parc de Tourvoie, 92160 ANTONY, FRANCE

(**) INPL-ENSAIA, 38 rue Sainte Catherine 54000 NANCY, FRANCE

The follow-up of experimental networks is required to understand better both the water movement and drainage mechanisms in heavy soils. In 1969, ENSAIA and CEMAGREF set up such a network on "*LA BOUZULE*" farm near Nancy, in Lorraine (Eastern part of France).

The environment and the experimental lay-out are first briefly described and then some drainage hydrographs are analysed with respcct to the water movement in a heavy soil and in a leached brown soil. The statistical survey of the *drainage outflow coefficient* and of the *peak flows* carried out since summer 1974 will allow to establish the relationship between the drainage hydraulic and hydrological operating and the constituents of soil.

1 The experimental lay-out

The field under study shows a characteristic catena of Lorraine soils, the physical-chemical properties of which are given in table 1. The experimental lay-out is described in table 2 (CROS and JACQUIN 1972). On the drained pelosol, the damaged meadow was superseded in 1972 by a maize monoculture. The monitoring system basically includes a rainfall gauge and a discharge recorder which have been operating since 1972 for each plot.

2 Analysis of some drainage hydrographs (fig. 1)

Two characteristic hydrographs (FLORENTIN, 1982) are analysed here (HARRIS, 1977 ; HERVE et al., 1984 ; JACQUIN and FLORENTIN, 1977 ; LESAFFRE et LAURENT, 1983 a) :

Table 1. Physical and Chemical properties of both soil types of La Bouzule (From Florentin, 1982; Waleed, 1983)

1st lign: brown pelosol; 2nd lign: leached brown soil with pseudo-gleyed layer

Depth in cm	Soil texture					pH	Base exchange capacity me/100 g					Bulk specific gravity	Permeability (Vergiere) cm/s
	Clay	Fine silt	Coarse silt	Fine sand	Coarse sand		Ca**	Mg**	K*	S	T		
0-20	51.5	29.0	9.5	3.0	2.0	7.0	Sat	1.8	0.78	-	18.4	1.32	1.2×10^{-4}
	24.2	32.5	15.5	5.3	15.0	6.6	10.5	1.75	0.78	13.0	13.7	1.47	2.6×10^{-3}
20-40	53.0	31.0	11.0	2.0	1.0	7.9	Sat	2.7	0.53	-	18.7	1.46	1.8×10^{-5}
	29.8	31.9	17.9	4.7	11.3	6.5	9.5	1.68	0.73	12.0	12.8	1.46	-
40-60	-	-	-	-	-	-	-	-	-	-	-	-	-
	34.7	34.8	10.6	4.1	8.9	6.6	5.5	1.2	0.38	7.1	16.7	1.61	1.2×10^{-3}
60-100	53.0	35.0	8.0	2.0	2.0	8.0	Sat	4.8	0.48	-	21.7	1.57	2×10^{-5}
	38.3	33.4	15.1	3.2	5.6	6.3	15	3.03	0.61	18.6	18.0	1.59	5.7×10^{-4}

- in winter, the response times to normal rainfalls are short ; the peak-flow stage is followed by a discharge recession stage, shorter in the pelosol. The latter period corresponds to the drawdown of the water-table under the tilled layer. During the former period, water mainly flows along the surface and within the tilled layer, so as to reach the drainage trench ;
- the drainage response to summer storms is instantaneous (high peaks, very short response time). The maximum flow rate is higher in the pelosol than in the brown leached soil. The water either passes down the soil profile through the cracks, numerous and deep in the pelosol, or flows from the surface to the drain pipes through the trench backfill.

Table 2. Some characteristics of the experiment location in 'La Bouzule' Farm

• Available measuring periods (a period begins in september and finishes in august)	8 periods (1974/75 till 1981/82) - Rainfull data and hydrographs are recorded on magnetic files. Record time step: one hour.	
• Climate	Semi-continental	
- Average yearly rainfall	710 mm (Nancy-Tomblaine, 1931-1960)	
- Three days' rainfall of return period one year.	42 mm (average 14 mm per day)	
• Parent material	Clay marl with Sinemurian Hippopodium	Clay marl blanketed by silt
• Soil	Brown pelosol	Leached brown soil
- Clay mineral	Mainly swelling clay (montmorillonites)	
• Drainage layout		
- Plot size	1,85 ha	2,83 ha
- Pipes spacing	8 meters	12 meters
- Installation technique	Unenveloped clay pipes installed at .9 meter depth by a trench machine in 1969	
- Design drainage rate	1,0 l/s/ha	1,0 l/s/ha
- Average ground slope	6 %	1%

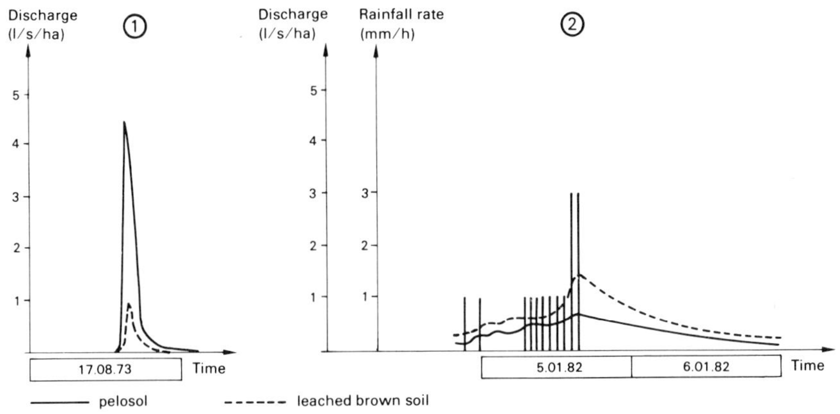

Fig. 1 Drainage response during two periods ① SUMMER - 27 mm rainfall in one hour
② WINTER - 15 mm rainfall in one day

3 Statistical survey of the drainage hydraulic operating

3.1 Double-mass curves

The double-mass method, used in hydrology (OBERLIN, 1971), consists in comparing concomitant and sufficently correlated data sequences :

- on both data sequences X et Y, summations are carried out, so as to obtain two series S_i et T_i ($S_i = \Sigma X_j = S_{i-1} + X_i$; $T_i = \Sigma Y_j = T_{i-1} + Y_i$, X_i et Y_i being the values for each sequence during a given time step);
- the couples (S_i, T_i) are then reported on an arithmetic graph. If the curve is roughly rectilinear, data are considered as homogeneous ; if there is a change in the trend, a rupture in the homogeneity must be suspected, dated and related to a known phenomenon.

Two applications of the method have been carried out through data processings.

3.1.1 Definition of drainage seasons

As far as a given year is concerned, the comparison of rainfall and discharge data of each plot enables to accurately shape the drainage seasons. From the example of the 1977-78 period, figure 2 shows that the curves are made of three straight segments, the slopes of which are equal to the season *outflow coefficients* (defined as the *ratio of actual drainflow to rainfall*) :

- the seasons of drainage beginning (until mid-December) and of drainage ending (from the end of March on) correspond to low coefficients during respectively the moistening (in autumn) and the drying up (in spring) of the soil. The coefficient is higher in the pelosol, because of the presence of numerous cracks ;

- during the intense drainage season, the outflow coefficient is definitely higher because of the soil saturation (reaching 80 % in the pelosol and 60 % in the leached brown soil). The higher coefficient of the pelosol is likely due to its position in the slope.

This analysis corroborates the results obtained from other networks through the implementation of the same method or others (KINJO et al., 1984).

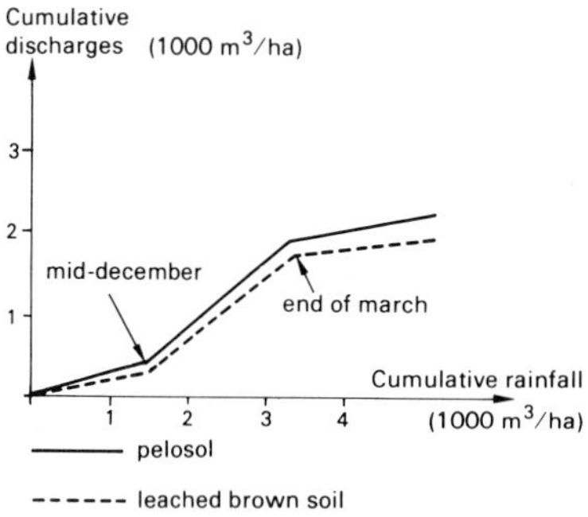

Fig. 2 Drainage operating curves in 1977-78

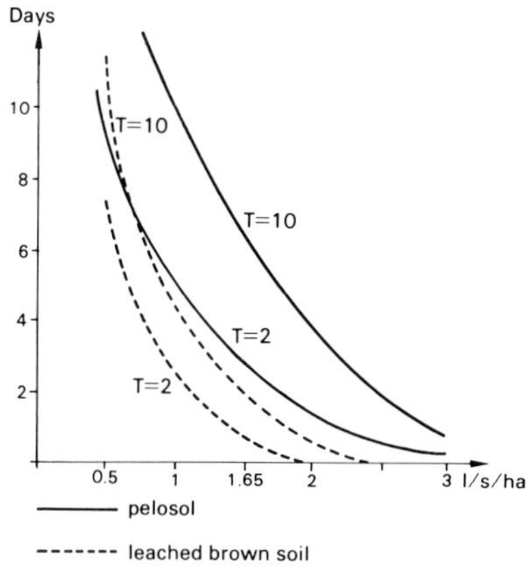

Fig. 3 Discharge exceeding duration (return period: T = 2 ans or mean; T = 10 ans)

3.1.2 Comparison of both plots operating

The comparaison of flowrate data of both plots, either together, or with the rainfall data, enables to assess the trends of the hydraulic operating over the eight available years. No trend change appears when the flowrate of the leached brown soil is compared with rainfall data. But the application of the method to flowrate data of both plots mainly shows a significant, maybe irreversible, decrease in the coefficient of the pelosol from the 1980 summer on (KINJO et al., 1984), resulting from maize harvested under wet conditions and, therefore, damage to the physical properties of the trench backfill (FLORENTIN, 1982 ; KINJO, 1982 ; WALEED, 1983). The pelosol seems to be underdrained.

3.2 Value and duration of peak flows

The retained method, previously applied to two field experiments (LESAFFRE and LAURENT, 1983b), consists in sampling independant data, which are classified and to which a mathematical curve is fitted. Quantiles may then be determined for any frequency (or return period), with a confidence interval of prediction (here: 70 %).

The method was applied to the two following types of variables : annual or seasonal maximum instantaneous flowrates ; durations of choosen flowrates, during a whole year or season (fig. 3).

The results emphasize the flowrate exceeds the design drainage rate (1 l/s/ha) about six days a year in the pelosol. Taking a three days rainfall once per year (quite usual in France) as a basis for the network design drainage rate (1.65 l/s/ha at LA BOUZULE) results in a network overloading sixty hours a year and is likely to avoid the network damage.

4. Conclusion on the experimental results

The statistical analysis of rainfall and flowrate discharge data over a period of eight years shows simultaneously the similarity of the drainage operating in any soil temporarily waterlogged, and the differences according to the soil type nature and constituents :

- the drainage outflow coefficient is higher in the pelosol than in the leached brown soil, because of its swelling and shrinking property and its position in the catena ;
- the peak flows are higher in the pelosol, because of its low permeability and its slope ; the amount of surface flow is therefore bigger ; so seems the yield of solutes linked to the exchange complex (CATILLON, 1979 ; JACQUIN, 1983).

REFERENCES

CATILLON J. (1979) : Entraînement d'éléments fertilisants et d'herbicides par les eaux de drainage des terres agricoles. ENSAIA - SRAEL.

CROS P., JACQUIN F. (1972) : Expérimentation de l'ENSAIA de Nancy et du CEMAGREF. Ferme de la Bouzule. Bulletin technique d'information n° 273-274, pp 947-953.

FLORENTIN L. (1982) : Contribution à la connaissance des sols hydromorphes et apparentés de Lorraine et de leurs réponses au drainage. Thèse INPL Nancy.

HARRIS G.L. (1977) : An analysis of the hydrological data from the Langabeare experiment. FDELL - Technical bulletin n° 77/4.

HERVE J.J., LESAFFRE B., ALDANONDO J.C., LAURENT F. (1984) : Restitution et débits de pointe d'un réseau de drainage en sols limoneux lessivés hydromorphes battants peu perméables. XIIème Congrès International des Irrigations et du drainage, Fort-Collins (USA).

JACQUIN F. (1983) : Transfert de nitrates dans deux sols lorrains drainés. Compte-rendu Académie d'Agriculture de France, pp 804-813.

JACQUIN F., FLORENTIN L. (1977) : Possibilité de drainage dans les sols lourds lorrains. Effet tranchée dans un pélosol. Académie d'Agriculture de France, pp 907-914.

KINJO A. (1982) : Contribution à l'étude hydrodynamique des sols lourds lorrains à partir du protocole expérimental de drainage de la BOUZULE Thèse INPL Nancy.

KINJO A., LESAFFRE B., MOREL R. (1984) : Méthodes d'analyse de la restitution du drainage agricole en sols à excès d'eau temporaire. Académie d'Agriculture de France (1er Février).

LESAFFRE B., LAURENT F. (1983a) : Le fonctionnement hydraulique des réseaux de drainage agricole : débits de pointe et de tarissement non influencé : rôle de la tranchée de drainage. Académie d'Agriculture de France (26 Octobre).

LESAFFRE B., LAURENT F. (1983b) : Valeur et durée des débits de pointe du drainage agricole en sols à pseudo-gley. Académie d'Agriculture de France (9 Novembre).

OBERLIN G (1971) : Généralités sur les exigences et contrôles de qualité des données hydrologiques de base. Note interne CEMAGREF.

WALEED J. (1983) : Comportement hydrodynamique des sols lourds lorrains drainés en fonction des technologies de drainage. Thèse INPL Nancy.

Tritiated water movement in clay soils of a small catchment under tropical rainforest in North-East Queensland

M. Bonell
James Cook University, Townsville,
Australia
D.S. Cassells
University of New England, Armidale, Australia
D.A. Gilmour
Queensland Forestry Department,
Gympie, Australia

1 Introduction

Tritiated water was used to trace soil water movement in kaolinitic clays below the active surface and subsurface stormflow layer, >0.2 m, (Bonell et al. 1981) at two contrasting sites in a tropical rainforest catchment ($17^0 20$'S, $145^0 58$'E) in north-east Queensland. Mean annual rainfall is 4239 mm with a marked concentration (63.5 percent) in the summer months December to March. The deep (c. 6 m) Ultisols-Inceptisols range from light to medium clay (Northcote 1979) with a fine to moderate blocky structure and the largely kaolinitic clay content increases to a maximum of 51 percent at the 0.4-0.5 m depth.

The lower slope (26.5^0) tracing site 1 was characterised by hydraulic conductivity, K values (log mean K = 0.18 md^{-1}, n = 10) (Reynolds et al. 1983) somewhat higher than the upper slope (23.5^0) site 2 (log mean K = 0.11 md^{-1}, n = 10) between 0.20-0.90 m depth. However, within plot K variability was much higher at site 1 with a maximum point estimate of 0.94 md^{-1} cf. 0.28 md^{-1}, site 2. These differences were attributed to the large amount of weathered rock pieces within the soil matrix of site 1. Each experimental plot had lines of soil water extractors (Talsma et al. 1979) inserted 0.10 m apart and located at depths ranging from 0.25-1.50 m. Two transects, 2 m apart, were installed at site 1 to monitor any rapid lateral transfer of tracer. In contrast three lines of extractors were located closer together at site 2 because of the lower hydraulic conductivity, being separated by a gap of 0.30 m between each transect. A line injection ($2.7\mu Ci\ell^{-1}$) at 0.2 m depth was carried out on 22 February 1980 slightly above each of the top extractor transects and the pulse's progress monitored until July 1980. Further details of the site physical properties and experimental procedure are reported elsewhere (Bonell et al. 1982, 1983).

2 Results and Conclusions

The tensiometers at both sites indicated near-saturated conditions for most of the experiment. The possible flow lines of soil water at site 2 (Figure 1) show typical conditions following rainfall when positive mat-

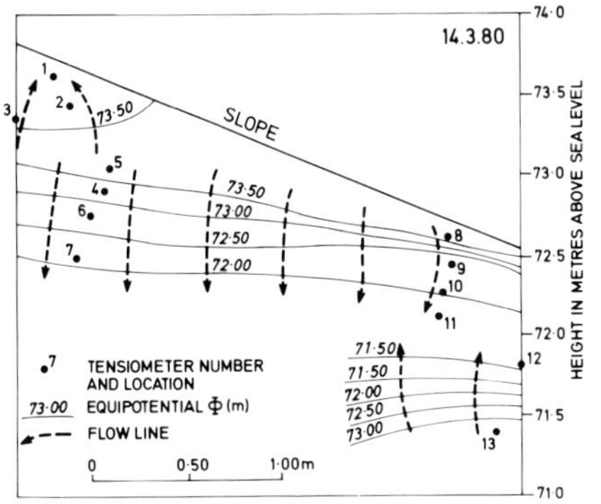

Figure 1. The equipotential and flow lines on 14.3.80 at site 2. Notes: The hydraulic potential, Φ is given as: $\Phi=\Psi_M+Z$ where Z is the elevation above sea level (upward is positive) and Ψ_M is the matric potential at a particular point.

ric potentials prevailed. A complicated three dimensional flow pattern is indicated with convergence of soil water through lateral movement into or out of the cross-section. A similar pattern was also observed at site 1 (Bonell et al. 1983) but with downslope flow more favoured. At both sites the translation laterally and vertically of the tritiated pulse appeared to be dominated by interstitial piston flow (Zimmermann et al. 1967) though moving at a faster rate when compared with results from temperate areas and could be attributed to the high frequency of moderate-heavy rainfall in this environment (Figure 2). Tracer downslope movement at site 2, however, was established as being more significant than previously indicated in Figure 1. In addition, the high initial activity in soil water extractors 1-25, 1-35 and 2-35 (Figure 2) and the deeper centre of mass at site 1 on 3.3.80 cf. site 2 were also indicative of short circuiting or preferential macropore flow (Table 1). The latter introduces the effects of the higher hydraulic conductivity

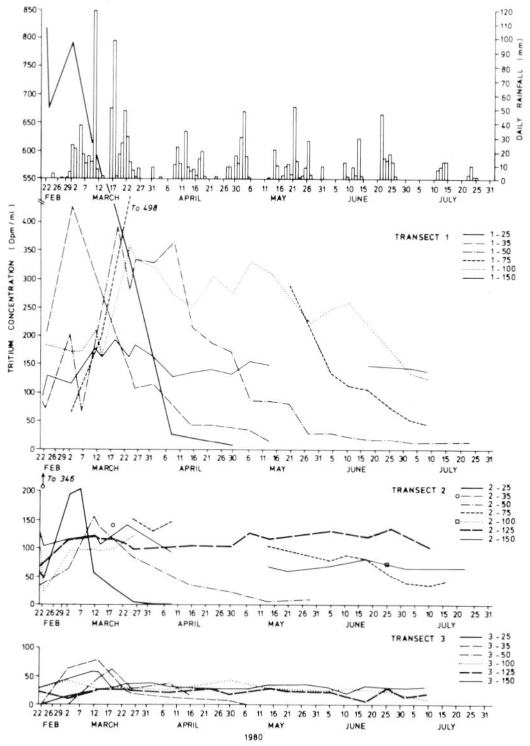

Figure 2. The time series of rainfall and tritium concentrations at site 2. Notes: i) 1-25 is extractor point: top (injection) line 1-0.25 m depth; ii) discontinuities in graphs denote breaks in record due to malfunctioning of soil water extractors. Isolated data are shown as unconnected points, eg 2-35; iii) tritium concentrations expressed in disintegrations per minute per millilitre (dpm/ml).

Table 1. The vertical displacement of tracer along the injection line and recharge on selected dates

Date	Site	Centre of mass z, (m)	Average volumetric water content, $\overline{\theta}$	Accumulated inputs from beginning of experiment: Rainfall (mm)	Throughfall[1] (mm)	Apparent vertical recharge, R_V (mm)	of throughfall[1]
3.3.80	1	0.595	0.405	34.90	22.04	159.97	725.82[2]
	2	0.270	0.390			27.30	121.88[2]
26.3.80	1	0.815	0.464	574.40	399.95	285.36	71.35
	2	0.635	0.515			224.03	56.01
14.5.80	1	1.060	0.457	892.80	690.89	393.02	56.89
	2	0.780	0.525			304.50	49.92
26.6.80	1	1.140	0.483	1237.50	839.50	454.02	54.08
	2	0.920	0.545			392.40	46.74

[1] Throughfall calculated from daily rainfall by equation $y = 0.7 + 0.725x$ where y = throughfall (mm), x = rainfall (mm) (Gilmour 1975)

[2] These high values are the result of short-circuiting

at site 1. There was a faster vertical displacement in this plot and a higher proportion of soil water recharge (Table 1) using Zimmermann et al. (1967) relation:

$$R_V = 1000\ (z - 0.20)\ \overline{\theta}$$

where

R_v = vertical recharge in mm in a unit area of profile between 0.20 and z

z = centre of mass of tracer pulse (m)

Θ = average volumetric water content ($cm^3\ cm^{-3}$) in a unit area of profile on given date, estimated from the soil moisture characteristic curve

Also similar amounts of lateral recharge were required under monsoonal intensities to push soil water downslope and cause a peak response at 0.35 m and 0.50 m depth on the lowest transect (Table 2) despite differ-

Table 2. Lateral recharge from injection line to peak response in selected extractors on lowest transect

Extractor depth (site no.)	Date of peak response	Accumulated rainfall (mm)	Accumulated throughfall (mm)	$\bar{\Theta}$[1]	Lateral recharge R_c = Accumulated Throughfall x $\bar{\Theta}$[1] (mm)
0.35 (1)	12.3.80	279.4	193.49	0.445	86.10
0.35 (2)	14.3.80	282.0	194.45	0.480	93.34
0.50 (1)	9.4.80	603.0	417.67	0.403	168.32
0.50 (2)	19.3.80	433.1	301.32	0.535	161.20
1.50 (1)	9.4.80	603.0	417.67	0.450	187.95
1.50 (2)	12.3.80	279.4	193.49	0.540	104.48

[1] Average volumetric water content on the given date in a unit area of profile between 0.20 m on top transect and the depth of each extractor on lowest transect

ences in distance between the points of injection viz. 2.0 m site 1, 0.60 m site 2. Greater amounts of recharge were, however, required at 1.50 m for site 1. In contrast no tritium was detected downslope on site 1 in an earlier experiment when lighter rainfalls prevailed (post-monsoon) (Bonell et al. 1983).

Below 1 m the profile remained labelled with tritium at both sites and indicated similar prolonged soil water transit times (eg Figure 2, site 2). A factor delaying the vertical advance of the tritium pulse was upward movement of soil water in various sections of the profile at both sites suggested from the distribution of hydraulic potential.

The experiment shows that both interstitial piston flow and preferential flow can occur simultaneously in these heterogeneous clays under low matric potentials. The effect of spatial variability of hydraulic properties has also been demonstrated. However the long soil water transit times especially below 1.0 m have management implications. Large areas of former tropical rainforest have been cleared for sugar cane farming and high chemical inputs are required to maintain productivity levels to offset intense leaching of the surface layers by lateral stormflow. These results indicate however the potential for accumulation of agricultural chemicals in the lower layers despite the high rainfalls.

References

Bonell, M., D.A. Gilmour and D.F. Sinclair 1981. Soil hydraulic properties and their effect on surface and subsurface water transfer in a tropical rainforest catchment. Hydrol. Sci. Bull., 26: 1-18.

Bonell, M., D.S. Cassells and D.A. Gilmour 1982. Vertical and lateral soil water movement in a tropical rainforest catchment. In: E.M. O'Loughlin and L.J. Bren (ed.), May 11-13, 1983, Melbourne, Instit. Eng., Aust., Canberra, pp 30-38.

Bonell, M., D.S. Cassells and D.A. Gilmour 1983. Vertical soil water movement in a tropical rainforest catchment in north-east Queensland. Earth Surf. Processes, 8: 253-272.

Gilmour, D.A. 1975. Catchment water balance studies on the wet tropical coast of north Queensland, Thesis (Ph.D.), James Cook University of North Queensland, Townsville, Australia, 254 pp.

Northcote, K.H. 1979. A factual key for the recognition of Australian soils, 4th ed. Rellim Tech. Publ., Glenside SA, 124 pp.

Reynolds, W.D., D. Elrick and G.C. Topp 1983. A re-examination of the constant head well permeameter method for measuring saturated hydraulic conductivity above the water table. Soil Sci., 136: 250-268.

Talsma, T., P.M. Hallam and R.S. Mansell 1979. Evaluation of porous cup soil-water extractors: physical factors. Aust. J. Soil. Res., 17: 417-422.

Zimmermann, U., K.O. Munnich and W. Roether 1967. Downward movement of soil moisture traced by means of hydrogen isotopes. In: G.E. Stout (ed.), Isotope techniques in the hydrological cycle, Geophys. Mon. Ser., No. 11, Amer. Geophys. Un., Washington D.C., pp 28-38.

Factors conditioning the surface waterlogging of leached clay chernozems in Bulgaria

Dr. M. Pencov, Dr. B. Djuninski,
Dr. T. Palaveev

Leached clay chernozems in Bulgaria have seasonal surface waterlogging due to high clay content (Table 1), humus-accumulation in the A12 horizon and a transition horizon (A13) with a depth of 50-70 below surface which has a low permeability. The latter is due to its extremely compact prismatic structure. The structural aggregates have slanting slickensides. The C horizon is calcareous. The soil reaction of the A horizon is moderately acid (pH 6,2-6,6) and the B horizon is neutral to moderately alkaline. These chernozems have a high potential fertility and are located in flat areas, which make them suitable for mechanized cultivation. However, during agricultural practice, these soils become waterlogged and acquire unfavourable agrotechnical characteristics almost every year during the autumn and winter but particularly during the spring season (2, 3, 4).

The climatic environment in which these soils occur is characterised by an excess of rain over evapotranspiration during the autumn-winter and the spring season. Waterlogging is caused by a low water permeability of the soil and the small slope of its surface, which does not allow run-off. The permeability, measured according to the method of De Boodt, of the A12 horizon and the A13 horizon varied from 0,026 to 0,060 m/day while the permeability of the B horizon varied between 0,010 and 0,015 m/day (Table 1), (2).

The low water permeability is thought to be due to the following factors: (1) There is a high clay content (Table 1), which in the A horizons varies between 48% and 60% and in the B1 horizon between 60% and 62%. Among the secondary clay minerals montmorillonite predominates (2) In part of the soils the percentage of exchangeable Mg is relatively high (20% to 40% of adsorbed cations).

Table 1 - LABORATORY PARAMETRES FOR THE LEACHED CLAY CHERNOZEMS WITH SEASONAL SURFACE WATERLOGGING

Horizons (cm)		Particle size distribution (%)			Permeability m/day	Particle density (g/cm^3)	Bulk density (at pF2,5) (g/cm^3)	Total porosity (pF2,5) (%)	Field water capacity (pF2,5) (%)	Aeration of soil at field water capacity (%)
		0,2-0,02	0,02-0,002 (mm)	<0,002						
A11	0- 30	24,8	15,0	56,7	0,765	2,70	1,18	56	28	23
A12	30- 50	24,8	14,6	57,1	0,054	2,72	1,35	50	30	9
A13	50- 72	23,8	13,8	59,6	0,026	2,73	1,40	49	31	5
B1	72- 96	23,9	13,6	59,8	0,013	2,73	1,46	46	31	1
B2	96-120	25,4	14,1	57,7	0,010	2,74	1,45	47	28	3
Ck	120-150	25,4	14,1	57,7	0,110	2,74	1,45	47	28	6

The soils being discussed have, in addition, the following properties (Table 1): (1) The water content at field capacity is high, (2) The A and B horizons are compact. Bulk density: (1,40 g to 1,50 g/cm^3 at field water capacity: 1,90 g/cm^3 in dry soil) (3). The porosity is high (determined at pF2,5-2,7). In correspondence with the above-mentioned soil properties, the aeration at field water capacity is lower than 10% in the A12 and the B horizons. The above metioned results, have been checked under laboratory conditions with model experiments. When sand is added to the soil sample (without drying or wetting) the water-permeability in the column increases. After a period of waterlogging, the quantity of exchangeable NH_4^+, Mn^{2+} and Fe^{2+} slightly increases. The increase is, however, insufficient to cause a significant increase in dispersion. Treatment with gympsum results in a decrease of dispersion and in desorption of exchangeable Mg. After a double suspension in water, the initial degree of dispersion is reestablished. Laboratory treatment with gypsum increased the permeability with 20-50% in a soil sample that had been waterlogged for seven days. Other factors, controlling the low water-permeability are the relief and the agrotechnical activity of man. The relief is flat, with a slope of 0-6°/₀₀, which makes surface water run-off impossible or very difficult.

The negative influence of the agrotechnical activity of man has been evident by tillage with big agricultural machines, which is sometimes performed at high moisture contents. This leads to compaction of the soil, decrease in the humus content of the fallow land and soil structure degradation. Ploughing at a constant depth has caused the formation of slowly permeable plowpans.

REFERENCES

1. Mamaeva, L. A colloide-chemical method, predicting the rates of melioration material for solonetz soils. Proceedings of the Soil Science Insitute "V.V. Dokootchaev", Moscow, 1956, Vol.II.
2. Pencov, M. and B. Djuninski. Measures for the improving of the heavy clay surface waterlogged soils. Agriculture N 3, Sofia, 1982.
3. Pencov, M. and B. Djuninski. Problems of the amelioration of heavy surface waterlogged soils in Bulgaria. Pedology and agrochemistry, N 6, Sofia, 1982.
4. Pencov, M. Soils in Bulgaria - Preserving and improvement, Sofia, 1983.

Water and solute movement in a heavy clay soil

D.E. Smiles and
W.J. Bond
CSIRO Division of Soils
GPO Box 639, Canberra, ACT, 2601
Australia

Abstract

Solute movement in soil is conventionally described in terms of diffusive forces that affect the distribution of the salt relative to the water; and convective effects due to soil water flow. We describe a series of experiments that illustrate the application of these ideas to $CaCl_2$ movement during non-steady, unsaturated water flow in sand and in a structurally stable heavy clay soil. The results are discussed in terms of simple concepts in soil water theory, and surface physical chemistry. The paper concludes by identifying possible areas for further research.

1 Introduction

The distribution of water soluble salt in soil in the natural environment depends on the local soil mineralogy and hydrology. Any perturbation of this system by, for example, irrigation or drainage, change in land management, or addition of fertilizer must be reflected to an extent in a change in this natural distribution. As a consequence, we may observe enhanced soil salinity, or leaching, or the appearance of various water soluble salts in the groundwater, or streams. We may also observe substantial change in soil physical properties because of the effect of change of the soil solution composition. Often these effects are deleterious and their amelioration may be difficult. It is therefore critical that we develop management strategies that are not potentially deleterious. It is also important that we identify reliable methods for ameliorating areas already affected by high concentrations of water soluble salts. This challenge is complicated by problems related to scale and to field variability in addition to the conceptual difficulties presented by the basic mechanisms of salt and water transport during unsteady, unsaturated soil water flow.

In this paper we describe a program that attempts to define a set of general 'tools' that might help resolve particular problems of salt movement in soils. These tools are, to an extent, expressed in mathematical terms, so it is important that we set out the role of theory, and its mathematical formulation, in the study of properties and behaviour as complicated as those of soil. We then formulate a theory based on processes involved in the interaction of soluble salts with water and soil during unsteady, unsaturated flow and illustrate it with experiments. Finally, we discuss some areas for further research.

2 Role of theory

A rationale for the development of theory is the assumption that well-posed field problems will be tackled most efficiently and fruitfully using strategies based on an understanding of the mechanisms of the processes involved. The formalism of mathematics is then used to set out laws necessary to describe, for example, water, heat and solute movement in soil. The theory derives its authority from critical and well-conceived experiments and its validity must be continually tested and deficiencies in the theory identified in this way must be explored and corrected. The mathematics provides not only a language, but also a logic that often permits us to explore situations more efficiently and perceptively than is possible by experiment alone. Furthermore, solutions to flow equations for many practically important initial and boundary conditions have a valuable illustrative and didactic role. An exposition of the role of mathematics in natural science is presented by Philip (1957).

The theory should be based on a suitably small set of measurable properties of the system. The material properties should be macroscopic because the generation of space averages from many microscopic measurements in a geometrically complicated body like a soil is a problem of great difficulty. In fact, the scale, variability, and complexity of field soils all conspire to make the application of theory in terms of the solution of equations most unlikely. Indeed, field problems may defy formal solution using scientific methods even though they may be defined in a scientific way. Nevertheless, scientific understanding of basic mechanisms must be central to our problem-solving strategies if only to 'inject some intellectual discipline into the republic of trans-science' (Weinberg, 1972).

3 Theory of solute movement in unsaturated water flow in soil

The dependence of the salt flow on water makes it necessary to write continuity equations for both the water and the soluble salt. In one dimension these equations are

$$\left(\frac{\partial \theta}{\partial t}\right)_x = -\left(\frac{\partial Fw}{\partial x}\right)_t \qquad (1)$$

$$\left(\frac{\partial (\theta C)}{\partial t}\right)_x = -\left(\frac{\partial Fs}{\partial x}\right)_t \qquad (2)$$

where θ is the volume fraction of water in the soil, C is the solution salt concentration, Fw and Fs are the volume flux of water and the mass flux of salt respectively, and x and t are distance and time.
The mass flux of the salt is given by

$$Fs = Fs_r + Fs_c = Fs_r + CFw \qquad (3)$$

where the subscripts r and c refer respectively to salt movement relative to, and convected with, the water.
Substitution of Equation 3 in Equation 2, differentiation of the composite terms, substitution of Equation 1 and rearrangement then yield

$$\left(\frac{\partial C}{\partial t}\right)_x + \frac{Fw}{\theta}\left(\frac{\partial C}{\partial x}\right)_t = -\frac{1}{\theta}\left(\frac{\partial Fs_r}{\partial x}\right)_t \qquad (4)$$

in which the second term on the left-hand side represents flow of salt with the water of average velocity $u = Fw/\theta$. In fact, the left-hand side is the differential of C following the motion of the water (Bird et al., 1960) whence Equation 4 may be written

$$\left(\frac{\partial C}{\partial t}\right)_q = -\left(\frac{\partial Fs_r}{\partial q}\right)_t \qquad (5)$$

where $q(x,t)$ is a material coordinate based on the distribution and flow of the water so that

$$\frac{\partial q}{\partial x} = \theta \quad \text{and} \quad \frac{\partial q}{\partial t} = -Fw \qquad (6)$$

The q coordinate is discussed by Smiles et al. (1981) and also by Wilson and Gelhar (1981). It is important to note that q satisfies the continuity of water requirement of Equation 1.

Equation 5 is a statement of continuity of salt presented in terms of a material coordinate based on the distribution of the water. Its further development requires definition of Fs_r. This requirement is generally met by the equation

$$Fs_r = -\theta D_s(u)\ \partial C/\partial x \qquad (7)$$

in which D_s is the dispersion coefficient. The nature of D_s is discussed elsewhere (e.g. Fried and Combarnous, 1971) but in essence it has both diffusional and hydrodynamic elements represented, for example, by the equation (Rose, 1977)

$$D_s = D_o(\Gamma + aPe^b) \qquad (8)$$

In Equation 8, Γ, a and b are empirical constants, D_o is the diffusion coefficient of the solute in bulk solution, and Pe is the Péclet number given by

$$Pe = u\ell/D_o \tag{9}$$

The Péclet number relates convective and diffusive components of transport using a characteristic length, ℓ (often identified with an average particle size). In general, as Equations 8 and 9 imply, D_s is velocity dependent. When the value of u is small enough, however, D_s may be regarded as velocity-independent.

Substitution of Equation 7 in Equation 5 using Equation 6 leads to the equation

$$\frac{\partial C}{\partial t} = \frac{\partial}{\partial q}\left[\theta^2 D_s(u)\,\frac{\partial C}{\partial q}\right] \tag{10}$$

Equation 10 provides the theoretical framework against which we develop insights into salt movement in soils.

Since our concern is primarily to investigate salt movement, we have worked with the simplest cases of water flow that reveal substantial evolution, in space and time, of well-defined water content profiles. Constant flux, or constant potential absorption provide two such simple systems. For both cases the material coordinate q(x,t) may be found by integrating the total differential of q using Equation 6. The detail is presented, for example, by Smiles et al. (1981). For experiments subject to conditions defined by

$$\theta = \theta_n \,,\quad C = C_n \,,\quad x > 0 \,,\quad t = 0 \tag{11}$$

and either

$$\theta = \theta_o \,,\quad C = C_o \,,\quad x = 0 \,,\quad t > 0 \tag{12}$$

or

$$Fw = V_o \,,\quad C = C_o \,,\quad x = 0 \,,\quad t > 0 \tag{13}$$

the material coordinate is conveniently defined by the equation

$$q = \int_o^x \theta_{t=t}\, dx - \int_o^t V_o\, dt \tag{14}$$

As Smiles et al. (1981) show, the origin of the q coordinate according to this definition coincides with the notional piston-front of the water, i.e. the interface between the initial and the invading water which would exist if the former was completely displaced by the latter.

Thus, for the initial and boundary conditions represented by Equations 11-13, salt movement studies basically centre on examination of the location of a step change in solution concentration (from C_o to C_n) relative to q = 0, and the way the step change is 'smeared' with increasing time.

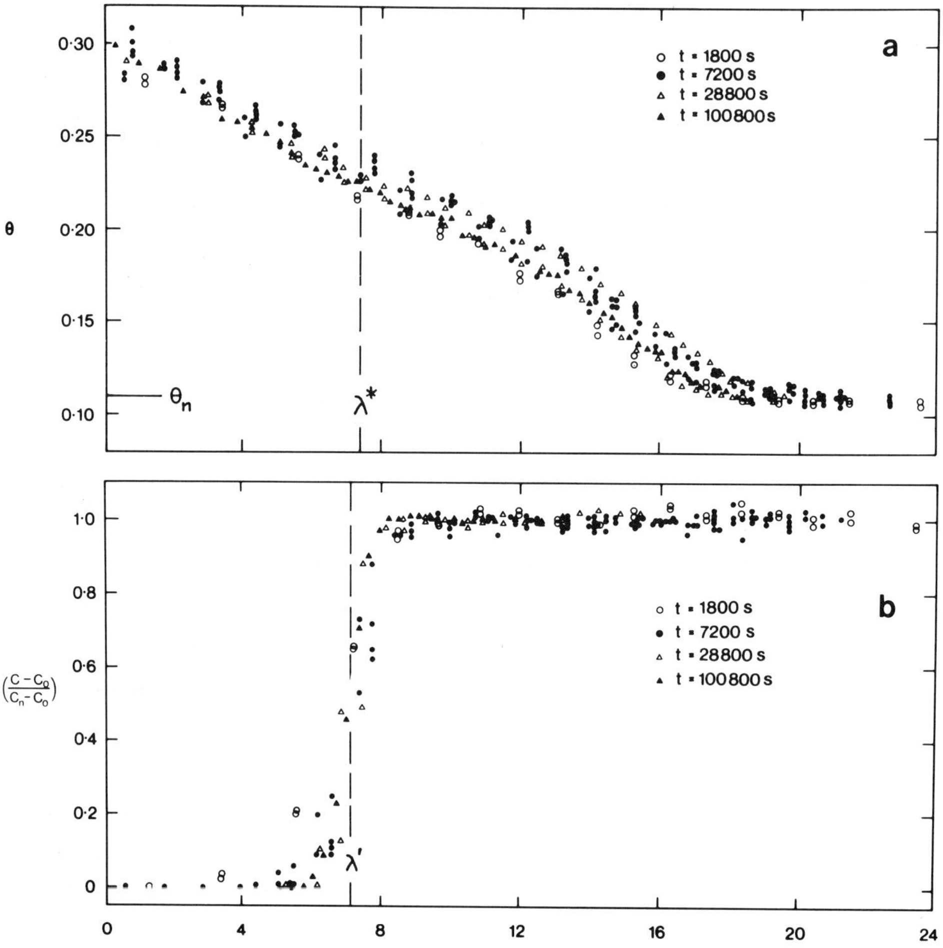

Figure 1. Profiles of (a) water content and (b) normalized solution salt concentration at four times during constant potential horizontal absorption in a mixture of fine sand and kaolinite. The low initial water content θ_n containing 8.5 kg m^{-3} KCl (C_n) was invaded by a solution of 1 kg m^{-3} KCl (C_o) supplied to the soil surface at a zero head. λ^* is the position where g = 0 and λ' is the mean position of the salt front (data from Smiles and Philip, 1978)

4 Illustrative experiments

The simplest experiments that demonstrate the basic aspects of this approach are performed in a chemically unreactive material. When the conditions of Equations 11 and 12 are imposed and it is assumed that the dispersion coefficient is velocity (Péclet number) independent, both the water flow equation and the salt flow equation reduce to ordinary non-linear diffusion equations in terms of the variable $\lambda = xt^{-\frac{1}{2}}$. In this space the piston front is identified by $g = 0$ where $g = qt^{-\frac{1}{2}}$.

Results of a series of such experiments are presented in Figure 1. It is observed that both the water content and the salt concentration profiles preserve similarity in terms of λ. The uniqueness of the salt concentration profiles indicates that D_s was effectively velocity independent in these experiments (Smiles and Philip, 1978). It is also noted that the salt front corresponds closely with the position where $g = 0$, as calculated from the water content profiles using an equation derived from Equation 14. Consequently the piston flow model is valid for this chemically unreactive sand at low velocities.

Further experiments applying constant flux of water (i.e. realizing Equations 11 and 13) confirm these observations (Smiles et al., 1981). When these conditions were applied to a non-swelling heavy clay soil, however, the movement of the salt was substantially altered as is demonstrated in Figure 2.

Theory predicts that for constant flux absorption, water content profiles in terms of $X(= V_o x)$ should be unique for constant values of $T(= V_o^2 t)$. In addition, if D_s is velocity independent $C(X)$ should be unique. In these coordinates, the piston front is defined by $Q = 0$, where $Q = V_o q$.

It will be observed that the tritium front corresponds with the plane $Q = 0$ but the chloride profile is different. Two features should be noted:

a) the amount of both chloride and tritium recovered from the soil agreed with the sum of the amount known to be present initially and the product of C_o with the volume of water which had entered the soil. In other words, the entry of both substances appears to be unhindered.

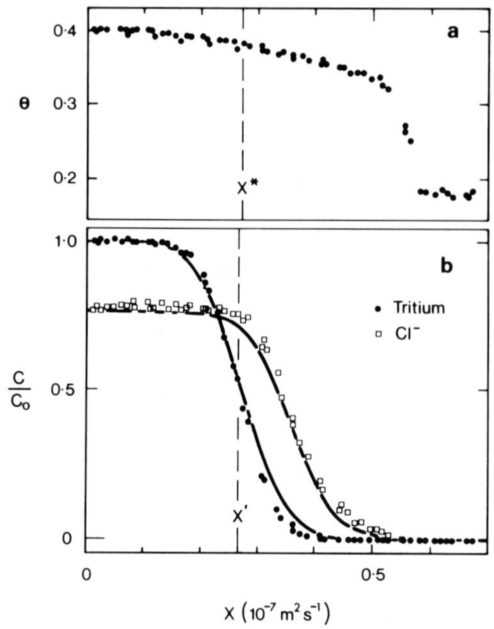

Figure 2. Profiles of (a) water content and (b) normalized concentration for $T = 1.1 \times 10^{-8}\ m^2s^{-1}$ and values of V_o between 3.3×10^{-7} and $6.6 \times 10^{-7}\ m\ s^{-1}$ during constant flux horizontal absorbtion of a tritiated calcium chloride solution by a clay soil. The low initial water content θ_n, which had a small concentration of tritium and chloride, was displaced by a solution with a specific tritium activity of 0.14 GBq kg^{-1} and a chloride concentration of 200 equiv m^{-3}. X* is the position where Q = 0 and X' is the mean position of the tritium front (data from Bond et al., 1982)

b) the apparent concentration of chloride close to X = 0 is about 0.8 C_o and the chloride front is proportionately in advance of Q = 0.

The effect shown in Figure 2 on the chloride profile was discussed for an analogous situation by Smiles and Gardiner (1982) who attributed it to anion exclusion from a volume of soil water equal to the external specific surface of the soil ($1.02 \times 10^5\ m^2\ kg^{-1}$ based on continuous flow N_2 absorption measurements; K.G. Tiller, personal communication) multiplied by a 'depth' of 0.87 nm. This depth corresponds with calculations of Quirk (1968) for the depth of chloride exclusion at the surface of clay in equilibrium with 100 equiv m^{-3} $CaCl_2$ where there is no double-layer interaction. The depth is also consistent with the layer of partially hydrated Ca ions adsorbed at the clay surface with a characteristic decay length of about 0.9 nm observed by Israelachvili and Adams, 1978. Such a layer would be inaccessible to chloride ions. Tritium, of course, is not affected by either of the above mechanisms. At this stage, the most important observation is that in a soil with a net negative charge and a substantial specific surface there will, almost inevitably, be water close to the colloid surface that is inaccessible to anions even in quite concentrated divalent salt solutions. The thickness of this layer may be only 2 to 3 water molecules; nevertheless, the actual volume of water involved may be quite great relative to the total water content.

Further experiments have shown that the thickness of the layer appears to depend on solution concentration, as Figure 3 illustrates.

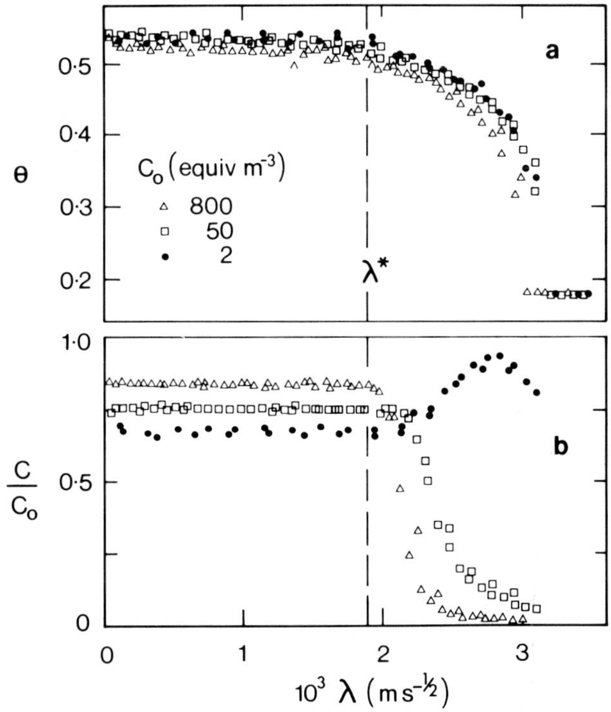

Figure 3. Profiles of (a) water content and (b) normalized chloride concentration during constant potential horizontal absorption of $CaCl_2$ solutions by a clay soil for three different inflowing solution concentrations C_o, λ* is the position where g = 0 (data from Bond et al., 1984)

In this figure we observe that as the solution concentration decreases, the amount of inaccessible water increases, as does the displacement of the salt front from g = 0. It should be noted with regard to these data, that for the least concentrated invading solution, the equilibrium solution (that 'external' to the double layer) initially present was more concentrated than C_o. These data are explored in more detail by Bond et al. (1984), who concluded that:

a) in chemically reactive clay soils, the exclusion of anions from the diffuse double layer at the clay surface significantly affects the shape of the salt concentration profile and, in particular, the position of the salt front;
b) this double layer effect depends on the concentration of the ambient equilibrium salt solution but, because of double layer interaction associated with soil structure, the effect cannot be predicted *a priori*, although it can be simply measured;
c) the position of the salt front is determined by both the initial and the inflowing solution concentrations, but control of this position is dominated by the highest concentration solution present.

5 Quantitative prediction of dispersion

For chemically non-reactive soils, dispersion occurs in a region close to q = 0 where θ and $D_s(u)$ do not change very much in space, although they may change in time. Equation 10 may then be written

$$\left(\frac{\partial C}{\partial \tau}\right)_q = \left(\frac{\partial^2 C}{\partial q^2}\right)_\tau \qquad (15)$$

where the time-like variable τ is given by

$$\tau = \int_0^t (\theta^*)^2 D_s(u^*)\, dt \qquad (16)$$

In Equation 16 the * indicates values of variables at q = 0, and the variable τ takes into account the evolution of θ^* and $D_s(u^*)$ up to time t.

Solutions of Equation 15 are available for a variety of initial and boundary conditions (see, for example, Carslaw and Jaeger, 1959). In particular, the conditions defined by Equations 11 and 12 or 13, which imply a step-change in solution concentration at q = 0, lead to the solution (Smiles et al., 1981)

$$\frac{C - C_n}{C_o - C_n} = 1/2 \text{ erfc } \left(\frac{q}{2\sqrt{\tau}}\right) \tag{17}$$

This equation describes the diffusion of solute about the moving plane q = 0 with a diffusivity given by τ/t.

Equation 17 has been tested and found to describe dispersion reasonably well (Smiles et al., 1981; Bond et al., 1982). It is important to note that this solution applies to material space and the water content distribution distorts the salt distribution in real-space.

When we extend the model represented by Equation 10 to chemically reactive soils, the situation is complicated because the inaccessible water content changes across any solution concentration front, and an iterative analysis of the problem appears obligatory. Bond et al. (1982) and Smiles and Bond (1982) pursue the problem in $CaCl_2$ systems, and discuss the interpretation of salinity profiles observed in the field.

6 Areas for further research

Solute movement is often more complicated than in the examples described here. It is therefore necessary to understand the limitations of the above approach in such situations and to modify it where necessary. Some areas where further experimental and theoretical work is possible are briefly outlined below.

a) Mixed cation systems. Although there have been a number of studies of solute movement incorporating cation exchange (predominantly for steady flow), there remains uncertainty. There is a need, for example, to clarify the nature of the exchange process, particularly the rate of attainment of equilibrium, and to examine the consequences of the effect of exchange on θ_i, through the effect of cation type on the double layer.

b) Strongly structured soils. As for mixed cation systems, there has been much work done for steady flow, particularly relating to the treatment of 'immobile' or, more correctly, 'slowly accessible' water. Study should be extended to unsteady flow and should consider interactions, of the type described by Blackmore (1976) for example, between structure and chemical/physico-chemical processes.

c) Structurally unstable soils. The structure and hydraulic properties of many soils depend on cation type and concentration (Collis-George and Smiles, 1963; Quirk and Schofield, 1955). There is scope for quantitative study of the interaction between soil solution, soil structure and water flow during miscible displacement.

References

Bird, R.B., W.E. Stewart, and E.N. Lightfoot 1960. Transport phenomena. John Wiley & Sons, New York, 780 pp.

Blackmore, A.V. 1976. Salt sieving within clay soil aggregates. Aust. J. Soil Res., 14: 149-158.

Bond, W.J., B.N. Gardiner, and D.E. Smiles 1982. Constant-flux absorption of a tritiated calcium chloride solution by a clay soil with anion exclusion. Soil Sci. Soc. Am. J., 46: 1133-1137.

Bond, W.J., B.N. Gardiner, and D.E. Smiles 1984. Movement of calcium chloride solutions in an unsaturated clay soil: the effect of solution concentration. Aust. J. Soil Res. (in press).

Carslaw, H.S., and J.C. Jaeger 1959. Conduction of heat in solids. (2nd Edition) Oxford University Press, Oxford, 510 pp.

Collis-George, N., and D.E. Smiles 1963. An examination of cation balance and moisture characteristic methods of determining the stability of soil aggregates. J. Soil Sci., 14: 21-32.

Fried, J.J., and M.A. Combarnous 1971. Dispersion in porous media. Adv. Hydrosci., 7: 170-282.

Israelachvilli, J.N., and G.E. Adams 1978. Measurement of forces between two mica surfaces in aqueous electrolyte solutions in the range 0-100 nm. J. Chem. Soc. Faraday Trans. I., 74: 975-1001.

Philip, J.R. 1957. The role of mathematics in soil physics. J. Aust. Inst. Agric. Sci., 23: 293-301.

Quirk, J.P. 1968. Particle interaction and soil swelling. Israel J. Chem., 6: 213-234.

Quirk, J.P., and R.K. Schofield 1955. The effect of electrolyte concentration on soil permeability. J. Soil Sci., 6: 163-178.

Rose, D.A. 1977. Hydrodynamic dispersion in porous materials. Soil Sci., 123: 277-283.

Smiles, D.E., and W.J. Bond 1982. An approach to solute movement during unsteady, unsaturated water flow in soils. In: Prediction in water quality, E.M. O'Loughlin and P. Cullin (eds.), Proc. Symp. Aust. Acad. Sci., Canberra, December 1982, pp. 265-287.

Smiles, D.E., and B.N. Gardiner 1982. Hydrodynamic dispersion during unsteady, unsaturated water flow in a clay soil. Soil Sci. Soc. Am. J., 46: 9-14.

Smiles, D.E., K.M. Perroux, S.J. Zegelin, and P.A.C. Raats 1981. Hydrodynamic dispersion during constant rate absorption of water by soil. Soil Sci. Soc. Am. J., 45: 453-458.

Smiles, D.E., and J.R. Philip 1978. Solute transport during absorption of water by soil: laboratory studies and their practical implications. Soil Sci. Soc. Am. J., 42: 537-544.

Weinberg, A.M. 1972. Science and trans-science. Minerva, 10: 209-222.

Wilson, J.L., and L.W. Gelhar 1981. Analysis of longitudinal dispersion in unsaturated flow: 1. The analytical method. Water Resour. Res., 17: 122-130.

Discussion

G.H. Bolt:

It appears to me that the use of a material coordinate (like your q in Eq. (5)) in solving the partial differential equation describing mixed convection/dispersion transport of solutes was already apparent in the early works on bed-exchange by deVault, Thomas, Glueckauf, and Hiester and Vermeulen of the forties and early fifties (for a review, see Chapter 9 in Bolt, G.H.,ed., 1982. Soil chemistry B. Physico-chemical models. Elsevier, Amsterdam, 527 pp.). In fact, in those days the name solution throughput parameter was used for minus q,

negative values of your q referring to regions that had experienced a positive throughput of solution, while the positive values then designate the region that has not yet been 'touched' by the incoming solution.
A question that arises in my mind in relation to this parameter is the merit of attaching the material coordinate to the carrier solution, as compared with its attachment onto the solute described with the differential equation. Although of not too much consequence in the case treated here, it is of importance in case of solute retardation due to adsorption. Even in the present case the decision has to be made if, in the presence of a stagnant liquid phase Θ_s, one uses θ or $(\Theta - \Theta_s)$ to define the material coordinate. Personally, I would favour the latter and would be interested in your comment.

Author:

We do not claim to have introduced the material coordinate as a new concept, merely to have applied it to the more general situation, where θ and u vary with position and time. In this situation the material coordinate is a nonlinear function of distance and time. The exact form of the material coordinate used must certainly take account of the properties of the solute and the presence of any 'stagnant' or inaccessible water. As alluded to in part 5 of our paper, such a material coordinate is described by Bond et al. (1982).

G.H. Bolt:

The penetration of Cl ions exceeding your 'chosen' (see also my earlier comment) zero point of the material coordinate q is interpreted as being associated with anion exclusion. Isn't is necessary, though, to also assume the existence of a relatively immobile layer of solvent in the system? Thus, if all the solvent were fully mobile, the carrier influx would reach up to your point g = 0, and the mean Cl penetration could not reach beyond that point. The fact that it does so then must imply that part of the water is (relatively) immobile, so $\Theta_m < \Theta$.
If now this immobile zone (presumably next to the solid surface) is also an anion exclusion zone, one would expect the fairly symmetric diffusion front (as found) located around the point $x = Fw.t/\Theta_m$ and not at $x = Fw.t/\Theta$. However, immobility and inaccessibility because of a low transfer coefficient would give the same effect. Finally, exclusion of

Cl without immobility of the liquid phase in that zone could not give the pattern as found. Instead, one would enter into some 'salt sieving' - a term I missed in your story - giving rise to increased concentrations.

Summarizing, I venture to state that a) your S-shaped concentration profiles extending beyond g = 0 point to the existence of an immobile liquid layer, and only secondarily hint at the possibility of Cl exclusion from that zone (temporary exclusion due to low transfer would already suffice!); b) your low-concentration 'bubble' curve points to the existence of an exclusion zone extending outside the above immobile zone and gives rise to some salt sieving with accompanying local concentration increase. Any comment?

Author:

Figure 2 in the paper shows that when both tritiated water and chloride are present in the inflowing solution, the position of the front of tritiated water is found at the notional 'piston front' of the water, while the chloride front is ahead of this position. From this we conclude that there is no appreciable volume of soil water that is inaccessible to the tritium (i.e. no stagnant zone, as it is usually referred to) and that, regardless of the mobility of the water, there must be a volume of soil water that is inaccessible to the chloride, probably as an result of anion exclusion. Undoubtedly, the water that is inaccessible to the chloride is far less mobile than the inflowing water by virtue of its proximity to the clay surface.

However, we prefer to consider that there is a continuous distribution of water velocities in this soil and no 'immobile zone' as such.

As explained in more detail by Bond et al.(1984), the low concentration 'bubble' in Figure 3 is a consequence of the fact that in that experiment the concentration of the inflowing solution was less than the concentration of the initial solution. There is no need to invoke 'salt sieving' to explain it.

Analysis of solute movement in structured soils[1]

J. Skopp
Department of Agronomy,
University of Nebraska
Lincoln, NE 68583 U.S.A.

Abstract

The movement of solutes is analyzed using moment techniques. Moments are an empirical procedure for interpreting data without bias towards a specific theory. Data for miscible displacement through packed, saturated soil columns at different flow rates is presented. Application of moment techniques to unsaturated flow conditions as well as theoretically based moments are also discussed.

1 Introduction

The analysis of solute movement consists of several stages including:

- conceptualization of process,
- development of descriptive equations, and
- measurement of solute movement.

At some point the measurements must be reconciled with the descriptive equations. One of the difficulties with analyses of solute movement is the inherent uncertainty of our concepts, and hence, our governing equations. This has combined with a lack of readily available tools to directly characterize data sets. Instead, experiments in solute movements are characterized by model parameters. Unfortunately as the model changes, so do the estimates of model parameters. This detracts from our confidence in developing a physical interpretation of these parameters.

Moment techniques provide a powerful means of overcoming these problems and allowing analysis of experiments independent of any particular model. Moments can be used to characterize and describe any data set which can be interpreted as a distribution. The output of a classical experiment on miscible displacement readily fits this requirement (Biggar and Nielsen, 1962). Examples abound in the Chemical Engineering and Petroleum Engineering literature of the use of moment techniques for similar problems. However, only one publication in the soils literature exists applying this technique to solute movement (Calvet et al., 1978). Some of the properties of moments will be described in this paper. Moment analysis will then be used to describe solute movement through structured soils.

2 Properties of moments

Distributions representing solute movement may be either a function of space or time. The concentration is interpreted as the density of the distribution. Classical experiments on miscible displacement provide a data set of concentrations varying with time. Studies of solute migration 'in situ' tend to produce data set of concentrations varying with position. Either data set can be analyzed with regard to moments. Although spatial and temporal moments may be related, they are not equivalent. Here the temporal distribution of concentrations will be emphasized.

Ordinary moments (M_n) of concentration (C) with respect to time (t) can be defined as:

$$M_n = \int_0^\infty t^n C \, dt$$

The subscript n is the order of the moment. A second type of moment is called central moment (M_n^*) is frequently more convenient for n greater than 1:

$$M^*_n = \int_0^\infty (t - M_1)^n \, C \, dt \quad ; n \geq 2$$

The physical interpretation of the moments follow:

- M_0 or zeroth ordinary moment represents the total amount of solute within a column for spatial moments or total solute eluted for temporal moments.
- M_1 or first ordinary moment describes the mean position of the solute concentration. Note the mean position may not identify the peak or modal position, but is a measure of central tendency. For temporal moments, this is the mean time for elution.
- M_2 or M_2^* are measures of the width in the solute distribution with M_2^* being directly interpreted as a variance.
- M_3 or M_3^* describes the asymmetry in the solute distribution and M_3^* can be used to calculate a skewness. Skew = $M_3^*/M_2^{*3/2}$.
- M_4 or M_4^* are measures of flatness relative to a normal distribution and extreme values are indicative of multiple peaks. M_4^* can be used to calculate the kurtosis. Kurtosis = $(M_4^*/M_2^{*2}) - 3$.

Moments possess the following characteristics:

a) Moments are model independent. They do not invoke any assumptions regarding the mechanisms controlling the movement of solutes.

b) Moments are orthogonal. The definition insures that each moment is independent of any other moments of different order. This is convenient for analysis of variance and regression.

c) Moments define a distribution. Not only does the data define a set of moments, but a set of moments unambiguously defines a distribution. Hence, moments summarize the data. It is possible to generate a distribution from the moments using a Gram-Charlier expansion.

d) Moments are portable. Data sets are cumbersome and rarely accessible in the literature. Moments provide an objective expression of the data which can be presented in a manuscript. This allows an experiment to be analyzed using models not available to the scientist when the experiment was performed.

e) Moments can be related to model parameters. This is most readily done for linear models by using Laplace transforms. The n'th derivative of the Laplace transform defines the n'th order ordinary moment.

3 Application of moments to solute movement

Prior to analyzing experimental data on solute movement, a data set based on a specific model was generated to evaluate the behavior of the moments. The classical solution to the dispersion equation (Ogata and Banks, 1961) was used to generate breakthrough curves from which moments were calculated. The equation used was:

$$C = (1/2)\ \{\mathrm{erfc}[(1 - T)/\sqrt{T/P}] + \exp(P)\ \mathrm{erfc}[(1 + T)/\sqrt{T/P}]\}$$

where

C and T are dimensionless concentration and time respectively. Time is expressed as pore volumes or number of displacements of soil solution. $T = Vt/L$

where

V is the convective velocity, L the sampling position and t the time. P is a Peclet number defined as:

$P = VL/D$

where

D is the dispersion coefficient. For a Peclet number of 100, the moments were:

$M_1 = 1.000$, $M_2^* = .02005$, skew = .425 and kurtosis = .25 (1)

The value of M_1 describes the rate of movement of the mean solute position. A value of 1.0 indicates the mean passes the position L after displacing exactly one pore volume. M_2^* describes the width in time of the solute history. The width depends directly on the dispersion coefficient in a simple model such as this. This model allows the theoretical calculation of the moments as:

$$M_1 = 1.0 \tag{2}$$

$$M_2^* = 2/P \tag{3}$$

$$M_3^* = 12/P^2 \text{ and skew} = 4.24/P^{\frac{1}{2}} \tag{4}$$

$$M_4^* = (12/P^2) + 120/P^3 \text{ and kurtosis} = 30/P \tag{5}$$

The skew is always positive which indicates tailing towards longer times or pore volumes. This tailing also increases as the dispersion coefficient increases.

The theoretical moments are a useful means of calculating dispersion coefficients. This is done by equating the theoretical and empirical moments. However, their use is predicated on the validity of the model upon which they are based. The empirical moments do not suffer from this defect and hence can be used to characterize the process of solute transfer. The empirical moments do not require the explicit description of a model or its validation.

Estimation of moments is not completely immune to all problems. The most significant difficulty is the decrease in precision as the order of moment increases. One consequence is that moments higher than fourth are rarely calculated. Even the third and fourth moments may be difficult to estimate. This is reflected in the example by the variation in estimates of the Peclet number. From the second moment, P = 99.75, from the third moment P = 99.53, while from the fourth moment P = 121.46 all in response to the same data set and P = 100.

4 Application to structured soils

The techniques of moment analysis will now be applied to studies of solute movement in packed columns of soil aggregates. The work of Biggar and Nielsen (1962) provides a sequence of experiments performed at two different flow rates and three different aggregate sizes. The calculation of the moments based on their data is given in Table 1.

The first moment increases as particle size decreases, while the second central moment decreases as particle size decreases. This trend is maintained regardless of whether the velocity is high or low. However, the higher velocity consistently shows an increased first moment and decreased second central moment when compared to the lower velocity. These interpretations are consistent with the observations of Calvet et al. (1978).

Calvet et al. (1978), however, did not determine the appropriateness of the solution to the classical dispersion equation. Since the Peclet number is unknown, a direct comparison of theoretical (Equation 2-5) and experimental (Equation 1) moments is not possible. However, a check for consistency is possible by comparing different order moments. M_1 from Equation 2 is always equal to 1 and this relation can be used by itself. However, M_2^* and M_3^* from Equations 3 and 4 are related in a fixed manner if the dispersion equation applies to the data. Eliminating P from Equations 3 and 4 gives the relation skew = $3\sqrt{M_2^*}$. This holds for the data set represented by the moments of Equation 1 generated earlier. This test for consistency fails for the data of Table 1. The lack of agreement suggests that this solution to the dispersion equation is inappropriate despite its visual goodness-of-fit. Negative values for the skew and kurtosis are also indications of this inappropriateness since Equations 4 and 5 do not allow negative values. This discrepancy may lie with either the governing partial differential equation or the boundary conditions.

A physical interpretation of the transport phenomena described by the moments of Table 1 can be developed using a mixing model suggested by Skopp et al. (1981). At low velocities, mixing between intra- and interaggregate pores is decreased. The soil behaves more like a bimodal pore system with M_2^* and skew increasing with aggregate size as the disparity in pore sizes (between intra- and interaggregate pores) increases.

At the higher velocity, mixing is increased. With the large aggregates the tendency is reduced for solute to spend time in only one of the two dominant pore sizes. M_1 is larger, $M_2^*$is smaller, although the skew is negative. As aggregate size decreases, mixing becomes more effective and the soil behaves more like a homogeneous material. In other words, at high flow rates and small aggregate sizes, the moments are closest to that predicted for the classical dispersion equation. However, the negative skew is still indicative that effective homogeneity has not been achieved.

More detailed models can be developed, although most models tend to give complete concentration profile histories. Frequently it is only M_1, the speed with which the solute moves that is desired. Models such as that of Rose et al. (1982) determine the appearance of the peak. The peak migration, in turn, depends on soil properties as well as water inputs. The simplest use of moments in this connection would be regression of M_1 against water velocity and aggregate size. The data of Table 1 is not sufficiently extensive to consider this form of regression analysis.

5 Application to unsaturated soils

The concept of moment analysis is based on the assumption that a concentration profile or history can be interpreted as a distribution. Unsaturated flow does not preclude the use of moments. Moreover, the uncertainty regarding the best description of the dispersion process lends additional support to the use of moment techniques. Parameter estimation is dependent on model formulation. Consequently, any uncertainty in the accuracy of the model must result in uncertainty in the accuracy of the estimated parameters. The use of moments provides an objective way of characterizing the data which is of increased importance when unsaturated conditions are studied.

6 Summary

Moments have been introduced as a convenient tool to analyze solute distribution data. Moments are defined, and their advantages listed. An example of the application of moments to solute movement through structured soils is given. The observed moments are then interpreted in terms of the physical processes controlling solute movement.

Table 1. Summary statistics for data of Biggar and Nielsen (1962)

Aggregate size (mm)	M_1	M_2^*	Skew	Kurtosis
V = .04 cm/hr				
1 - 2	.7406	.3476	.843	-.18
.5 - 1	.8425	.3091	.689	-.31
.25 - .50	.9175	.1710	.296	-.46
V = 2.0 cm/hr				
1 - 2	.8121	.1715	-.377	-.070
.5 - 1	.9692	.0609	-.148	4.9
.25 - .50	1.001	.0275	-1.47	12.3

Notes

[1] Published as Paper Number 7430, Journal Series, Nebraska Agricultural Experiment Station.

References

Biggar, J.W. and D.R. Nielsen 1962. Miscible displacement II. behavior of tracers. SSSA. Proc., 26: 125-128.

Calvet, R., J. LeRenard, C. Tournier and A. Hubert 1978. Hydrodynamic dispersion in columns of pumice particles. J. Soil Sci., 29: 463-474.

Ogata, A. and R.B. Banks 1961. A solution of the differential equation of longitudinal dispersion in porous media. U.S.G.S. Prof., paper 411-A.

Rose, C.W., F.W. Chichester, J.R. Williams and J.R. Ritchie 1982. A contribution to simplified models of field solute transport. J. Environ. Qual. 11: 146-150.

Skopp, J., W.R. Gardner and E.J. Tyler 1981. Solute movement in structured soils: two-region model with small interaction. SSSA J. 45: 837-842.

Discussion

G.H. Bolt:

Would you follow me with the statement, 'One of the main advantages of the use of moments for solute-distribution curve description is its sobering action on the model builders'? The higher moments stress the weight of the extreme tail and head sections of the curve, while the model builder too often is so satisfied with reasonable agreement between experiment and model in the central section of the curve that he erroneously concludes that his model has been proved correct: any model is likely to produce some type of S-shape. At the same time, I regret the sensitivity of the higher moments to the boundary conditions (if this is indeed the case), because the latter are precisely the conditions that are so difficult to define in reality.

Author:

If moments induce model builders to think more carefully about critical tests of their hypotheses, then well and good. However, this represents a subjective advantage of the technique rather than an objective advantage in addition to those listed in the paper.

P.A.C. Raats:

You show that for the linear convection/dispersion equation the dispersion coefficient can be inferred from the moments. Have similar parameter identification techniques been worked out for various types of linear models involving mobile and stagnant phases?

Author:

Solutions are given in the chemistry and chemical engineering literature. In many cases reaction terms are also included. The two earliest reports of moment techniques were given separately in papers by Kubin and Kucera.

Salt transport in heavy clay soil

J.W.van Hoorn
Department of Land and Water Use,
Agricultural University, Wageningen

Abstract

Salt transport in heavy clay soil is analysed, using field data of saline clay soils.

1 Introduction

Salt movement is linked together with water movement which depends on the hydraulic conductivity of the soil profile and its structure. In heavy clay soils one can usually distinguish a top layer with a rather high hydraulic conductivity consisting of the tilled layer in arable land or the turf layer in grassland, below which the hydraulic conductivity is much lower and often decreases with depth. In basin clay soils in river areas and in marine clay soils one may encounter at a depth of about 1 m again a layer having a moderate to high hydraulic conductivity and consisting of clay with iron concretions, organic matter, peat or soil material with a coarser texture.
The discharge of water in such soil profiles depends on the ratio between the rainfall (or irrigation) rate and the infiltration rate of the second layer of low permeability. Under Dutch climatic conditions the rainfal rate during autumn and winter is rather low, ranging between 1 and 2 mm/hour for rainfall amounts up to 10-15 mm/day. Observations in tile-drained fields have shown that most of the water is drained off by infiltration into the heavy clay layer and flows through the permeable subsoil towards the drains (Figure 1a). Horizontal flow through the highly permeable top layer towards the drain trench only occurs during short periods after heavy rainfall (Van Hoorn, 1960).

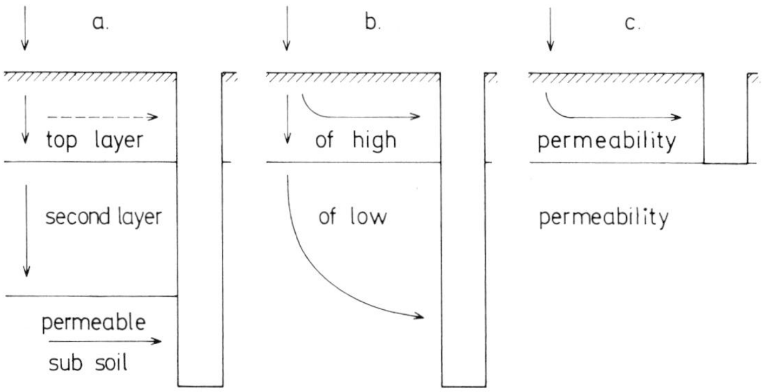

Figure 1. Water movement in heavy clay soil towards subsurface and surface drains

In the absence of a highly permeable subsoil, the water table will rise faster causing a decrease of the infiltration rate. If moreover the rainfall rate is higher than under Dutch conditions, part of the water will move downward through the second layer of low permeability, but an important part of it will flow through the top layer towards the drain trench (Figure 1b).
In the case of a surface drainage system consisting of furrows, water will infiltrate at the start of the rainy season into the layer of low permeability causing a rise of the water table to the top layer; afterwards all rainfall will be discharged by horizontal flow through the top layer (Figure 1c).

The structure of heavy clay soils is characterized by rather big soil aggregates separated by large pores formed as root or worm holes by biological activity or as cracks during dry periods. Owing to the biological activity and the alternation of crack formation and swelling the hydraulic conductivity generally increases in summer and decreases in winter, a reversible process with short and long term trends.
Better drainage leads on long term to a clear increase of the hydraulic conductivity, whereas bad drainage and shallow groundwater tables cause a decrease in hydraulic conductivity in soils which originally had a low water table (Van Hoorn, 1981).
The large pores are essential for the removal of surplus water by gravity force and determine the saturated hydraulic conductivity.
Inside the soil aggregates pores are too small for water movement by

gravity force and water can only be removed by the evapotranspiration process.

2. Natural conditions of the experimental sites

The salt movement in heavy clay soils will be illustrated by examples from two saline areas:

- the Marismas in the Quadalquivir delta in Southern Spain (Van Hoorn et al., 1976),
- the Leziria Grande near Lisbon between the rivers Tague and Soraya (Mann et al., 1982).

The soils in both areas can be classified as silty clay, containing in the Marismas about 20% $CaCO_3$ and $MgCO_3$ and in the Leziria less than 1.5% $CaCO_3$.

Both areas are characterized by rainfall from September through May, the average precipitation being about 550 mm in the Marismas and 650 mm in the Leziria.

Table 1 presents the salinity at the start of reclamation of an experimental station in the Marismas and of three experimental fields in the Leziria. The salinity is expressed as electrical conductivity in mS/cm of an extract of 200 g of water per 100 g of dry soil ($EC_{1:2}$). Since the saturated paste contains about 80 g of water per 100 g of dry soil, $EC_{1:2}$ can be converted into EC_e by multiplication by 2.5.

Table 1. $EC_{1:2}$ in mS/cm at the start of reclamation

Layer in cm	0-25	25-50	50-75	75-100	100-125	125-150	150-200	0-125
Marismas	12.0	20.0	26.0	28.0	30.0	32.0	34.0	23.2
Leziria 1	7.6	9.6	15.6	19.3	25.8	33.0	34.2	15.6
2	6.5	7.7	10.9	14.1	19.7	24.2	28.2	11.8
3	3.3	5.3	7.2	9.7	13.0	18.1	23.6	7.7

Before the installation of the experimental station the soil in the Marismas was inundated every year during autumn and winter and water was removed by natural surface flow to depressions and by evaporation. The Leziria was already protected by dikes and water was drained off by a system of shallow surface drains. This difference in drainage between the two areas is reflected on the one hand in the salinity, as shown in table 1, and on the other hand in the hydraulic conductivity of the upper meter of the soil profile.Whereas the hydraulic conductivity of the Marismas soil to a depth of about 1 m ranged around 0.1 m/day, the Leziria soil showed a value of about 0.3 m/day.

3. Horizontal flow versus vertical flow

Table 2 presents the relation between drain spacing and the average salinity of drain water from the experimental station in the Marismas. A wider spacing leads to a higher water table (Figure 2) and an increase of horizontal flow through the top layer.

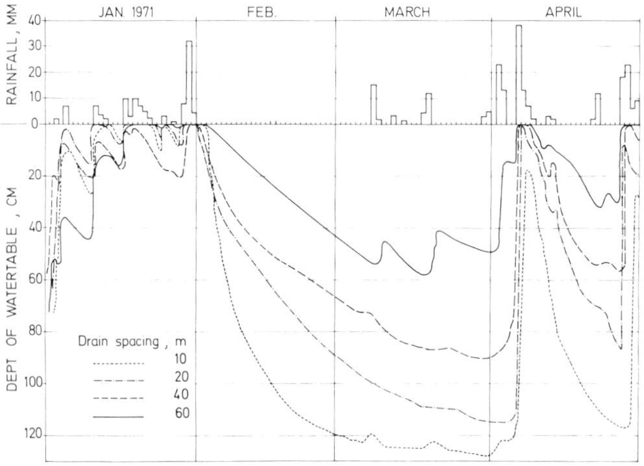

Figure 2. Drainspacing and depth of water table

Table 2. Drain spacing versus average salinity of drain water, winterperiod 1970-74, Marismas

Spacing , m	10	20	40	60
Salinity , g/l	60	55	48	41

Table 3 shows the change in salinity of drain water from surface and subsurface drains during a winter period. It presents a second example of the increase of horizontal flow through the top layer due to a higher water table, in this case as a consequence of more precipitation.

Table 3. Salinity of drain water from surface and subsurface drains, winterperiod 1970-71, Marismas

Month	Nov	Dec	Jan	Feb	M	Apr	May	Aver.	Min.
Rainfall ,mm	53	108	104	2	42	168	62		
Surface ,g/l	58	27	19	9	48	8	27	17	5
Subsurface,g/l	91	39	31	19	75	29	72	34	18

Table 4 presents the salinity of drain water for three successive winter periods from surface and subsurface drains, the latter one expressed as average for the whole winter period and also at low discharge. The salinity of drain water at low discharge corresponds with that of soil water at a depth of about 1 m.

Table 4. Salinity of drain water from surface and subsurface drains, average of 3 experimental plots, Leziria

Winter period		1976-77	1977-78	1978-79
Surface drain	,g/l	10.5	5.8	4.8
Subsurface drain, average	,g/l	30.8	16.4	10.4
, at low discharge	,g/l	40.3	29.7	23.0

These examples clearly show that part of the water is drained off by vertical flow through the soil profile and part of it as horizontal flow through the top layer. The ratio between vertical and horizontal flow changes with the depth of the water table, the latter one depending on spacing and precipitation.

4. Salt movement from soil aggregates to large pores

Table 5 presents for six succesive years the precipitation, the drain discharge and the average salinity of the subsurface drains of the three experimental fields in the Leziria

Table 5. Salinity of subsurface drains during 6 winter periods,Leziria

		1976-77	1977-78	1978-79	1979-80	1980-81	1981-82
Rainfall	,mm	620	712	748	415	308	554
Discharge	,mm	269	210	230	0	0	113
Salinity,	field 1,g/l	41.3	19.5	12.4	-	-	23.7
	field 2,g/l	23.8	19.3	12.6	-	-	21.9
	field 3,g/l	27.4	10.5	6.3	-	-	17.7

During the first 3 years the salinity of the soil decreased, owing to the leaching by the rainfall. The lower the salinity of the soil, the lower the salinity of the drainage water, which is quite normal.During the first year, however, field 2 showed a lower salinity of the drainage water than field 3 despite a higher soil salinity (Table 1). This points to a lower leaching efficiency coefficient on field 2, that means a larger part of the water is passing through cracks without mixing with soil water. During the third year, fields 1 and 2, which in spring 1978 attained a soil salinity level equal to that of field 3 in 1976, showed a lower drainwater salinity than field 3 during the first year. Such a decrease of the leaching efficiency coefficient with time is often observed. This phenomenon can be ascribed to rapid leaching of easily attainable salts in cracks and large pores, whereas the removal of salt inside the soil aggregates takes much longer.

After two dry years without drain discharge the drain water salinity increased strongly. During the dry period salt moved from the interior of the structural elements towards the large pores and new cracks were probably formed, providing a fresh oppurtunity for leaching by the rain water.

Assuming a yearly precipitation of 650 mm in the period 1979-81, a drain discharge of about 200 mm per winter and a continuing decrease of the drain water salinity, this higher rainfall during two winters would not have leached more salt from the soil than the rainfall in the winter 1981-82, that was even below the average.

5. Leaching by applying irrigation water

Trying to accelarate the leaching process of heavy clay soils by large water applications does not seem an interesting and useful operation. An irrigation test carried out on the experimental farm of the Marismas points in a similar direction (Table 6).
During this test two amounts of irrigation water were compared, spread over three applications during a period of about one month. Part of the water was used for filling up the soil water deficit and for evapotranspiration, so that in the case of the small water application of about 200 mm the amount of drain water remained very low. However, this low amount of drain water was more or less compensated by the high salinity of the water,whereas in the case of the large water application the drain water salinity decreased considerably. This decrease may be ascribed to an increase of horizontal flow through the top layer and to a closing of cracks.

Table 6. Irrigation test for desalinization, Marismas

Irrigation	Drainage	Salinity drain water	Salt removed
197 mm	9.0 mm	82 g/l	7.4 t/ha
462	90.3	19	16.9

In stead of applying large amounts of irrigation water in summer it seems more interesting to apply a modest amount of water in early autumn in order to fill up the soil water deficit and to promote more drainage from winter rainfall.

6. Resalinization of the top layer by capillary rise

During summer the salinity of the top layer increases again by capillary rise of saline water from underlying layers. Figure 3 shows an example of such a redistribution of the salinity, observed in the Marismas. According to calculations based on the amount of salt transported to the top layer capillary rise from the underlying layer (15-75 cm) ranged in the order from 25 to 30 mm. This rather low amount may be ascribed to horizontal cracks that are formed when the soil is drying out and that impede the vertical capillary movement.

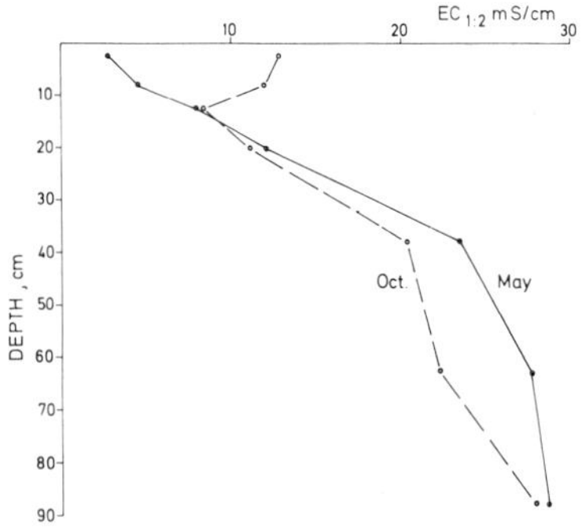

Figure 3. Redistribution of salinity due to capillary movement

Although the amount of water moving upward in summer is rather small, it can bring a considerable amount of salt into the top layer. For example in the case of the Marismas the average value of $EC_{1:2}$ was 20 mS/cm in the layer 25-50 cm (Table 1), corresponding with about 65 g/l for the soil water at field capacity. Capillary rise of 25 mm corresponds then with 16 t of salt per ha.
At the start of the rainy season part of this salt will move downward again and part of it will be removed by horizontal flow through the top layer. Owing to this combination of capillary rise during the dry season and surface flow during the rainy season a surface drainage system can also contribute on long term to the desalinization of heavy clay soils, although the process will take more time than desalinization by subsurface drains and may not attain the same depth.

7. Model for prediction of desalinization

Desalinization of heavy clay soil is characterized by three typical phenomena:

- the combination of horizontal flow through the top layer and vertical water movement through the deeper layers;
- the leaching efficiency coefficient, expressing the percentage of rain or irrigation water mixing with the soil solution, can either

decrease or increase with time, depending on the amount of water applied and the alternation of wet and dry periods;
- the resalinization of the top layer by capillary rise during the dry season.

Owing to these phenomena it is not possible to predict desalinization of heavy clay soil by using leaching formulas that assume the same vertical flow through the entire rootzone, e.g. over a depth of at least 1 m.

For the Marismas the desalinization process by rainfall was calculated by a numerical method, assuming a decrease of the amount of percolation water with depth: 100 % of the water percolates the first layer (0-25cm), 80 % the second layer (25-50cm), 60 % the third layer (50-75cm),40 % the fourth layer /75-100cm) and 20 % the fifth layer (100-125cm).

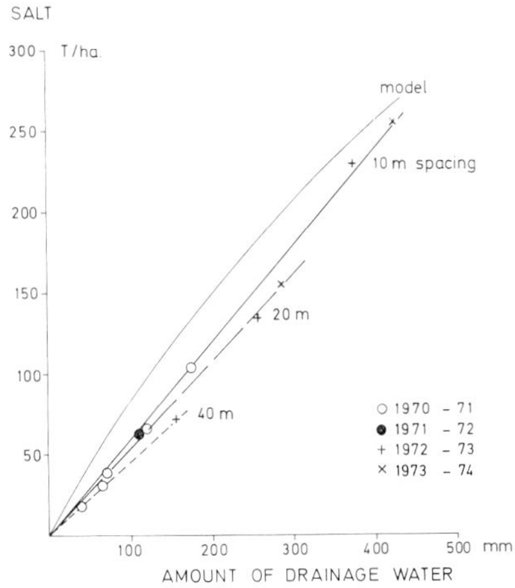

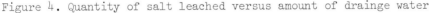

Figure 4. Quantity of salt leached versus amount of drainge water

Figure 4 shows the relation between the amount of drainage water and the amount of salt removed, on the one hand calculated according to the model described above and on the other hand observed for three different drain spacings. The observations for a spacing of 10 m are rather close to the calculated curve; the observations for the wider spacings deviate more and more, probably because more water is drained off by surface flow.

Although Figure 4 indicates a rather good agreement between the model used for the calculation and the observations, the model has a very limited value for the prediction of salinization, since it does not take into account the resalinization of the top layer by capillary rise and the change of the leaching efficiency coefficient with time. In reality a part of the desalinization may have been caused by the combination of capillary rise in summer and surface flow in winter and vertical flow may have been less than assumed in the model, both phenomena compensating each other more or less. The change of the leaching efficiency coefficient owing to the alternation of wet and dry periods is however unpredictable.

8. Desodification

In order to maintain or improve the structure and the infiltration rate of heavy clay soil, gypsum can be applied if the $CaCO_3$ content is low. The soil of the Marismas experimental farm contained about 20 % $CaCO_3$ and $MgCO_3$ and did not show any response to gypsum application. Since the desalinization process is slow, $CaCO_3$ notwithstandig its low solubility provides sufficient calcium for the desodification of the soil.

The soil in the Leziria only contained between 0.2 and 1.5 % $CaCO_3$ in in the upper 50 cm. Gypsum application at a rate of 10 t/ha per year during 4 years lowered the E.S.P., improved the infiltration rate and the amount of drainage water. However, a comparison between the amount of calcium applied as gypsum and the amount exchanged in the soil showed that only part of the calcium applied as gypsum was used in the exchange process. Also the decrease in ESP between 1978 and 1980 was smaller than between 1976 and 1978. This points in the same direction as the decrease of the leaching efficiency coefficient between 1976 and 1979. The exchange of sodium by calcium inside the structural elements of heavy clay soil is a slow process. In order to avoid losses, small amounts of gypsum should be applied with intervals of several years.

References

Mann, M., A.Pisarra and J.W. van Hoorn, 1982. Drainage and desalinization of heavy clay soil in Portugal. Agric.Water Manage., 5:227-240.

Van Hoorn, J.W., 1960. Groundwater flow in basis clay soil and the determination of some hydrological factors in relation with the drainage system. PUDOC, Wageningen. Versl.Landbk.Onderz. no.66. 10, 136 pp (in Dutch with English summary).

Van Hoorn, J.W., 1981. Drainage of heavy clay soils. In: Land Drainage. A Seminar in the EC Programme of Coordination of Research on Land Use and Rural Resources. Cambridge, U.K., July 1981. M.J.Gardiner (ed), A.A.Balkema, Rotterdam, pp 57-72.

Van Hoorn, J.W., I.Risseeuw and J.M.Jurado Prieto, 1976. Desalinization of a highly saline soil, prediction and field observation. In: Managing Saline Water for Irrigation. Proc.Int.Sal.Conf., Tex. Tech.Univ., Lubbock, Texas, August 1976. H.E. Dregne (ed),pp 558:574

Discussion

M.G.M. Bruggenwert:

Can you give some information concerning your statement, 'The Marismas soil did not show any response to gypsum application, and $CaCO_3$ provides sufficient calcium for the desodification'? What were the total salt concentration and composition of the soil solution during the desodification process, in particular the initial and final E.S.P.?

Author:

In contrast to the Leziria soil, the Marismas soil did not show improvement of soil structure and infiltration rate in the case of gypsum application. The water discharged by the tile drains was always clear and free of dispersed soil particles, whereas in the Leziria the water was loaded with fine clay particles, tiles had to be cleaned regularly and, finally, gypsum had to be applied to the drain trench.

The total salt concentration in the top layer of the Marismas was at the start about 300/meq/l in the saturated paste and, in the case of the Leziria, about 150 meq/l, decreasing to 60-70 meq/l within a few years. For these marine clay soils the composition of the soil solution resembles that of seawater, sodium about 80%, calcium plus magnesium about 20%, chloride 85%, and sulphate 10-15%. The initial E.S.P. in both cases ranged between 25% and 30% and decreased to about 15% in 4 years. In a soil containing 20% $CaCO_3$ and $MgCO_3$, sufficient calcium and magnesium are available to replace sodium if $CaCO_3$ and $MgCO_3$ are well distributed and not occurring in concretions, and if the desalinization process is slow.

P.A.C. Raats:

What is the basis for the assumed decrease of the amount of percolation water, with depth in the model, mentioned in paragraph 7 (figure 4)?

Author:

In the case of the Marismas the hydraulic conductivity of the soil profile decreases with depth, and the drains at a depth of 1.25 m are located on a layer that can be considered as impervious. In the top layer we may assume that the soil is unsaturated and the flow is vertical for most of the time; in the underlying, predominantly saturated layers the flow decreases with depth owing to the decreasing hydraulic conductivity. The data assumed for the successive layers (100%, 80%, etc.) are a guess.

Effect of anion exclusion on solute transport in soil

W.J. Bond
CSIRO Division of Soils
Canberra, A.C.T., Australia

The fate of solutes during water movement in soils is of considerable interest and importance and it is therefore essential to understand all factors which affect it. Of particular interest is the location and rate of movement of a solute front when a solution with one concentration and composition moves through soil initially containing a different solution. This has relevance to, for example, studies of leaching and of the use of poor quality irrigation water. The position of such a solute front is determined principally by the movement of the water, but may be modified by many factors, including the presence of physically or chemically inaccessible water, and adsorption and solubility reactions.

This paper examines some recent findings (Bond et al., 1984) concerning the effects of inaccessible water, and in particular concentration-dependent anion exclusion, on the location and movement of a solute front for the general case of uni-dimensional, uni-directional water movement. Although several authors have shown that the presence of anion exclusion enhances the movement of solutes, few have explored the consequences of the concentration-dependence of anion exclusion. Bond et al. (1984) examined the effects of different solution concentrations on the movement of calcium chloride during unsteady, unsaturated, miscible displacement experiments with a calcium-saturated, structurally stable heavy clay soil. From data independent of the salt front they calculated the water content that was inaccessible to chloride and found that this was concentration-dependent.

Because this water was permanently inaccessible to chloride, for a given concentration, but accessible to tritiated water they suggested it was largely a result of anion exclusion.

The position of the solute front may be defined by its centre of mass, i.e. the position where there would be a step change in concentration if the solute moved as a plug in the accessible water. The position of the solute front x' at any time t can therefore be found from a mass balance equating the mass of solute which should be in the soil with the mass which would be found on either side of the notional step change, viz.

$$\int_0^\infty [\theta_n - \theta_i(C_n)]\, C_n\, dx + C_o\, V(t)$$

$$= \int_0^{x'} [\theta(x,t) - \theta_i(C_o)]\, C_o\, dx + \int_{x'}^\infty [\theta(x,t) - \theta_i(C_n)]\, C_n\, dx \qquad (1)$$

where

x = distance from the soil surface

θ = volumetric water content

θ_n = initial water content

$\theta_i(C)$ = inaccessible water content corresponding to the concentration C in the accessible solution

C_n = initial concentration in the accessible solution

C_o = concentration of the inflowing solution

$V(t)$ = volume of solution per unit cross-sectional area, which has entered the soil up to time t

Manipulation of Equation 1 leads to the expression

$$q(x',t) - \alpha x' = 0 \qquad (2)$$

where

$$q(x,t) = \int_0^x \theta(x,t)\, dx - V(t) \qquad (3)$$

and

$$\alpha = \frac{C_o\, \theta_i(C_o) - C_n\, \theta_i(C_n)}{C_o - C_n} \qquad (4)$$

The function q is a material coordinate based on the distribution of water and has its origin at the notional piston-front of the water, i.e. the interface between the initial and the invading water which would exist if the former was completely displaced by the latter.

These expressions are completely general for uni-directional water movement; they apply equally to transient water flow, such as infiltration and redistribution, and to steady flow.

The parameter α describes the displacement of the solute front relative to the piston-front of the water; as α increases the displacement increases. It is useful to consider various possibilities for the value of α.

a) If there is no inaccessible water, $\alpha = 0$ and the solute front is coincident with the piston front.

b) If θ_i is a constant, $\alpha = \theta_i$ and the solute front is displaced ahead of the piston front. For a given value of θ_i the magnitude of this displacement depends on $q(x,t)$ and therefore on the water content distribution.

c) For concentration-dependent θ_i, if $C_o >> C_n$, $\alpha \simeq \theta_i(C_o)$, while if $C_o << C_n$, $\alpha \simeq \theta_i(C_n)$. It is therefore evident that the displacement of the solute front relative to the piston front depends largely on the more concentrated solution in the soil, whether it be the initial or the invading solution.

d) When the difference between C_o and C_n approaches zero, the value of α may be obtained by taking the limit of Equation 4.

$$\lim_{C_n \to C_o} \alpha = d[C\,\theta_i(C)]/dC \,\Big|_{C=C_o}$$

$$= \theta_i(C_o) + C_o\, d\theta_i(C)/dC \,\Big|_{C=C_o} \qquad (5)$$

The value of α depends on the slope of the $\theta_i(C)$ relationship, but because the slope is negative (for anion exclusion) $\alpha \leqslant \theta_i(C_o)$.

e) For values of C_o and C_n between the limiting cases (c) and (d), α must be calculated from Equation 4. As an example, this has been done using the $\theta_i(C)$ data of Bond et al. (1984). The results are shown in Figure 1, where α is expressed as a fraction of $\theta_i(C_o)$ and graphed against C_n/C_o for several values of C_o. Note that it has been assumed that $C_o > C_n$, but that if the values of C_o and C_n are interchanged α remains the same. It is apparent that when the range of concentrations in the soil is small, the value of α may be considerably different from $\theta_i(C_o)$.

It is concluded that when there is anion exclusion, satisfactory calculation of the location of the solute front relies on considering the properties of both the initial and the invading solutions.

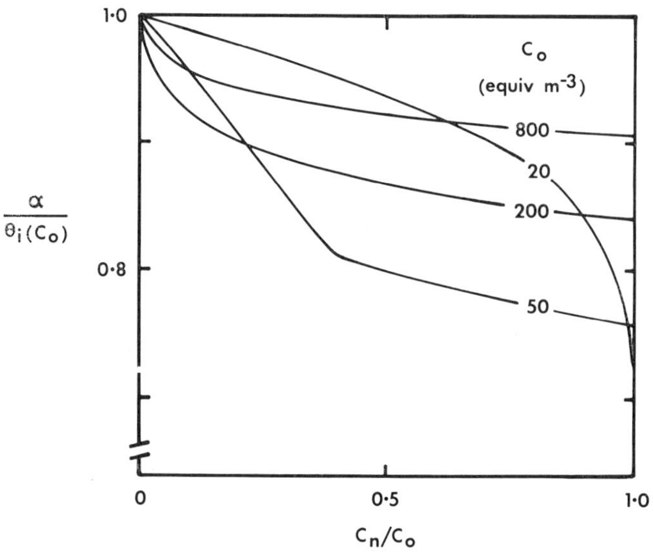

Figure 1. The dependence of α on C_n for various values of C_o, calculated from the data of Bond et al. (1984).

Acknowledgements

Helpful discussions with Dr D.R. Scotter of Massey University, N.Z., and Dr J.H. Knight of CSIRO Division of Environmental Mechanics are gratefully acknowledged.

Reference

Bond, W.J., B.N. Gardiner, and D.E. Smiles 1984. Movement of $CaCl_2$ solutions in an unsaturated clay soil: the effect of solution concentration. Aust. J. Soil Res. (in press).

Discussion

G.H. Bolt:

While deferring till later the outcome of a discussion on our differences of opinion signalled already yesterday, I might attempt to summarize a main aspect. If you use Θ_i to indicate 'inaccessible' water, I would rather refer to your Θ_i as 'immobile' water. We could then probably state that if the immobile water was also inaccessible for anions (via exclusion) we would fully agree on the effect to be expected. Our differences of opinion come about when the inaccessible water was fully mobile, or if the immobile water was fully accessible to the solute.

Author:

As pointed out in the reply to yesterday's question, the 'inaccessible water' is undoubtedly much less mobile than the accessible water. Our difference of opinion would seem to be whether or not there is a discrete 'immobile zone'.

Evaluating a model for nitrate leaching in clay soils with macropores

R.E. White
Soil Science Laboratory,
Department of Agricultural Science,
University of Oxford, Oxford,
OX1 3PF, England

Introduction

Rapid breakthrough of surface-applied solutes due to preferential flow of water down macropores in cracking clay soils is well documented (Thomas and Phillips 1979, Bouma 1981). The effects of this flow on the leaching of indigenous soil nitrate are more complex, depending on the ratio of macropore to matrix flow, the spatial distribution of nitrate and its rate of diffusion between the two flow domains (White 1984). As yet, attempts to predict patterns of NO_3^- redistribution and quantities of N leached in structured soils under field conditions have proved disappointing (Addiscott and Cox 1976, Cameron and Wild 1982, Barraclough et al. 1983). This paper outlines a model for nitrate leaching under unsteady flow conditions and indicates its predictive value using data from large, undisturbed soil cores.

Methods

Duplicate cores of cracking clay soil (Evesham series), 23 cm dia. x 20-24 cm long, were taken from a grassland and an arable site using the method of Smith et al. (1984). Volumetric water contents (θ) were 0.40 to 0.49 $cm^3\ cm^{-3}$ (< "field capacity"). The cores were irrigated at a rate of 7 mm h^{-1} for 13-15 h with $10^{-2}M\ CaCl_2$ solution (Cl^- concentration, C_o). Flow was always unsaturated and the soils absorbed water throughout the experiment. Drainage was collected at frequent intervals and analysed for Cl^- and NO_3^- (concentration C). Initial soil Cl^- and NO_3^- concentrations (C_i) were measured in 8-10 small cores taken next to each large core. Cumulative drainage volume (V) and total pore volume (V_o) were calculated. Nitrification rates were measured in separate incubation experiments.

Model development

Breakthrough curves of C/C_o (Cl) vs V/V_o suggested extensive bypass flow occurred in these cores (Anderson and Bouma 1977). Effluent NO_3^- concentrations changed with pore volume and differed markedly between replicate cores. To model NO_3^- leaching, it was assumed that the applied water mixes with a variable fraction of soil water to produce a miscible volume V_m. Drainage occurs from the miscible volume only, and Cl^- and NO_3^- can diffuse across a contact area A between the miscible and immiscible volumes. For any time interval t_2-t_1 ($=\Delta t$), V_m was calculated from the equations:

$$V_i = V_m - (V_a+V_s) \quad (1)$$

$$V_aC_o + V_sC_{t_1} + V_iC_i = V_mC_m \quad (2)$$

$$M = 2A\theta\ (C_m-C_i)\ (D\Delta t/\pi)^{\frac{1}{2}} \quad (3)$$

$$M/V_m = C_m-C_{t_2} \quad (4)$$

where

C_m = Cl^- concentration in the miscible volume

V_a = volume applied during time Δt

V_i = volume of soil solution with which V_a mixes in time Δt

V_s = miscible volume remaining in the soil at time t_1

M = quantity of Cl^- diffusing in time Δt

D = effective diffusion coefficient for Cl^- in soil

C_{t_1}, C_{t_2} = Cl^- concentration in the effluent at times t_1 and t_2 respectively. For $t_1 = 0$, $V_s = 0$ and $V_i = 0$.

Knowing V_m, an analogous set of equations can be written in which NO_3^- concentrations are substituted for Cl^-. When solved simultaneously, these give a value for C, the predicted NO_3^- concentration in the effluent, for each interval Δt.

Values for A and D (for Cl^- and NO_3^-) were selected from the literature (Bouma and Wosten 1979, White et al. 1984). Provided V_i in equation (1) was > 0, the predicted NO_3^- concentrations were insensitive to changes, in A and D within reasonable limits. The equation

$$\Delta C_s = k \, \Delta t \qquad (5)$$

was used to allow for nitrification in the cores, where

C_s = soil NO_3^- concentration

k = 0.92 ± 0.13 ng N cm^{-3} soil min^{-1}

Model output

The mean fractional miscible volume (θ_m) in these experiments ranged from 0.12 to 0.29 compared to mean θ values of 0.45 to 0.53 respectively. θ_m values were determined from Cl^- breakthrough data and not arbitrarily chosen as in other models (Addiscott 1977, Addiscott et al. 1978). In Fig. 1a and b, mean NO_3^- concentrations for pairs of grassland and arable cores are plotted against V/V_o. The change in NO_3^- concentration with V/V_o was well simulated (r^2 for the regression of observed on predicted NO_3^- concentration = 0.92 in both cases). In contrast, Barraclough et al. (1983) reported poor simulation of the change in NO_3^- concentration with cumulative drainage from a field site, using the models of Burns (1974) and Addiscott (1977).

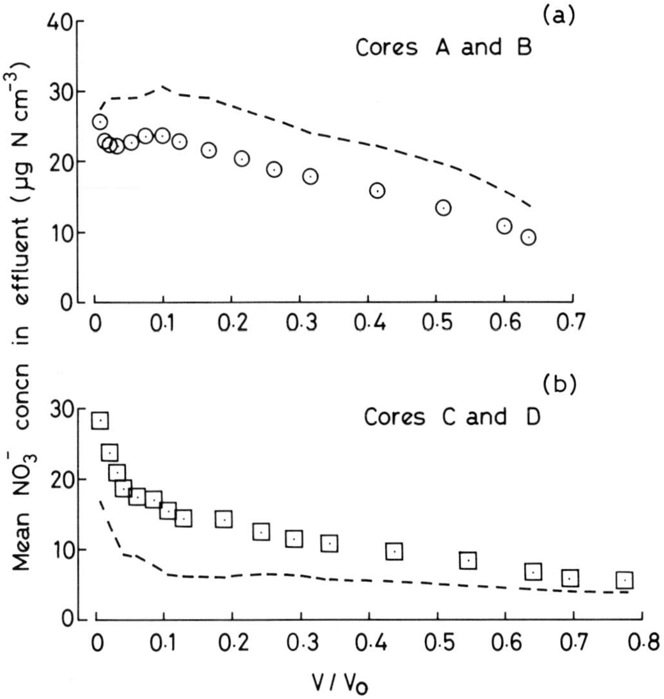

Figure 1. Observed (□,○) and predicted (--) NO_3^- concentrations in the effluent from undisturbed soil cores

Prediction of the total N leached was less accurate, being in the one case too high and in the other too low (Table 1).

Table 1. Actual and predicted fractions of original soil N leached in paired cores of grassland and arable soils

Soil core	Actual fraction leached	Predicted fraction leached
Grassland A	0.14	0.39
Grassland B	0.36	0.28
Mean	0.25	0.34
Arable C	0.03	0.22
Arable D	0.57	0.11
Mean	0.30	0.17

The problem, common to other models, probably lies in the measurement of the mean NO_3^- concentration in the soil initially. As well as between-core variation, exemplified in Table 1, NO_3^- concentration may vary by as much as 50% between the interior and exterior of peds. This aspect is being investigated further.

References

Addiscott, T.M. and D. Cox 1976. Winter leaching of nitrate from autumn-applied calcium nitrate, ammonium sulphate, urea and sulphur coated urea in bare soil. J. Agric. Sci., Camb. 87: 381-389.

Addiscott, T.M. 1977. A simple computer model for leaching in structured soils. J. Soil Sci. 28: 554-563.

Addiscott, T.M., D.A. Rose, and J. Bolton 1978. Chloride leaching in the Rothamsted drain gauges: influence of rainfall pattern and soil structure. J. Soil Sci. 29: 305-314.

Anderson, J.L. and J. Bouma 1977. Water movement through pedal soils. II. Unsaturated flow. Soil Sci. Soc. Am. J. 41: 419-423.

Barraclough, D., M.J. Hyden, and G.P. Davies 1983. Fate of fertilizer nitrogen applied to grassland. I. Field leaching results. J. Soil Sci. 34: 483-497.

Bouma, J. and J.H.M. Wosten 1979. Flow patterns during extended saturated flow in two undisturbed swelling clay soils with different macrostructures. Soil Sci. Soc. Am. J. 43: 16-22.

Bouma, J. 1981. Soil morphology and preferential flow along macropores. Agric. Water Manage. 3: 235-250.

Burns, I.G. 1974. A model for predicting the redistribution of salts applied to fallow soils after excess rainfall or evaporation. J. Soil Sci. 25: 165-178.

Cameron, K.C. and A. Wild 1982. Prediction of solute leaching under field conditions: an appraisal of three methods. J. Soil Sci. 33: 659-669.

Smith, M.S., G.W. Thomas, R.E. White, and D. Ritonga 1984. Transport of Escherichia coli through intact and disturbed columns of soil. J. Environ. Qual. (submitted).

Thomas, G.W. and R.E. Phillips 1979. Consequences of water movement in macropores. J. Environ. Qual. 8: 149-152.

White, R.E. 1984. The influence of macropores on the transport of dissolved and suspended matter through soil. Adv. in Soil Sci. 2: (in press).

White, R.E., G.W. Thomas, and M.S. Smith 1984. Modelling water flow through undisturbed soil cores using a transfer function model derived from ^{3}HOH and Cl transport. J. Soil Sci. 35:(159-168).

Discussion

A.P. Whitmore:

Bob, I feel I should point out that the model used by Barraclough et al. (1983) was not the Addiscott model, but an 'Addiscott type' model, in fact a BASIC version written by Dedan Barraclough from the original FORTRAN version. Also, do you think it is fair to compare your own validations on your own model with validations made by Barraclough and by Camoron and Wild on models developed by Addiscott and by Burns. We hope to attempt our own simulations on these data sets and publish them soon.

Author:

Thank you for pointing out that Barraclough et al. used an Addiscott model - would the fact it was written in BASIC rather than FORTRAN matter? I was not trying to compare models directly in this paper - merely making a passing reference to Barraclough's comparatively unsuccessful results using Burns' and Addiscott's models on his data.

Soils and solute patterns in reclaimed estuarine marshland in South-East England

P.J. Loveland, R.G. Sturdy and J. Hazelden,
Soil Survey of England and Wales,
Rothamsted Experimental Station,
Harpenden, Herts. U.K.

1. Introduction

The heavy clay soils of embanked marshes bordering the Thames Estuary in Kent were last flooded by the sea in 1953. Some land has been drained in the last 10-15 years to allow cereal production, but blocking of the pipe drains with dispersed clay, followed by structural collapse and crop failure is widespread. The soils are Typic Fluvaquents with a saline horizon (EC > 4 mmho cm^{-1}) within 1 m of the soil surface. Like the 'knip' or 'knick' soils of The Netherlands and North Germany, the subsoils have small permeability, large exchangeable sodium percentage and Ca:Mg <3. This paper reports the preliminary findings of an investigation of relationships between soil chemical and physical properties, microrelief and land use history.

2. Methods

Air photos show patterns of creeks and pools, the extent (Figure 1) of waterlogging, and crop damage on arable land, and were used to locate six transects on Cliffe (1-4) and St. Mary's (5-6) marshes, N. Kent, chosen to include undrained grassland and arable land drained at different times. All sampling sites were classified into creeks, pools, levees or, particularly on levelled land, 'neutral' ground, and sampled at 0-15cm, 35-50cm and 70-85cm during August 1983. Samples were air dried, crushed <2mm and clay content, organic carbon content, $CaCO_3$ equivalent, (Avery and Bascomb, 1974), exchangeable sodium percentage (ESP), electrical conductivity (EC 1:5 extract) (USDA, 1954), and dispersion ratios (DR) (MAFF, 1982) determined.

3. Results and Discussion

Analytical data for 43 sites are grouped by land use (Table 1) and microrelief (Table 2). Clay contents are large and relatively uniform.

Table 1. Soil physical and chemical properties in relation to land use history

Field and transect (n)	Land use history	Sample depth cm.	Clay %	Org. Carbon %	$CaCO_3$ %	ESP %	EC (1:5) $mScm^{-1}$	DR %
Cliffe	Undrained	0-15	61	16	0	14	1.5	12
Trans.2+3	Perm.grass	35-50	61	1.4	0.5	24	1.1	12
(11)	No gypsum	70-85	58	0.9	0.9	29	1.4	83
Cliffe	Drained 3yr	0-15	64	6.5	1.0	10	0.9	14
Trans.1	Winter Wheat	35-50	65	1.4	0.6	22	1.3	32
(16)	No gypsum	70-85	63	1.0	1.2	26	1.6	49
Cliffe	Drained 3yr	0-15	67	5.1	0.2	7	1.5	13
Trans.4	Fallow	35-50	68	1.6	0.2	17	0.8	37
(6)	Gypsum	70-85	66	1.8	0.3	25	1.3	52
St.Mary's	Drained 12yr	0-15	57	4.0	0.7	3	0.4	11
Trans.5+6	Spring barley	35-50	59	1.9	2.6	11	0.4	24
(10)	Gypsum	70-85	61	0.9	2.9	21	0.8	54

(n) = number of observations

Topsoil organic carbon contents are large, particularly in grassland where there is often a thin surface root mat. $CaCO_3$ values are generally small, exceptions being the heavily limed topsoils at Cliffe (Transect 1) and the subsoils on St. Mary's Marsh. ESP increases with depth at all sites being largest at Cliffe on grassland, exceeding 15% everywhere below 35 cm except in the calcareous soils on St. Mary's Marsh. Smaller ESP values on arable land at the latter sites and at Cliffe undoubtedly reflect the addition of gypsum. There is also clearly a management effect in that ESP, EC and DR are all smaller in the upper horizons of the St. Mary's soils which have been drained and gypsum treated for the longest period. Dispersion ratios parallel ESP values in subsoils, but in topsoils are ca. 12% irrespective of land use. EC values show a less clear pattern, largest values being in topsoils of Transects 2, 3 and 4. In the latter case this perhaps being due to the recent application of gypsum (Shanmuganathan and Oades, 1983).

ESP, EC and DR show the same trends with depth when sites are grouped according to microrelief (Table 2). The ranking pools > creeks > 'neutral' > levees in terms of EC (i.e. 'salinity') is confirmed by a further set of 65 measurements made with a conductivity probe over a wider area.

Table 2. Relationship of some soil chemical properties to microrelief

Microrelief (n)	Exch.Na %			EC (1:5) $mScm^{-1}$			DR %		
	0-15	35-50 cm	70-85	0-15	35-50 cm	70-85	0-15	35-50 cm	70-85
Pools (14)	13	28	37	1.5	1.5	1.8	13	50	75
Creeks (9)	12	19	22	1.6	1.0	1.5	13	39	45
'Neutral' (14)	9	19	26	0.9	0.9	1.1	12	42	68
Levees (6)	9	17	24	0.9	0.7	1.0	10	32	58

(n) = number of observations

The qualitative relationship between ESP, DR and organic carbon content is shown in Figure 2. Correlation coefficients >0.7 ($p<0.001$) were found between DR and ESP for both subsoils and topsoils. There was little correlation between EC or organic carbon and DR ($r \ll 0.5$).

Multiple regression analysis gave the equation:

$$DR = 15.4 - 0.97\ Org.C. + 2.86\ ESP - 12.1\ EC$$

for all the transect data ($r^2 = 0.737$). Although a good correlation, the regression analysis suggests that if DR is seen as the best measurement of stability in these soils, it would be better measured directly rather than inferred from other, more conventional analytical data.

Dispersion ratios for other nonsaline arable soils suggest that increasing structural instability is likely with DR > 15. Fullstone and Watson (1981) showed that saline soils with ESP > 3 were virtually impermeable probably because of clay dispersion. The generally large ESP values found in this study, and their large contribution to the regression equation which reflects their substantial contribution to the DR values, suggests that structural instability is likely at DR <<15 in these soils. All of the latter in this area may be unsuitable therefore for prolonged arable cropping.

References

Avery, B.W. and Bascomb, C.L. (Eds.) 1974. Soil Survey Laboratory Methods. Technical Monograph of the Soil Survey No. 6, Harpenden, U.K. pp. 83.

Fullstone, M.J. and Watson, J.P. 1981. The saturated hydraulic conductivity of sodium-influenced soils. Proc. 10th National Congress (1981), Soil Science Society of Southern Africa, Tech. Comm. 180, pp. 58-61, Dept. Agric. Pretoria, S. Africa.

MAFF 1982. Techniques for measuring soil physical properties. Ministry of Agriculture, Fisheries and Food Ref. Book 441. H.M.S.O., London, U.K. pp. 116.

Shanmuganathan, R.T. and Oades, J.M. 1983. Modification of soil physical properties by addition of calcium compounds. Austr. J. Soil Res. 21, 285-300.

USDA. 1954. Diagnosis and improvement of saline and alkali soils. Agric. Handbk. 60, U.S. Dept. Agric. Washington D.C., U.S.A., pp. 160.

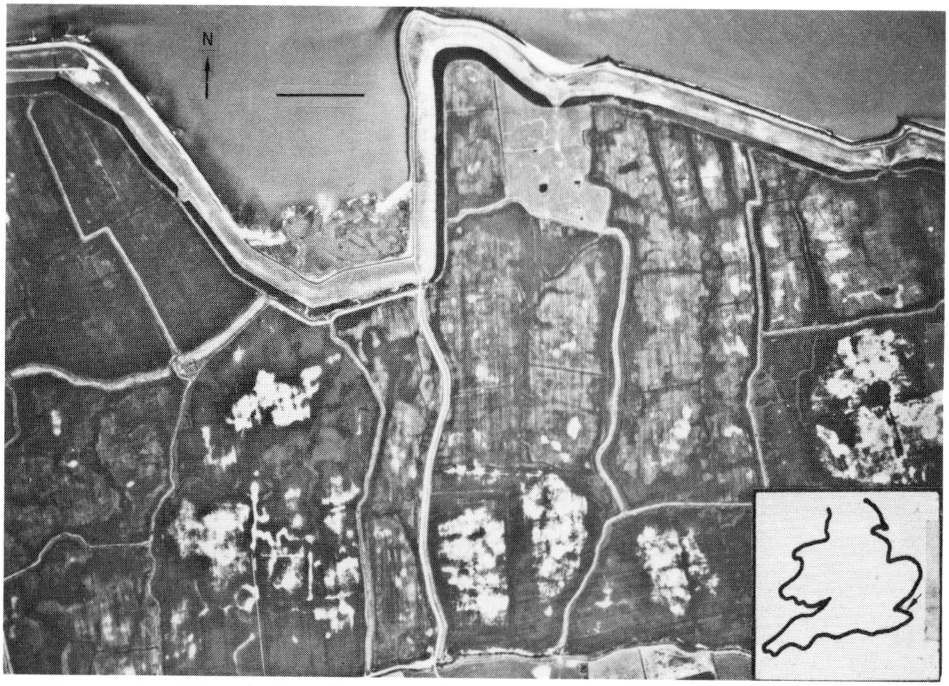

Figure 1. Air photo of part of study area showing creek pattern and areas of crop failure. Bar = 200m

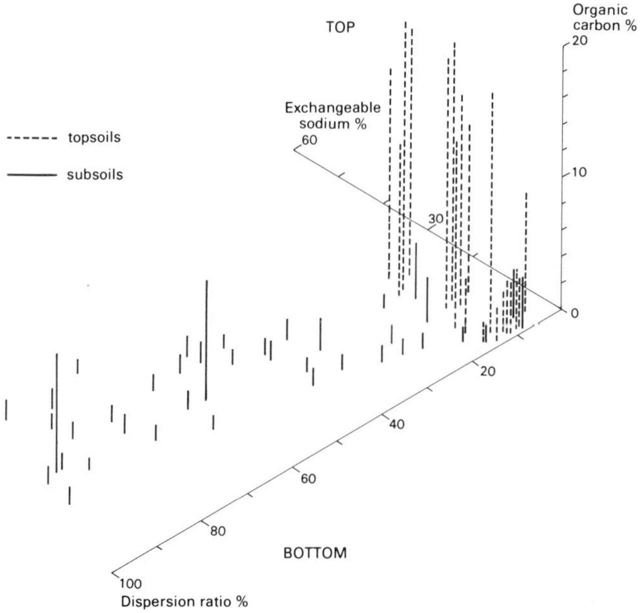

Figure 2. Relationship between ESP, DR and organic carbon; broken lines - topsoils, solid lines - subsoils

Discussion

<u>L.P. Wilding</u>:

Could you, in fact, have predicted the solute patterns observed on your maps from aerial photography tonal patterns before you conducted your survey?

<u>Author</u>:

For grassland, possibly yes, but for arable land, no. The two problems are: a) the area was flooded by seawater in 1953 and the residual effects of this are unknown, as is the distribution of ameliorative gypsum placed soon after the floods, and b) when the land is prepared for arable cultivation and for some improved grassland, many of the levees are bulldozed into the creeks. This changes both the distribution of the topsoil and subsoil and related chemistry in an unknown and unpredictable manner.

Transport of solutes in highly structured soils

M. Loxham
Delft Soil Mechanics Laboratory
Delft, The Netherlands

Peat soils form a class of highly structured soils whose fabric is a result of the build-up and decay of macro organic matter such as leaves, stems and even tree trunks.
The prediction of the movement of solutes in the peat is important in the whole discussion of the possibilities of utilisation of the vast but marginal wet lands of the world.
Peat contains about as much organic matter as many mineral soils. However the mineral fraction is absent and the peat has a water content of up to 98% of the sample volume. Photomicrographs of undisturbed samples and comparison of pF data with particle size distributions of the (dispersed) organic matter show that most of this water is locked up in volumes largely inaccessible to the normal hydraulic flows in the profile. The water flows in large multiply connected pores (white in Fig 1.) and air bubbling studies show that they are continuous for at least tens of centimeters.

Theory

Based upon the above observations a three phase (solid, dead zone, and active zone) model is proposed in which the active zone is assumed to consist of parallel cracks of (conceptual) width 2b, spaced 2B apart. Convection and dispersion are considered in the crack, and molecular diffusion only in the dead zone. Transfer between the zones is by diffusion over the crack wall controlled by the diffusion rate in the dead zone. Adsorption etc are not considered here but can be easily

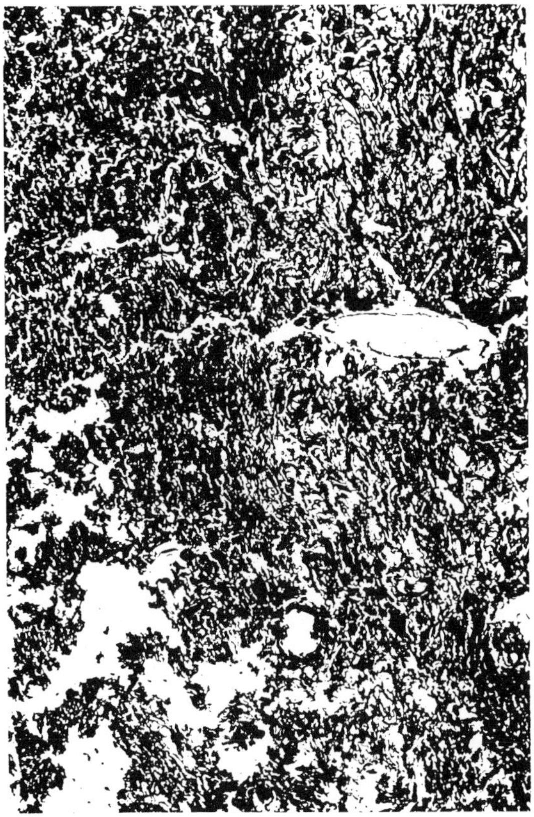

Fig. 1.

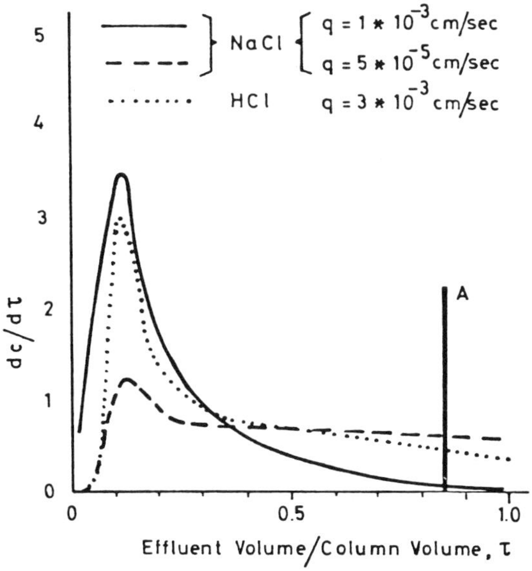

NaCl
q = 1 * 10^-3 cm/sec
q = 5 * 10^-5 cm/sec
HCl
q = 3 * 10^-3 cm/sec
A
dc/dτ
Effluent Volume/Column Volume, τ
0
1
2
3
4
5
0.5
1.0

Fig. 2.

introduced.

The model equations are then:

a) for the fissure:

$$\partial c/\partial t + v\ \partial c/\partial x - av\ \partial^2 c/\partial x^2 - \frac{\theta D}{b}\ \partial c/\partial z\Big|_{z = b} = 0 \qquad 0 < x < \infty$$

b) for the dead zone:

$$\partial c/\partial t - D\ \partial^2 c/\partial z^2 = 0 \qquad b < z < B$$

where 2b is the fissure width, 2B the distance between fissures, v the in-pore convective velocity, a the dispersion length, D the molecular diffusion coefficient in the dead zone, t the time and x, z the distance coordinates parallel and normal to the fissure.

There are many analytical solutions for this model depending upon what terms are ignored, see for example Skopp & Warrick (1974), Frind et al. (1982). In practice however the system is easily handled by commonly available numerical techniques.

In principle there are only two unknowns in the equations, a and b. (B is given from geometrical considerations once b is set and the active porosity is known from the pF data.). In fact there is also considerable uncertainty as to the real value of the in-pore molecular diffusion coefficient for these systems.

Experimental

In order to verify this model breakthrough curves were measured for undisturbed peat samples of various sizes but from one location in North Germany, see Loxham and Burghard (1983). Different water velocities were imposed and two tracers (NaC1 and HC1) were studied because of their differing molecular diffusion co-efficients. Some of the results are shown in Fig 2. for a pulse input. As would be expected from the model the modal travel times (dominated by the active zone) are the same but there are differences in the "tail" (controlled by the balance between the convective and diffusive processes). The modal travel times indicate an active porosity of about 14% of the total and this corresponds to pore sizes above 30 microns obtained from image analysis of the photomicrographs.

Discussion

The model was tested by fitting the values of a and b to the data at the higher water velocity for the NaC1 case, and using it to predict the NaC1 curve for lower velocities and the HC1 curve, (Fig 3). The agreement is suprisingly good for the NaC1 case. However for the acid case the data can only be explained by assuming either a higher effective diffusion coefficient than expected (Proton hopping) or protonation of the solid phase, both of which effects are known to occur.

It can be concluded that the crack models can account for the general trends in the data in a way that the classical convection-dispersion theory cannot and that the model gives a reasonable quantative description of the measured breakthrough curves.

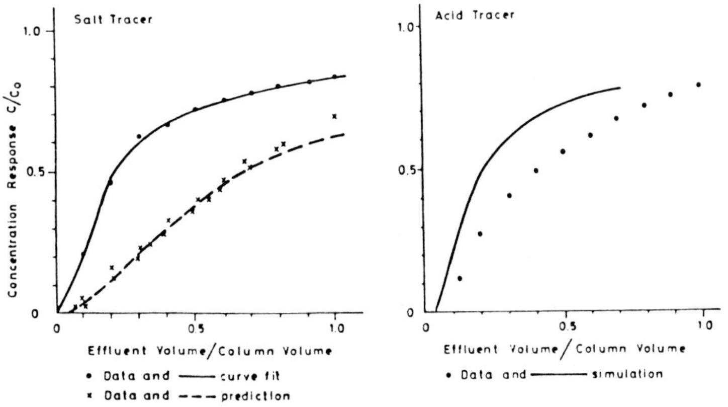

Fig 3.

References.

Skopp J. & Warrick A.W. (1974), A Two Phase Model for the Miscible Displacement of Reactive Solutes in Soils. Soil Soc. Amer. Proc. 38(4), 545-550.

Frind E.O. & Sudicky E.A. (1982), Contaminant Transport in Fractured Porous Media, Water Resour. Res. 18(6) 1634-1642.

Loxham M. & Burghard W. (1983), Peat as a Barrier to the Spread of Micro-contaminants to the Groundwater, Int. Peat Symp. Bemidji U.S.A.

Solute displacement through a Rendzina

R. Schulin
Swiss Federal Institute of Technology
Zurich, Switzerland
P.J. Wierenga
New Mexico State University
Las Cruces, NM, United States
H. Flühler
Swiss Federal Institute of Technology
Zurich, Switzerland

1 Introduction

Solute transport through porous media is commonly described by a convective dispersion type equation that employs a Fick-type concept of solute dispersion. This concept proved to be successful in the case of homogeneous porous media like laboratory sand packings. It failed, however, to explain certain features of solute displacement through aggregated porous media such as the tailing of breakthrough curves (BTC) (Tyler and Thomas, 1981; Nkedi-Kizza et al., 1983). To account for these assymmetries, the original model has been extended by partitioning the liquid phase into a mobile and an immobile region. According to this model, convective-dispersive transport is restricted to the mobile region, whereas in the immobile region solute transport takes place by diffusion only (Coats and Smith, 1964; van Genuchten and Wierenga, 1976; DeSmedt and Wierenga, 1979). In this paper, the implications of a high skeleton content especially of coarse fragments, on the characteristics of tracer displacement are investigated.

2 Models

The two models evaluated in this paper can be written as the one-dimensional convective-dispersion equation

$$(\theta - \theta_{ex}) \frac{\partial c}{\partial t} = \theta D \frac{\partial^2 c}{\partial z^2} - q \frac{\partial c}{\partial z} \qquad \text{(one-region model)}$$

and the one-dimensional non-equilibrium exchange equations

$$\theta_m \frac{\partial c}{\partial t} + \theta_a \frac{\partial c_a}{\partial t} = \theta D \frac{\partial c^2}{\partial z^2} - q \frac{\partial c}{\partial z}$$

and (two-region model)

$$\theta_a \frac{\partial c_a}{\partial t} = \alpha (c - c_a)$$

where

- θ = total volumetric water content
- θ_m = volumetric water content in the mobile region
- θ_{ex} = volumetric content of water excluded from solute transport (inaccessible part of the immobile region)
- θ_a = volumetric water content in the accessible part of the immobile region ($\theta_a = \theta - \theta_m - \theta_{ex}$)
- q = Darcy flux
- D = Fick-type hydrodynamic dispersion coefficient
- c = solute concentration (in the mobile region)
- c_a = solute concentration in the accessible part of the immobile region
- α = mass transfer coefficient

Both models account for the presence of immobile water inaccessible for solutes either because of anion exclusion or because of physically excluded pore space. In the case of tritium, the exclusion effect of inaccessible water (θ_{ex}) will be underestimated because adsorption is neglected. However, comparing the θ_{ex} estimates of different experimental runs (varying infiltration rates) the influence of adsorption is negligible.

3 Methods

BTC of tritium and bromide were recorded using two undisturbed large-scale columns of a Rendzina soil (Weissenstein, Swiss Jura). The two models were evaluated by fitting analytical solutions of the model equations to the experimental data, using the least square curve fitting program CFITM of van Genuchten (1981).

4 Results

As shown in Figure 1, the two-region model yields an adequate fit of the data. The one-region model fails to explain the pronounced tailing of the BTC. The fitted parameters of both models are presented in Table 1 for various infiltrations rates. These data support the concept of splitting the immobile water in two fractions, one in contact and the other excluded from the mobile water. Surprisingly, the mobile water content θ_m as well as the accessible immobile water content θ_a remains rather constant regardless of the total water content. Changes of θ due to variable infiltration rates seem to affect primarily the fraction of immobile water excluded from solute transport (θ_{ex}). Due to anion exclusion, this volume is larger for bromide than for tritium. The fact that also tritium is excluded from a noticeable fraction of the pore water may be attributed to a "roof effect" of the soil skeleton.

The dispersion coefficient D which expresses Fick-type hydrodynamic dispersion in the mobile phase exhibits a strong linear correlation

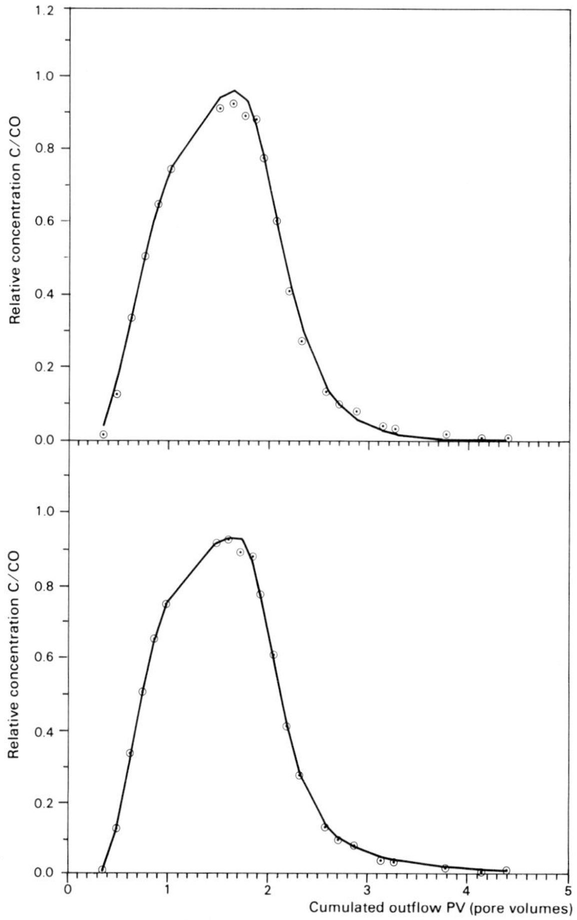

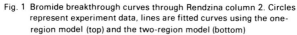

Fig. 1 Bromide breakthrough curves through Rendzina column 2. Circles represent experiment data, lines are fitted curves using the one-region model (top) and the two-region model (bottom)

Table 1. Measured and estimated model parameters

Experiment	measured		one-region model		two-region model				
	q [cm/d]	θ	D [cm^2/d]	θ_{ex}	D [cm^2/d]	α [d^{-1}]	θ_m	θ_a	θ_{ex}
Tritium									
1	2.13	0.147	36.9	-0.002	23.0	0.027	0.132	0.019	-0.004
2	6.04	0.153	129.2	0.004	84.8	0.040	0.136	0.018	0.001
3	16.97	0.164	399.5	0.008	276.7	0.047	0.147	0.020	-0.003
4	40.76	0.176	915.9	0.019	506.1	0.315	0.139	0.028	0.011
5	14.52	0.165	303.9	0.008	195.7	0.065	0.146	0.022	-0.003
6	5.26	0.160	129.1	0.009	80.7	0.018	0.142	0.035	-0.017
7	2.02	0.156	45.1	-0.001	21.2	0.034	0.129	0.031	-0.004
Bromide									
1	2.13	0.147	51.8	0.025	30.3	0.005	0.116	0.050	-0.019
2	6.04	0.153	164.5	0.028	101.7	0.027	0.115	0.019	0.019
3	-	-	-	-	-	-	-	-	-
4	40.76	0.176	945.2	0.054	541.2	0.221	0.110	0.018	0.048
5	14.52	0.165	328.9	0.035	214.0	0.045	0.123	0.021	0.021
6	5.26	0.160	123.3	0.037	68.8	0.025	0.112	0.018	0.030
7	2.02	0.156	58.6	0.024	34.7	0.007	0.122	0.023	0.011

Table 2. Parameters of linear regression between dispersion coefficient D and apparent pore water velocity v ($v = q/\theta$)

One-region model		
Tritium	$D = 4.003\,v - 19.08$	$r = 0.999$
Bromide	$D = 4.077\,v - 6.27$	$r = 0.999$
Two-region model		
Tritium	$D = 2.245\,v + 2.05$	$r = 0.994$
Bromide	$D = 2.340\,v + 1.54$	$r = 0.999$

with the pore water velocity. No similar simple relationship could be obtained to describe the variation of the mass transfer coefficient α.

References

Coats, K.H. and B.D. Smith 1964. Dead-end pore volume and dispersion in porous media. Soc. Pet. Eng. J., 4: 73-84.

DeSmedt, F. and P.J. Wierenga 1979. Mass transfer in porous media with immobile water. J. Hydrol., 41: 59-67.

Nkedi-Kizza, P., J.W. Biggar, M.Th. van Genuchten, P.J. Wierenga, H.M. Selim, J.M. Davidson, and D.R. Nielsen 1983. Modeling tritium and chloride-36 transport through an aggregated oxisol. Water Resour. Res., 19: 691-700.

Tyler, D.D. and G.W. Thomas 1981. Chloride movement in undisturbed soil columns. Soil Sci. Soc. Am. J., 45: 459-461.

van Genuchten, M.Th. 1981. Non-equilibrium solute transport parameters from miscible displacement experiments. Res. Rep. 119 U.S. Salinity Lab. and Dept. of Soil and Environ. Sci., Univ. of Calif., Riverside, 88 pp.

van Genuchten, M.Th. and P.J. Wierenga 1976. Mass transfer studies in sorbing porous media. 1. Analytical solutions. Soil Sci. Soc. Am. J., 41: 272-278.

Discussion

L.P. Wilding:

As I understand the porosity of an organic soil, there are three pore systems. They are the large macrovoids due to biological and desiccation cracks, the fine pores associated with the packing and compression of the organic fibers, and the very fine pores within the fiber tissues caused by cellular voids. The latter is by far responsible for most of the total porosity in an organic soil, but it is likewise less accessible. Does your model take this pore system (intra-tissue pores) into consideration?

Author:

I agree with you, but in our model we did not consider the porosity component associated with fiber tissues.

P.A.C. Raats:

Perhaps θD in the second equation on page 263 there should be $\theta_m D_m$. According to F. de Smedt (1981), θD in the first equation on page 263 is then the sum of $\theta_m D_m$ and a term proportional to q^2 and inversely proportional to α. ("Solute transfer through unsaturated porous media". In: Quality of Groundwater, Proceedings of an International Symposium, Noordwijkerhout, The Netherlands, 23-27 March 1981, W. van Duyvenboden, P. Glasbergen, and H. van Lelyveld, eds., p. 1011 - 1016.)

Some theoretical aspects of the influence of soil-root contact on uptake and transport of nutrients and water

P. de Willigen
Institute for Soil Fertility, Haren (Gr.), The Netherlands

Introduction

The importance of limited soil-root contact with respect to transport of soil water towards the plant root has been recognized for some time (Herkelrath et al., 1977; Weatherly, 1979). The probability of limited soil-root contact is greater the heavier the soil (Tinker, 1976). To investigate the consequences of limited soil-root contact for the uptake of water and nutrients the situation of Figure 1 was considered.

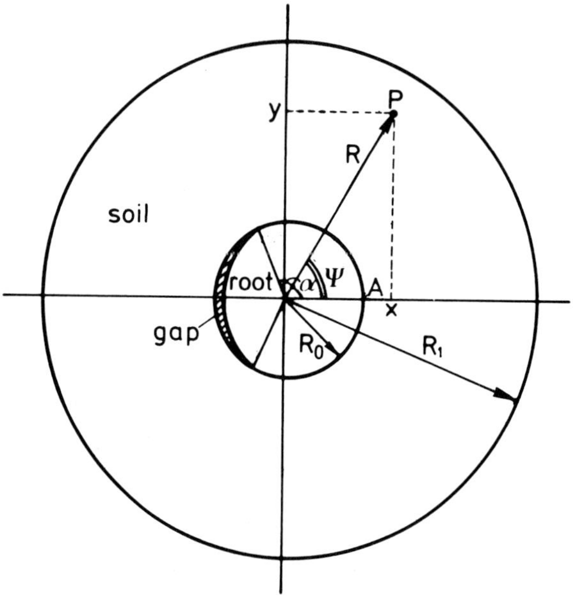

Figure 1. Schematic representation of the position of a root in a soil cylinder. The location of a point P in the cylinder is given either by rectangular (x,y) or polar (R,ψ) coordinates. The degree of soil-root contact is given by the angle α. The radius of the soil cylinder R_1 is related to the root density W by: $R_1 = (\pi W)^{-\frac{1}{2}}$

Such a situation could occur when a gap develops at part of the soil-root interface due to shrinkage of the soil and/or of the root (Altemüller und Haag, 1983). It is assumed that the root as depicted in Figure 1 is representative of a large set of roots, all under the same conditions and belonging to a crop taking up water and nutrients.

Transport and uptake of nutrients

The transport of a nutrient subject to linear adsorption and diffusive transport can be described by

$$\frac{\partial C}{\partial t} = D\nabla^2 C = \frac{D^1}{K + \Theta}\nabla^2 C \qquad (1)$$

where C is the concentration of the nutrient in mg ml^{-1}, t is time in days, D is the effective diffusion coefficient in cm^2 day^{-1}, D^1 is the diffusion coefficient of the nutrient in soil in cm^2 day^{-1}, K is the adsorption constant in ml cm^{-3}, Θ is the water content in ml cm^{-3}, and ∇^2 the Laplacian operator in cm^{-2}.

The boundary conditions chosen are: vanishing flux at the outer boundary of the soil cylinder and at part of the inner boundary, and constant flux at the remaining part of the inner boundary. The latter is a consequence of the assumption that uptake is determined by plant demand (Scott Russell, 1981), which is taken to be constant, rather than by the concentration, as long as the concentration exceeds a certain limiting value, which depends on uptake characteristics, root density and soil-root contact (de Willigen, 1981). If only part of the root surface is in contact with the soil (given by the angle α of Figure 1) the required flux through that part should be a factor π/α larger than the flux when the complete root surface is in contact with soil. Defining dimensionless variables and parameters as : $\tau = Dt/R_o^2$, $r = R/R_o$, $c = C/C_i$ (C_i is the initial concentration in mg ml^{-1}), $\nabla^{*2} = R_o^2\nabla^2$, $\phi = D^1C_i/(AR_o)$ (A is the required uptake rate in mg cm^{-2} day^{-1}, $\eta = h/R_o$ (h is the root length in cm), $\rho = R_1/R_0$, then equation (1) and the boundary conditions transform into :

$$\partial c/\partial \tau = \nabla^{*2} c \tag{2}$$

$$\tau>0 \quad r = \rho \quad 0<\psi<2\pi \quad \partial c/\partial r = 0 \tag{3}$$

$$\tau>0 \quad r = 1 \quad \left.\begin{matrix} 0<\psi<\alpha \\ 2\pi-\alpha<\psi<2\pi \end{matrix}\right\} \partial c/\partial r = \pi\rho^2/(2\alpha\eta\phi) \tag{4a}$$

$$\alpha<\psi<2\pi-\alpha \quad \partial c/\partial r = 0 \tag{4b}$$

$$\tau = 0 \quad \begin{matrix} 1<r<\rho \\ 0<\psi<2\pi \end{matrix} \quad c = 1 \tag{5}$$

Because of the boundary conditions (3), (4a) and (4b), ultimately a steady-rate situation may develop (de Willigen, 1981) where: $\partial c/\partial \tau$ = constant = $-\rho^2/\{\eta\phi(\rho^2-1)\}$
When a new variable U is defined as:

$$U = c + \rho^2 r^2/\{4\eta\phi(\rho^2-1)\} \tag{6}$$,

and this result is substituted in (2) - (4) one obtains:

$$\nabla^{*2} U = 0 \tag{7}$$

$$r = \rho \quad 0<\psi<2\pi \quad \partial U/\partial r = \rho^3/\{2\eta\phi(\rho^2-1)\} \tag{8}$$

$$r = 1 \quad \left.\begin{matrix} 0<\psi<\alpha \\ 2\pi-\alpha<\psi<2\pi \end{matrix}\right\} \partial U/\partial r = \{\pi\rho^2(\rho^2-1) + \alpha\rho^2\}/\{2\alpha\eta\phi(\rho^2-1)\} \tag{9a}$$

$$r = 1 \quad \alpha<\psi<2\pi-\alpha \quad \partial U/\partial r = \rho^2/\{2\phi\eta(\rho^2-1)\} \tag{9b}$$

The Laplace equation (7) with boundary conditions (8) - (9) can be solved so as to give U as a function of r and ψ. This can be most conveniently achieved by transforming the region of the hollow cylinder in the x,y-plane onto the half plane $\eta>0$ of another coordinate system ξ,η, by employing an appropriate mapping function. In the ξ,η-plane the solution of U can be found, save for an integration constant U_o, by application of a well-known integral formula from complex analysis

(Churchill et al., 1974), and so the steady-rate concentration profile can be calculated. The minimum value of C is to be found at the point $r = 1$, $\psi = 0$ (point A in Figure 1); as long as the concentration here exceeds the limiting value C_L, uptake can proceed according to the required rate. If the value of the integration constant U_o is chosen so as to make the concentration at A equal to C_L, the amount of nutrient taken up by the root at the required rate can be calculated by integration of $(K + \Theta)C$ over the hollow soil cylinder. Results of calculations are presented in Figures 2 and 3. In Table 1 the values of the parameters are given. Figure 2 shows an example of isoconcentration curves.

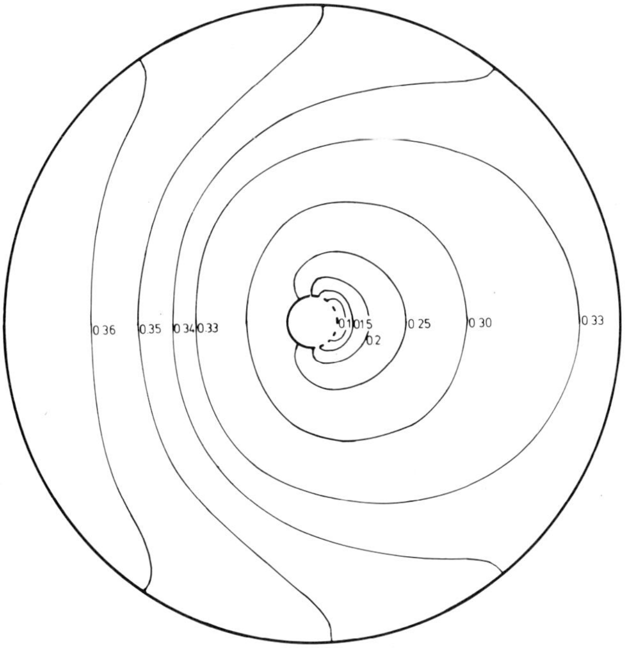

Figure 2. Isoconcentration lines in the soil cylinder when the minimum concentration at the root wall equals the limiting value. Parameters: $W = 5$ cm^{-2}, $c_L = 0.075$, $\alpha/\pi = \frac{1}{2}$, $K = 200$ ml cm^{-3}

Figures 3a and b give the depletion F_D of the nutrient relative to the initial amount of available nutrient as a function of the ratio α/π, and the root density. The degree of soil-root contact is more important as the nutrient is more strongly adsorbed by the soil.

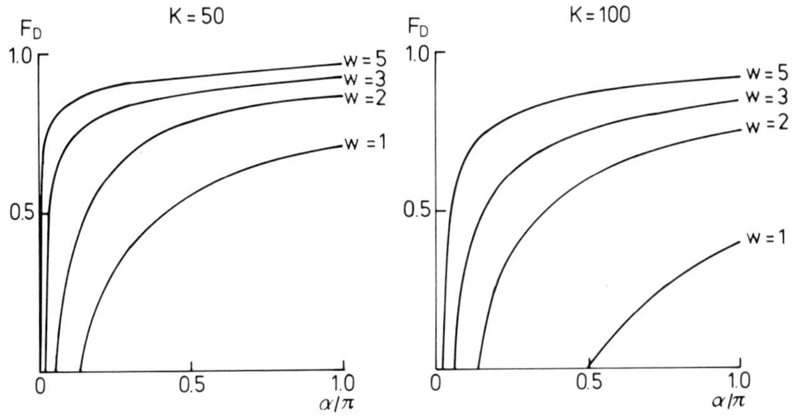

Figure 3a, b. Fractional depletion of the nutrient as a function of the ratio α/π, and root density

Table 2. Values of parameters

Variable	Symbol	Dimension	Value
Adsorption constant	K	ml cm^{-3}	50-200
Root density	W	cm cm^{-3}	1-3-5
Diffusion coefficient	D	$cm^2 day^{-1}$	0.25
Root radius	R_o	cm	0.02
Root length	h	cm	20
Uptake rate	A	mg $cm^{-2} day^{-1}$	$4.4*10^{-3}$
Initial concentration	C_i	mg ml^{-1}	$1.3*10^{-3}$(K=50)-$3.3*10^{-4}$(K=200)
	D_s	$cm^2 day^{-1}$	$4.3*10^3$(cl)*; $1.4*10^2$(scl)**
	β	-	66.8 (cl)*; 22.7 (scl)**
	Θ_s	ml cm^{-3}	0.4 (cl)*; 0.375 (scl)**
Transpiration	E	cm day^{-1}	0.5
Limiting water content	-	ml cm^{-3}	0.15 (cl)*; 0.12 (scl)**

(cl)* = clay loam

(scl)** = silty clay loam

Transport of water

The transport of water towards a root is described in a way similar to transport of nutrients:

$$\partial\Theta/\partial t = \underset{\sim}{\nabla}.D_w\underset{\sim}{\nabla}\Theta \tag{10}$$

The diffusivity D_w is a function of Θ. We shall here confine our attention to those soils where D_w is an exponential function of the water content (Stroosnijder, 1976).

$$D_w = D_s \exp\{\beta(\Theta-\Theta_s)\} \tag{11},$$

where D_s is the diffusivity in $cm^2\ day^{-1}$ at some reference water content Θ_s. When we define the dimensionless variables and parameters $\tau = D_s t/R_o^2$, $H = D/D_s$, $\phi_w = D_s/(ER_o)$ (E is the transpiration in cm day^{-1}), $r = R_1/R_o$, $\eta = h/R_o$, $\nabla^{*2} = R_o^2\nabla^2$, employing boundary conditions similar to (3) and (4a and b), we get:

$$\partial H/\partial\tau = H\ \nabla^{*2} H \tag{12}$$

$$\tau>0 \quad r = \rho \quad \partial H/\partial r = 0 \tag{13}$$

$$\tau>0 \quad r = 1 \quad \begin{matrix} 0<\psi<\alpha \\ 2\pi-\alpha<\psi<2\pi \end{matrix} \quad \partial H/\partial r = \eta\rho^2/(2\alpha\phi_w)\{\sin(2\pi D_s\tau/R_o^2)+|\sin(2\pi D_s\tau/R_o^2)|\}/2 \tag{14a}$$

$$\tau>0 \quad r = 1 \quad \alpha<\psi<2\pi-\alpha \quad \partial H/\partial r = 0 \tag{14b}$$

$$\tau = 1 \quad 1<r<\rho \quad 0<\psi<2\pi \quad H = 1 \tag{15}$$

Equation (14a) implies that during half of the day the uptake is sinusoidal, and during the other half it is zero. Again in analogy to nutrient uptake it is assumed that water uptake is independent of the water content up to a limiting value corresponding to a matric potential of 500 kPa (5 bar).

The solution of (12) - (15) was obtained by a numerical method (the alternating direction implicit method). Parameter values used can be found in Table 1. Results are shown in Figures 4a and 4b. Under the conditions chosen the effect of soil-root contact is only slight compared with the effect it has on the availability of nutrients (Figures 3 and 4).

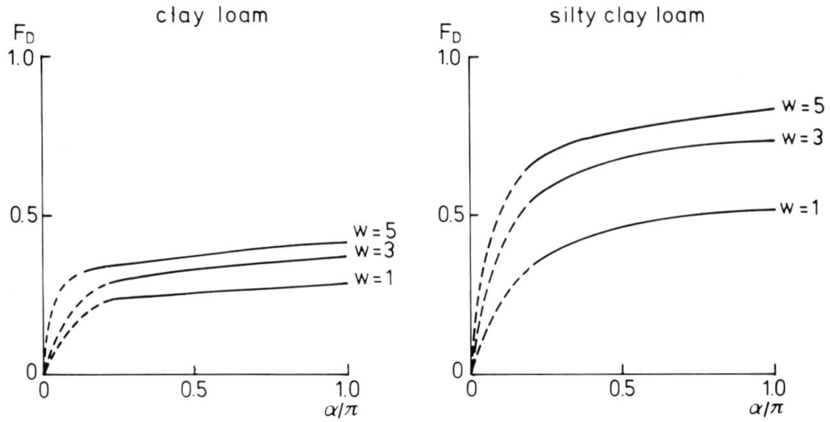

Figure 4a, b. Fractional depletion of water as a function of the ratio α/π, and root density for two soils

References

Altemüller, H.J. und Th. Haag 1983. Mikroskopische Untersuchungen an Maiswurzeln im ungestörten Bodenverband: Kali-Briefe (Büntehof) 16 (6): 349-363.

Churchill, R.V., J.W. Brown and R.F. Verhey 1974. Complex variables and applications, 3rd ed. McGraw-Hill-Kogakusha, Tokyo, 332 pp.

Herkelrath, W.N., E.E. Miller and W.R. Gardner 1977. Water uptake by plants: II The root contact model. Soil Sci. Soc. Am. Proc. 41: 1039-1043.

Scott Russell, R. 1981. Root growth in relation to maximizing yields. In: Proc. of the 16th Colloquium of the International Potash Institute, Bern, 295 pp.

Stroosnijder, L. 1976. Infiltratie en herverdeling van water in grond.

Thesis, Wageningen.
Tinker, P.B., 1976. Roots and water. Transport of water to plant roots in soil. Phil. Trans. R. Soc. London B. 273: 445-461.
Weatherly, P.E. 1979. The hydraulic resistance of the soil-root interface. A cause of water stress in plants. In: The soil-root interface, J. Harley and R. Scott Russell (ed.), Academic Press, London - New York - San Francisco, pp. 275-286.
Willigen, P. de 1981. Mathematical analysis of diffusion and mass flow of solutes to a root assuming constant uptake. Inst. Bodemvruchtbaarheid, Rapp. 6-81, 56 pp.

Discussion

L.P.Wilding:

I am not sure I understand the rationale of assuming that uptake will take place only over about that fraction of the root surface that is in contact with the soil. If water is present in the void, then diffusion or mass flow will take place surrounding the whole root, but contact exchange would only occur where the root contacts the soil. Perhaps you are assuming a much lower water content where these macrovoids are air filled. Would you expand on this?

Author:

Indeed. We have assumed the water content and the dimensions of the void at the root surface to be such that the void is air filled.

Improvement in leaching efficiency of a silty clay loam soil through application of sand

H.S. Sen and A.K. Bandyopadhyay
Central Soil Salinity Research
Institute, Regional Research
Station Canning, P.O. Canning Town,
24-Parganas West Bengal, India-
743329

Introduction

Desalination of heavy textured soils through leaching is limited by their low hydraulic conductivity. An attempt has been made to improve leaching efficiency by altering the pore size distribution at the surface through application of sand.

Materials and methods

Two masonry lysimeters of 0.75 × 0.75 m were filled with a disturbed silty clay-loam profile and the soil was allowed to settle for sufficient time. In one of the lysimeters, S_{30}, 30 per cent sand was added to the top 0.15 m. The other lysimeter, S_0, was used as a control. The entire soil profiles of 1.15 m were equilibrated with saline water, such that the EC of the soil prior to the start of the experiment varied between 10 and 11 mmhos/cm. The EC of the irrigation water was 1-1.5 mmhos/cm. A continuous water head of 4 ± 1 cm was maintained on the soil surface. The leachate was collected periodically at 1.15 m depth and analyzed for Cl^-. The breakthrough curves of Cl^- were expressed as a function of pore volume of effluent collected.

Results and discussion

Figure 1 shows the breakthrough (BT) curves.

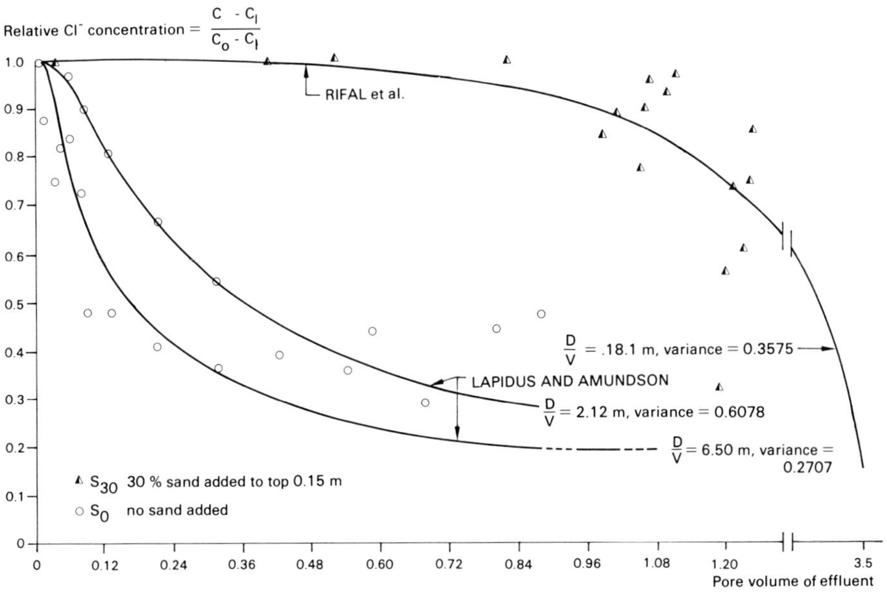

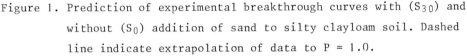

Figure 1. Prediction of experimental breakthrough curves with (S_{30}) and without (S_0) addition of sand to silty clayloam soil. Dashed line indicate extrapolation of data to P = 1.0.

The pore water velocities v in lysimeters S_0 and S_{30} were, respectively, 0.031 and 0.074 cm/h. In lysimeter S_0 some of the irrigation water appeared in the effluent as soon as 0.06 pore volume of effluent had been been collected. In lysimeter S_{30} irrigation water did not appear in the effluent until 0.80 pore volume of effluent had been collected, but thereafter the concentration decreased sharply. Evidently, in lysimeter S_0 a large fraction of the pores was not invaded by the leaching water. Adding 30 per cent sand to the top 15 cm in lysimeter S_{30} more than doubled the pore water velocity and resulted in far more efficient displacement of the soil solution. For S_{30}, after 0.2 pore volume had passed the Cl-concentration remained rather constant; it appears as if of the infiltrating water 40% displaces soil solution initially present and 60% bypasses.

An attempt was made to predict the BT curves. Two equations were used:
1. A solution of the convection/dispersion equation given by Lapidus and Amundson (1952; see also Bolt, 1982, p. 313);

2. An approximate solution of the simplest equation describing non-equilibrium exchange between a mobile phase and a stagnant phase (Rifai et al, 1956; see also Bolt, 1982, p. 328).
The results of the calculations are shown in Fig. 1. The parameter on the curves is the ratio of the dispersion coefficient D and the pore water velocity v. The solution based on the convection/dispersion model with dispersion length D/v = 2.12 or 6.50 m somewhat resembles the experimental data for the S_0 lysimeter. The data for the S_{30} lysimeter are approximated best by the solution based on the mobile/stagnant model.

The results clearly show a favourable influence of the sand mixed with the top soil upon the leaching efficiency.

References

Bolt, G.H. (Editor) 1982. Soil Chemistry. B. Physicochemical models. Second (Revised) Edition, Amsterdam, 527 pp.

Lapidus, L. and N.R. Amundson 1952. Methematics of adsorption in beds: VI. The effect of longstudinal diffusion in ion exchange and chromatographic columns. J. Phys. Chem. 56: 984-988.

Rifai, M.N.E., W.J. Kaufman, and D.K. Todd 1956. Dispersion phenomena in laminar flow through porous media. Sanit. Eng. Res. Lab., Divn. Civ. Eng., Rep. 3. Univ. of California, Berkeley: 157 p.

Modelling the interaction between solute leaching and intra-ped diffusion in clay soils

T.M. Addiscott,
Rothamsted Experimental Station,
Harpenden, Herts AL5 2JQ, U.K.

Abstract

A simple model is described that treats the soil profile in layers, each containing a mobile phase that can be displaced by rainfall and an immobile phase held in cubic peds, with inter-phase solute movement occurring by diffusion. This was tested against experimental analogues consisting of columns of chalk cubes and then used to simulate the effects of ped size and other variates on solute retention by clay soils undergoing leaching. The increase in solute retention due to the presence of an immobile phase is supplemented by retention arising from the fact that solute movement between the phases occurs by diffusion and never proceeds to equilibrium. The simulations showed this extra retention to become potentially more important as cumulative rainfall, and thence potential leaching, increased. The extent to which this extra retention occurs depends on the size and diffusional impedance of the peds, but was also strongly influenced in the simulations by the pattern of rainfall. Whether the ped sizes were uniform or normally or log-normally distributed also affected this retention appreciably. Simulations that took account of anion exclusion showed that salts became more strongly retained as their concentrations declined.

1 Introduction

Water percolating through heavy clay soils mainly passes round, rather than through, the peds. Solutes held within the peds are therefore protected to a large extent against leaching until they diffuse to the ped surface, so that intra-ped diffusion may have quite a strong effect on solute losses by leaching in these soils. Such effects are difficult to quantify by experiment and this paper presents an attempt to evaluate their importance using simple computer modelling techniques. The questions discussed include the following. How large do soil peds have to be to afford significant protection against leaching? How important is the variation in ped size, and does it matter whether the ped sizes are normally or log-normally distributed? How much effect does the pattern of rainfall have on solute losses? Do variations in the diffusional impedance factor in the peds much affect these losses? Is anion exclusion a significant factor in salt losses? The value of the answers given to the questions depends entirely on the validity of the model.

2 The model

The leaching model derives from a simple two-phase (mobile/immobile) layer model. In this, solute and water are displaced through the mobile phase of each layer by incoming rainfall, and displacement through several layers can occur when a large amount of rain falls. Solute movement between the phases then occurs. In the original version of the model (Addiscott, 1977) this solute movement was simply assumed to equalise the solute concentrations between the phases. In the version used here it occurs by, and is regulated by, diffusion.

To simulate diffusion in peds it is necessary to assign them a definite geometry. Since peds in clay soils are often described as 'blocky' a cube seems as appropriate as any shape and this choice is supported by the work of Bouma (p.298). Cubes also pack simply in model layers. Diffusion in a cube was simulated by dividing the cube into concentric volumes and computing the diffusional transfer of solute between each pair of volumes by direct application of Fick's first law (Addiscott, 1982). This method is not as exact as an analytical approach (e.g. Crank, 1956) but is more flexible. For these computations each cube was divided into ten concentric volumes.

Diffusion of bromide from mixed 20, 30 and 40 mm chalk cubes could be simulated better with separate computions for each size than by taking a cube root mean cube size (Addiscott et al, 1983), implying that this is important for soil peds, which have a much wider size range. Sizes representative of each ped size distribution were obtained by dividing the distribution into six sections each representing the same volume of peds plus associated mobile water and represented by its median size. This procedure was outlined earlier (Addiscott, 1981) and will be treated in greater detail elsewhere (Addiscott and Wagenet - in preparation). Provision was made for normal or log-normal distributions. Log-normality is likely when the peds occur as a result of breakage (Epstein 1948) and has been found in cultivated surface soils (Gardner, 1956; Smith, 1977) but may not occur in uncultivated subsoil clays.

3 Preliminary testing of the model

3.1 An experimental analogue

Chalk cubes from the batch used for earlier diffusion studies (Addiscott, 1982) were packed into a column about 40 cm high separated by thin layers of 1-2 mm quartz chips that enabled water to flow between the cubes. The solutions within and between the cubes were both initially 0.01 Molar in calcium bromide. Quantities of distilled water equivalent to 2-12 mm increments of 'rainfall' were applied at the top usually at 15 minute intervals but separated by longer periods overnight and at the weekend. Immediately after each addition, an equivalent volume of effluent was removed from the bottom of the column, giving practically instantaneous drainage, and analysed for bromide with a previously-calibrated bromide-specific ion electrode. Runs were made with identical rainfall regimes for (a) fifty-six 20 mm chalk cubes arranged in 14 layers separated by 0.5 mm of quartz chips, and (b) seven 40 mm cubes, each constituting a layer, separated by 1.25 mm of quartz chips.

In both runs there was a steady decline in effluent bromide concentration (Figure 1) that was not much influenced by the 13 hour overnight break in water addition (point A) but was reversed by the 42 and 31 hr breaks at the week-end (points B and C). The rate of decline was smaller and the reversal larger with the larger cubes.

3.2 Simulation of the experiment

The diffusion coefficient and impedance factor used in the previous studies with these cubes (1.2×10^{-5} cm^2 s^{-1} and 0.4) were taken again. The simulations for both cube sizes followed the general trend of the measured points quite well (Figure 1). The lack of exact agreement was partly because the mobile water was in horizontal as well as vertical gaps between cubes so that some moved much less than the rest. This distinction could not be made in the model. The small increase in effluent concentration A and the much larger increase at B were simulated with fair accuracy for both cube sizes. The simulation also reproduced some of the lesser inflections in the measured points, notably in the final section of the relationship for the 20 mm cube column. This degree of agreement between the model and the experimental analogue suggests that the model functioned as intended and that it can fairly be used to assess the likely effect of ped size on solute holdback.

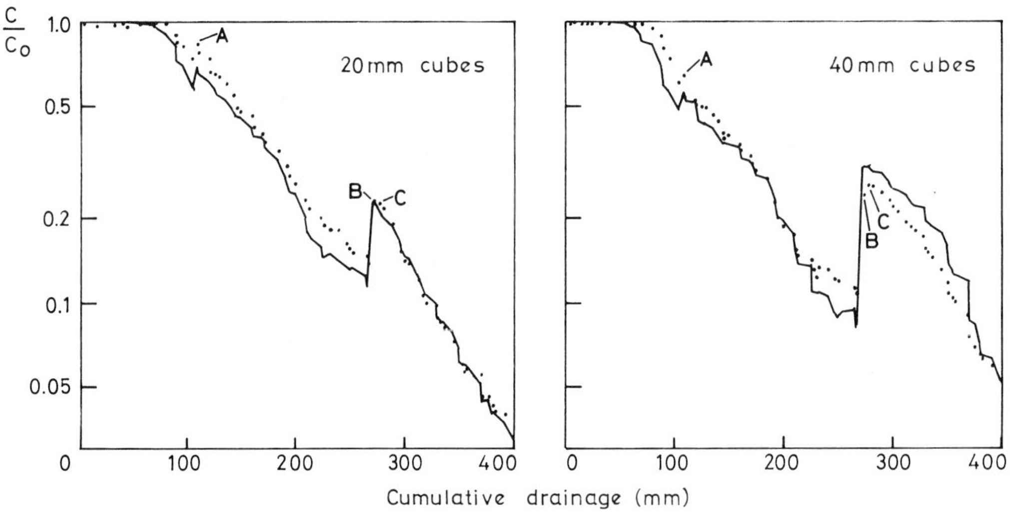

Figure 1. Ratio of bromide concentration in drainage to that initially in column (c/c_0). Measured (.) and simulated (——) values for 20 and 40 mm cubes plotted vs cumulative drainage.

4. Simulations of solute retention by peds of differing sizes and size distributions under various rainfall regimes.

4.1 Details of the simulations

Simulations were made of a 500 mm profile of cubic peds divided into ten layers, with each layer having the same mean ped size. The profile was assumed to be fully recharged and the peds not to shrink or swell. The ped porosity was 0.35 and the proportion of each layer occupied by air or non-porous material was 0.25. The computer program was written to make the amounts of mobile and immobile water (WM and WR) in each layer the same for all mean sizes, so that only the direct effects of ped size on solute retention were simulated. WM and WR were 3.654 and 11.846 mm respectively, corresponding to a volumetric moisture content (θ_v) of 0.31. The solute was assumed to be non-adsorbed and uniformly distributed throughout the profile at the start. Values of 1.2×10^{-5} $cm^2\ s^{-1}$ and 0.3, appropriate to calcium nitrate or a calcium halide, were taken for the diffusion coefficient and the impedance factor respectively. The effects of anion exclusion were assumed to be included in the impedance factor. The variate simulated was the percentage retention of the solute initially in the profile, and the inputs varied were as follows. (a) Mean ped size (i.e. cube side length): 1 mm, 5 mm, 10-100 mm in 10 mm increments. (b) Ped size distribution. Uniform or normally- or log-normally distributed. The mean (m) and variance (s^2) of the log-normal distribution were obtained from the mean (μ) and variance (σ^2) of the corresponding normal distribution as follows (Haan, 1977):

$$m = \ln\mu - 0.5 \ln(1 + \sigma^2/\mu^2) \qquad (1)$$

$$s^2 = \ln(1 + \sigma^2/\mu^2) \qquad (2)$$

For normally-distributed sizes σ was put equal to μ, because this gave an s-value in the corresponding log-normal distributions comparable with that in field top-soils (e.g. Smith, 1977; Figure 2). This resulted in some negative sizes in the normal distributions, which were replaced by sizes of 0.1 mm (negligible in terms of diffusional "hold-back").

(c) Rainfall pattern: Forty daily increments of 2.5 mm, ten of 10 mm, or eight cycles of 10 mm followed by 2.5 mm. (Total 100 mm).

4.2 Results of the simulations

The percentage of initial profile solute retained against leaching (R) was computed first with the simple two-phase model that makes no allowance for retention due to diffusion not proceeding to equilibrium. This gives a value (R_{SM}) which is the smallest possible R-value for the system simulated. The largest possible value (R_{max}) corresponds to a ped of infinite size, for which only solute in the mobile phase would be lost and is given by expressing the immobile water as a percentage of the total layer water, i.e. for the values used here:

$$R_{max} = 100 \times 11.846/(3.654 + 11.846) = 76.43\%$$

Simulations with the full leaching/diffusion model gave R-values for uniform 1, 5 and 10 mm peds that were practically indistinguishable from each other and from R_{SM}, and values for (hypothetical) 1000 mm peds that were close to R_{max} (Table 1).
Retention at intermediate sizes was influenced by ped size (cube side), because of the varying extents to which diffusion proceeded towards equilibrium, and also by the amount and pattern of rainfall. To quantify these effects, the size-dependent retention (R_{SD}) was computed for each ped size,

$$R_{SD} = R - R_{SM} \tag{3}$$

and expressed as a percentage (P_{SD}) of the maximum potential size-dependent retention (R_{SDP}) as follows.

$$R_{SDP} = R_{max} - R_{SM} \tag{4}$$

$$P_{SD} = 100 \times R_{SD}/R_{SDP} \tag{5}$$

The ped size needed to give discernible size-dependent retention (a_{SDR}) was defined as that at which P_{SD} first exceeded 0.005.

a_{SDR} depended on the amount and pattern of rainfall, being larger with 2.5 mm than with 10 mm rainfall increments and larger after 50 mm rain than after 100 mm with the smaller, but not the larger, increments (Table 1).

Table 1. Simulated values for R_{SM}, R_1, R_{20}, R_{40}, R_{60}, R_{80}, R_{100} (values for 1....100 mm uniformly sized peds). Also, values computed for a_{SDR}.

	Rainfall increment and cumulative amount (mm)					
	2.5		2.5/10		10	
	50	100	50	100	50	100
R_{SM}	67.74	36.29	68.29	40.74	68.42	41.37
R_1	67.74	36.29	68.29	40.74	68.42	41.38
R_{20}	67.74	36.30	68.29	40.74	68.43	41.40
R_{40}	67.74	36.45	68.34	41.01	68.54	41.97
R_{60}	67.75	37.13	68.60	42.28	69.07	44.25
R_{80}	67.81	38.36	69.16	44.60	69.97	47.77
R_{100}	67.97	40.00	69.94	47.47	71.00	51.49
R_{1000}	76.04	74.11	76.31	75.50	76.32	75.84
a_{SDR} (mm)	37	18	18	17	15	15

The simulations showed no potential size-dependent retention (R_{SDP}) until a certain amount of rain (37 mm in this instance) had occurred. After this, R_{SDP} increased with cumulative rainfall (i.e. potential leaching) but was relatively little affected by the rainfall pattern (Figure 2). The proportion (P_{SD}) of this retention that occurred depended on rainfall as well as ped-size, but, in contrast to R_{SDP}, P_{SD} depended more on the pattern than the cumulative amount. For example, after 50 or 100 mm of rain, P_{SD} values were appreciably

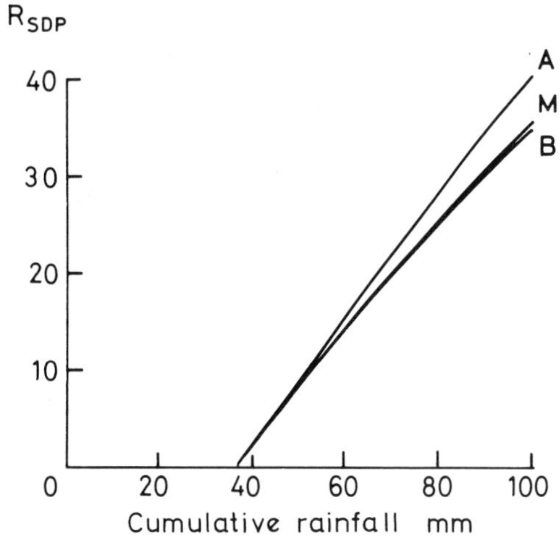

Figure 2. Simulated relationship between R_{SDP} and cumulative rainfall. A, 2.5 mm; B, 10 mm; M, mixed rainfall increments.

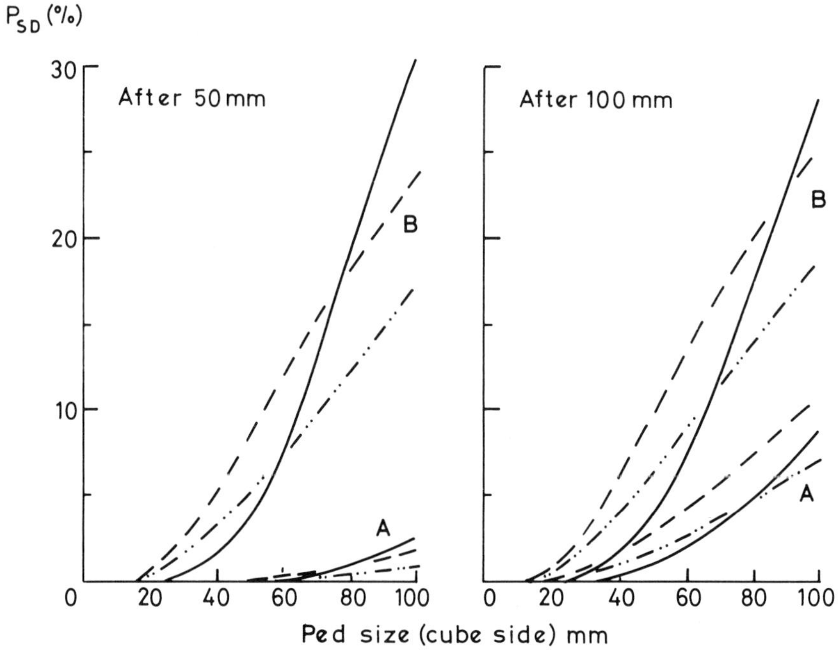

Figure 3. Simulated relationship between P_{SD} and ped size (after 50 or 100 mm rainfall). A, 2.5 mm; B, 10 mm; M, mixed rainfall increments. Uniform (——) or normally (— — —) or log-normally (—·—·—·) distributed ped sizes.

larger when the rain was in 10 mm increments (B in Figure 3) than when it was in 2.5 mm increments (A), with the mixed regime (not shown) intermediate. With normally and log-normally distributed ped sizes, P_{SD} became discernible at smaller mean sizes, but also reached smaller values at the larger mean sizes, than with uniformly-sized peds. P_{SD} was also larger for normally than for log-normally distributed peds. These differences must simply reflect the spread of ped sizes characteristic of each type of distribution. R_{SDP} was independent of the kind of distribution.

5. Simulating the effect of the diffusional impedance factor in the peds

The profiles simulated had the same dimensions and rainfall regimes as those in Section 4 and contained normally distributed ped sizes with means of 50 or 100 mm. The impedance factor (f), discussed in detail by Nye and Tinker (1977), takes account primarily of the tortuous pathway

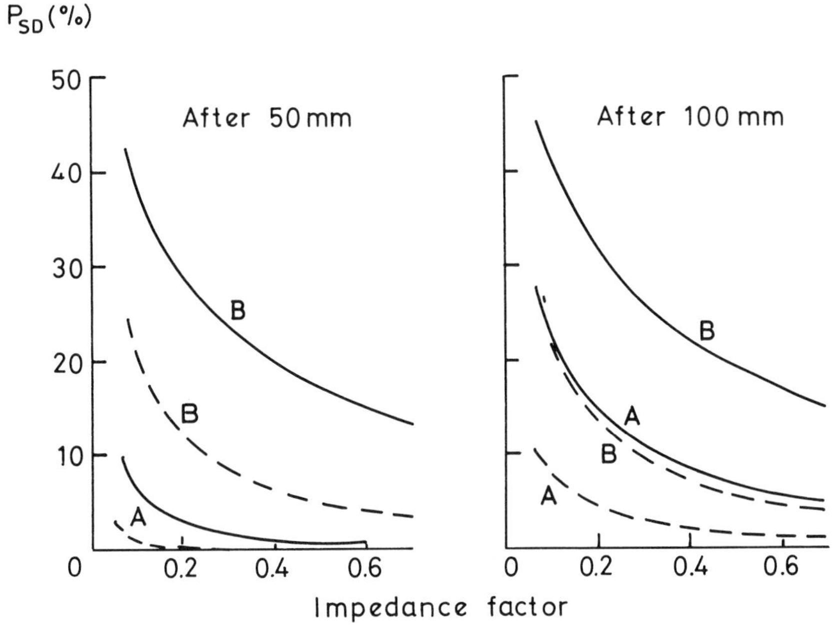

Figure 4. Simulated relationship between P_{SD} and impedance factor for 50 mm (---) or 100 mm (——) normally-distributed peds after 50 or 100 mm rainfall in (A) 2.5 mm or (B) 10 mm increments

taken by the solute in the ped pores and was varied between 0.1 and 0.7 in increments of 0.1. For all combinations of ped size and rainfall regime P_{SD} declined markedly as f increased with most of the decrease occurring at f-values up to 0.4 (Figure 4). Ped size, rainfall amount and rainfall regime all influenced this decrease. It is clear that making f appreciably different from 0.3 in the previous simulations would have changed the a_{SDR} values in Table 1 and given somewhat different curves in Figure 3. However, Pinner and Nye (1982) found an f-value of about 0.26 in a fully-wetted block cut from a clay loam ped, and Barraclough and Tinker (1982) found values around 0.3 in small undisturbed soil cores that were not necessarily single peds, so the value taken should be fairly appropriate.

6 Simulating the effects of anion exclusion on the retention of electrolytes against leaching

In simulating the retention of an electrolytic solute it is necessary to take account of the anion exclusion that occurs in part of the intra-ped water. This seems likely to have three relevant effects: 1) increasing the concentration in the non-excluding water and thence the concentration gradients within the ped and between the ped and the external solution, 2) restricting the volume and thence the cross-sectional area through which diffusion occurs, and 3) making diffusion pathways more tortuous (i.e. decreasing the impedance factor). The second and third effects should tend to counteract the first. Simulations were made for normally distributed peds with mean sizes of 50 and 100 mm in profiles and rainfall regimes identical with those in section 4 except that the mixed rainfall regime was omitted. The effects of intra-ped anion exclusion were incorporated as follows. At each (diffusion) time step and for each of the concentric volumes into which the model divided each ped, the fraction (F_e) of water that excluded anions was computed (after Schofield, 1947) from the relationship $F_e = X_1 + X_2 c^{-\frac{1}{2}}$, where X_1 and X_2 were constants and c was the mean anion molarity in that volume. This was used to compute the concentration in the non-excluding volume, and thence its gradient between adjacent volumes, and to adjust the cross-sectional area and impedance factor for diffusion by multiplying them by $(1-F_e)$ and $(1-F_e)^{\frac{1}{2}}$ respectively (Addiscott et al, 1983). Simulations were made for assumedly monovalent anions with X_1 and X_2 held constant at the values of 0.112 and 0.0395 found for a heavy sandy clay in the work cited above. The initial concentrations tested were derived from a 'baseline' concentration of 64.516 mM (used because it gave a convenient profile total of 10^4 m-moles m^{-2}), by dividing it by 1, $\sqrt{10}$, 10 and $10\sqrt{10}$ to give the concentrations shown at the head of Table 2. Solute retention after 100 mm rain, shown by R, increased appreciably with decreasing initial solute concentration. i.e. The postulated effects on anion exclusion on the volume available for diffusion and the impedance (2 and 3 above) outweighed that on the concentration gradient (1) in the simulations. The effects of ped size and rainfall pattern were greatly diminished by decreasing the initial concentration.

Table 2. Simulated anion exclusion effects on R resulting from various initial solute concentrations. After 100 mm rain.

Ped size (mm)	Rain inc. (mm)	R				
			Initial solute concentration (mM)			
		No AA*	64.516	20.402	6.5416	2.0402
50	2.5	37.45	37.13	52.92	70.50	73.06
	10.0	44.75	48.11	57.95	71.13	74.08
100	2.5	40.64	45.96	56.86	70.52	73.12
	10.0	50.45	54.99	63.22	71.24	74.13

* No anion exclusion

7 Discussion

The results of the simulations can be summarised simply as follows. Soil peds help to retain solutes against leaching by keeping part of the soil water relatively immobile, but also provide extra retention due to the fact that solute diffusion between water inside and outside the peds never proceeds to equilibrium. This extra retention becomes apparent only after a certain amount of rainfall (and potential leaching) have occurred. The potential extra retention then increases cumulatively with rainfall, but only a proportion of it actually occurs. This proportion depends on the size and diffusional impedance factor of the peds and is also strongly influenced by the pattern of rainfall. The ped size and impedance factor exert their influence in the inter-phase diffusion following each rainfall event, but the effect of rainfall pattern is of a more cumulative nature, analogous to solvent extraction procedures, in which a larger number of small volumes of the extracting solvent are more effective than a smaller number of large volumes comprising the same total volume. Anion exclusion enhances salt retention by the peds and becomes more important as the intra-ped solute concentration declines, i.e. as losses increase, but diminishes the effects of other factors.

Barraclough et al (1983) deduced from Crank's (1956) graphical solutions to the equation for diffusion in a sphere that diffusion-controlled solute hold-back was unlikely to be significant in nitrate leaching unless the immobile zones were equivalent to spheres of at least 40 mm diameter. This is a fair conclusion in respect of a single leaching event, but the simulations suggest that when intra-ped diffusion is considered in conjunction with varying amounts and patterns of leaching no such unique size can be defined.

One of the less satisfactory aspects of the simulations lies in the way in which the normal size distributions had to have negative ped sizes replaced by 0.1 mm (section 4.1). This will have slightly affected the shape of the "normal" curves in Figure 3. The underlying problem here is that normal and log-normal ped populations have differently structured size ranges. Transforming one type of distribution to the other is mathematically quite feasible but the result may not be physically satisfactory. The simulations were arranged to favour the log-normally distributed ped populations usually found in top-soils (Gardner, 1956; Smith, 1977) and therefore gave this problem with the hypothetical normal populations. There seem to have been few studies on ped size distributions in heavy clay soils and such studies would be difficult.

The simulations incorporating anion exclusion should also be treated with caution as they seem to have given rather exaggerated effects. The most relevant conclusion from these simulations is probably that such effects merit further attention, not least because the results imply that the effects of ped size, impedance and rainfall pattern could be less important in practice than the other simulations suggest. An interesting possibility arising from these simulations is that the diffusional uptake and release of salts by clay peds could show hysteresis. Diffusion in the ped peripheries should be more free during the uptake period, when the salt molarities in the peripheries and the external solution are relatively large, than in the release period when the molarities would be smaller.

The sensitivity of the model to changes in the impedance factor gives added relevance to work on this factor in clay soil cores and peds by Barraclough and Tinker (1982) and Pinner and Nye (1982). These measurements are usually difficult, often because of the problems of achieving a uniform initial solute concentration in a ped or core. Barraclough and Tinker tried several core conditioning methods but found none to be without problems or applicable to all soils. Core leaching methods were affected by preferential flow in some cores and impermeability in others. Long-period immersion was feasible, but only for very stable soils. Capillary uptake methods have the disadvantage that the soil must be dried and this may change its pore structure. The leaching and capillary uptake methods may also be affected by anion exclusion. One possibility is to measure the diffusion of solute already in the peds (Addiscott et al, 1983). This should be feasible if the peds come from a soil under a constant cropping and fertiliser regime for a long period and are close to having a uniform intra-ped solute concentration.

Acknowledgements

Grateful acknowledgement is made to P.B. Tinker, J.A. Currie and P.S.C.

Rao for helpful discussions, V.H. Thomas for practical help and the Rothamsted Experimental Station Computer Unit for computing facilities.

References

Addiscott, T.M. 1977. A simple computer model for leaching in structured soils. J. Soil Sci., 28: 554-563.

Addiscott, T.M. 1982. Simulating diffusion within soil aggregates: a simple model for cubic and other regularly shaped aggregates. J. Soil Sci., 33: 37-45.

Addiscott, T.M., V.H. Thomas, and M.A. Janjua 1983. Measurement and simulation of anion diffusion in natural soil aggregates and clods. J. Soil Sci., 34: 709-721.

Barraclough, D., M.J. Hyden, and G.P. Davies 1983. Fate of fertiliser nitrogen applied to grassland. I. Field leaching results. J. Soil Sci., 34: 483-497.

Barraclough, P.B., and P.B. Tinker 1982. The determination of ionic diffusion coefficients in field soils. II. Diffusion of bromide ions in undisturbed soil cores. J. Soil Sci., 33: 13-24.

Crank, J. 1956. The mathematics of diffusion. 1st Edn. Oxford University Press, Oxford, 347 pp.

Epstein, B. 1948. Logarithmico-normal distribution in breakage of solids. Ind. Eng. Chem., 40: 2289-2291.

Gardner, W.R. 1956. Representation of soil aggregate size distribution by a logarithmic-normal distribution. Soil Sci. Soc. Am. Proc., 20: 151-153.

Haan, C.T. 1977. Statistical methods in hydrology. Iowa State University Press, Ames.

Nye, P.H. and Tinker, P.B. 1977. Solute Movement in the Soil-Root System. Blackwell Scientific Publications, Oxford, p 78.

Pinner, A. and P.H. Nye 1982. A pulse method for studying effects of dead-end pores, slow equilibration and soil structure on diffusion of solutes in soil. J. Soil Sci., 33: 25-35.

Schofield, R.K. 1947. Calculations of surface areas from measurements of negative adsorption. Nature, 160: 408-410.

Smith, K.A. 1977. Soil aeration. Soil Sci., 123: 284-291.

Discussion

<u>G.H. Bolt:</u>

1. As a general remark, I would say that the phrase 'extra retention due to non-equilibrium diffusion' expresses some bias to me. Instead, I would favour stating, 'Instantaneous equilibration between a mobile and an immobile phase effectively prevents retention by the latter'. Admittedly one could expect retardation of a concentration front as compared to the mean liquid velocity in the mobile phase, but if the latter were not known, instanteneous equilibration would effectively hide the presence of an immobile phase.
2. Against the background of the above, I would be inclined to exclude the amount still necessarily present in your profile if only 50 mm of the 155 mm of liquid phase have been removed 'at the bottom' (equal to 105/155 = 0.677 !!) from the phrase retention. Similarly, 35.5% constitutes the amount necessarily still present upon passing through 100 mm, both numbers having nothing to do with the presence of any immobile phase. In line with this is also the conclusion that with a mobile phase constituting 23.6% of the total water, your set of assumptions makes it necessary that this percentage taken of the total of 155 mm water in the profile, i.e. 36.6 mm, must be leached out before one could ever expect that delayed diffusion would show up. No simulations are necessary for the above conclusions!
3. May I finally suggest that you take a look at the results obtained with the equations based on continuous flow of the mobile liquid in contact with a stagnant phase via some transfer coefficient. (See Passioura, J.B., 1971, 'Hydrodynamic dispersion in aggregated media: 1. Theory in Soil Sci. 111, pp. 339-344; Chapter 10 by M.Th van Genuchten and R.W.Cleary in Bolt, G.H., ed., 1982; Soil chemistry B; Physico-chemical models, Elsevier, Amsterdam, 527 pp.). Such equations give one a fair guessing procedure as to when it becomes worthwhile to run a computer simulation.

Author:

1. The only bias that I am able to see myself (not that it is easy to see one's own biases) is that I start from the simple model, to which reference is made, and proceed from there. Had I started from another simple model, the paper would presumably have been different.
2. There is a phrase missing in the text, for which I apologize. On page 285, seventh line from the bottom, please insert after (37 mm in this instance) 'equivalent to the amount of mobile water in the profile'. Obviously, Dr. Bolt is right in saying that you do not need simulations to realize that this is so, but the main point of the simulations comes after this point.
3. I am not sure that it is helpful to use a continuous flow model to simulate the effects of discontinuous rain on solute and water movement. These simulations are geared to the field rather than the laboratory.

G.H. Bolt:

I have some difficulty following how one could discuss the effect of anion exclusion in general terms, because such an effect would be critically dependent upon the pore size distribution within the aggregate. The so-termed exclusion distance (e.g. about $2/\kappa$ for your symmetric salt system) must be related to the pore size if its effect is to be estimated. Secondly, I fail to see how anion exclusion could lead to an increase in the concentration exterior to the exclusion zone, unless you dump dry aggregates into a solution without further equilibration. In that case it would seem fair to use the new equilibrium concentration as a point of departure; your retention as defined would then necessarily always increase, but you might as well study the effect of an increased tortuosity, i.e. a decreased D value! Finally, the above sentence should be interpreted as referring to the retention, taking into account the absence of salt in the exclusion zone. Reference is made to a paper by A.V. Blackmore on this issue ('Salt sieving within clay soil aggregates,' Amsterdam J. Soil Res., 1976:14, pp. 149-158).

Author:

I agree that the problem could be approached most fundamentally by considering the pore size distributions within the aggregate, but this would make this part of the model much more complex than the rest. The exclusion constants were inferred from simple measurements in which small (< 3 mm) crumbs of a chalky boulder clay soil were equilibrated with solutions containing chloride-36 and various concentrations of $CaCl_2$. This is not a fundamental approch, but it has been used before. I did not fully follow what Dr. Bolt was getting at in the middle section. One of the effects simulated is an increased tortuosity (or decreased impedance factor) resulting from anion exclusion.

W.J. Bond:

1. There is an apparent discrepancy between the effect of anion exclusion and salt movement adopted by Addiscott and that shown by Smiles and Bond. This arises, I believe, because Addiscott is determining salt movement within so-called dead pore space, where he postulates that there is no convective or advective transport of salt, only transport by diffusion. The Smiles and Bond case does have active advective transport of salt in addition to diffusion.
2. However, I would like to ask if there is experimental evidence to support your approach to the effect of anion exclusion on simple diffusion.

Author:

1. I agree with your basic assessment of the difference between our approaches.
2. I have been trying to do some experiments on the effects of anion exclusion on diffusion in small clods, but have had problems because the anion exclusion has been creating problems during the setting up of the experiments.

J. Bouma:

You are right in saying that there have been few studies on ped sizes in clay soils. However, I believe that we should move away from the ped concept here and focus on flow patterns during infiltration which involve only a fraction of the total vertical ped surface (see Bouma and Dekker, 1978, as referred to in my paper). Use of S values, as defined there, would be more relevant than ped sizes. The same is true for flow in saturated soils where staining has proved that only a fraction of all cracks are used during flow (figure 1 in my paper).

Author:

In general, I agree with Dr. Bouma. Using size distributions in the way that I have done is satisfactory only in a fairly limited range of circumstances. Where you have really massive peds, which are exposed to mobile water on only a very limited proportion of their surfaces, alternative approches are needed.

However, if we do move away from the ped concept in this context, we must be careful not to lose sight of the interactions between structural units of different sizes in the soil, e.g. the smallest units lose the largest proportion of their solute early in a period of diffusion. By doing so, they also lessen the concentration gradient not only for themselves, but also for the larger units. This means that the release curve for a distribution of aggregate sizes has a shape different from that for a collection of uniformly-sized aggregates. Also, small aggregates suffer most from the increased impedance of diffusion due to anion exclusion.

R.E. White:

Is it profitable to try and identify a particular geometry (e.g. a three dimensional array of solid cubes) for the immobile volume in soil? Firstly, in a real soil it is extremely difficult to describe quantitatively the ped geometry. Secondly, under intermittent rainfall in the field the immobile volume (and hence the mobile transport volume) will not be constant. Is it not preferable to define a variable transport volume (the mobile volume) which is the volume participating in solute transport by mass flow and diffusion?

Author:

I agree that cubes are not an entirely satisfactory representation of the immobile volume, but they do provide a means of assessing the possible effects of different sizes of structural units. They also provide the geometry necessary for diffusion calculations. Whether it is better to use cubes or to treat the soil as a semi-infinite medium, as you do, probably depends on the type of soil you wish to simulate.

Your comments on a variable mobile volume are also quite fair. In the most recent version of the simple model we do now have variable mobile volume. It can go above and below the 'field-capacity' value.

Using soil morphology to develop measurement methods and simulation techniques for water movement in heavy clay soils

J. Bouma
Netherlands Soil Survey Institute
Wageningen, The Netherlands

Abstract

Soil morphology data, in terms of number, size and length of large soil pores per unit surface area, are used to define: (1) optimal sample sizes for physical measurements and (2) computer simulation models for water movement in heavy clay soils. The simulation models use standard numerical procedures to predict vertical and horizontal fluxes, while morphology data provide boundary conditions for the flow system. The morphological techniques being used require not only a description but also application of staining and/or filling of large pores with gypsum. Examples are discussed for cracks (planar voids) and cylindrical worm-channels. Soil morphology provides data that cannot be obtained with physical methods. However, such data are only relevant for characterizing water movement when applied in a soil physical context.

1 Introduction

Many heavy clay soils have relatively large cracks or root and animal burrows which occur in a fine-porous soil matrix. This matrix has a very low hydraulic conductivity (K) and significant fluxes of water and solute through the entire soil are therefore only possible when continuous large pores are present. These large pores are unstable as their dimensions change upon swelling and shrinking of the soil following wetting and drying.

A particularly complex condition is found when free water infiltrates along vertical cracks into an unsaturated soil matrix. Such processes have widely been observed in the field (e.g. Thomas and Phillips, 1979; Beven and Germann, 1982), and the name "short-circuiting" has been used to describe the phenomenon (Bouma et al., 1978). The term "bypass flow" will be used hereafter.
Flow of water and solutes in heavy clay soils is quite different from flow in more sandy soils in which most soil pores contribute to water movement. The purpose of this paper is to review Dutch procedures which use soil morphological techniques for defining flow patterns in heavy clay soils. Attention will be focused on the measurement procedures themselves and on use of data in simulation models for water and solute transport.

2 Flow patterns

2.1 Introduction

Flow patterns in heavy clay soils are difficult to characterize with physical methods. Measurement of fluxes provides, of course, no clues. Breakthrough curves provide information on large-pore continuity by their point of initial breakthrough (e.g. Wösten and Bouma, 1979), but they do not indicate the functioning of various pores in the soil. It is sometimes important to know whether rapid breakthrough is due to flow along one large continuous pore or to flow along several pores. Also, flow along planar voids (cracks) has different dynamics than flow along root- and worm-channels. Obviously, physical methods can not be used to distinguish different types of pores.
Morphological staining techniques, to be discussed in the three following subchapters, have been used successfully to define flow patterns in heavy clay soils.

2.2 Saturated flow

Undisturbed samples of a clay soil that had been close to saturation for a period of several months, were percolated with a 0.1% solution of me-the samples were freeze-dried and thin sections were made in which stained pore walls were observed next to morphologically idential pores with unstained walls (Figure 1).

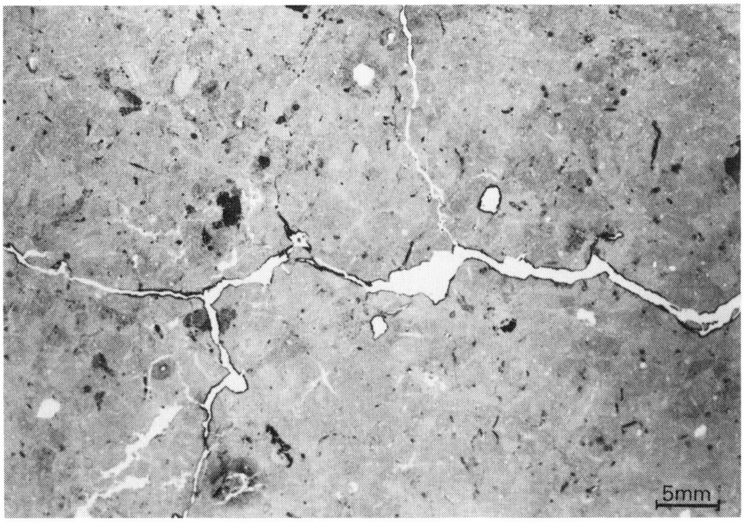

Figure 1. Thin section image of a wet clay soil in which water conducting planar voids are stained with methylene-blue (dark pore-walls). Only continuous voids are stained.

Staining indicates pore continuity which is more crucial to hydraulic conductivity than the often used pore size distribution. The studies, cited above, resulted in the following conclusions: (1) K_{sat} was governed by small pore necks in the flow system with diameters of approximately 30 !m. Small changes in pore-neck sizes had a large effect on K_{sat}. For example, a pore neck of 22 µm resulted in a K_{sat} of 5 cmd^{-1} and a neck of 30 µm in a K_{sat} of 25 cmd^{1}. (2) Water-conducting (stained) larger pores usually occupied a volume that was lower than 1%. Such pores should therefore be characterized in terms of numbers per unit area rather than in terms of relative volume. (3) Flow occurred mainly along planar voids (cracks). This contradicted the common assumption for Dutch clay soils that cracks close completely upon swelling. (4) Using morphological data, K_{sat} of six different clay soils could be calculated according to:

$$K_{sat} = \frac{\rho g}{\eta S} \left(\frac{d_n^3 l}{12} + \frac{r_n^4 n}{8} \right) \qquad (1)$$

in which: ρ = liquid density (kgm^{-3}); g = acceleration of gravity ($m\ s^{-2}$); η = viscosity ($kgm^{-1}s^{-1}$); S = cross-sectional area of soil (m^2) containing a length of l (m) of stained plane slits with neck width d_n (m) and n channels with neck radius r_n (m). Neck widths were calculated from the size distribution of stained pores, using a newly developed pore-continuity model (Bouma et al., 1979).

2.3 Bypass flow

Two case studies are discussed:

(1) A solution of methylene blue in water was used for sprinkling irrigation on a dry, cracked clay soil (Bouma and Dekker, 1978). Soil below the experimental plot (1.0 x 0.5 m) was excavated and visual observations were made of the infiltration patterns of the water, which consisted of 5 to 7 mm wide vertical bands on ped faces. The total number and surface area of bands were determined in soil below the 0.5 m^2 plot for each 10 cm depth increment down to 100 cm below surface. Five sprinkling intensities were tested in four different clay soils. The total surface area of bands, to be called "contact area" (S) hereafter, is an important characteristic as it defines the area which is available for lateral infiltration into the (dry) peds. The contact area increases up to 200 cm^2 as sprinkling intensity becomes higher (Bouma and Dekker, 1978), but the stained fraction of the total vertical surface area of cracks remains low in all cases. A coarse prismatic structure with peds of 10 cm cross-section has (per 10 cm thickness increment) a contact area of 20 000 cm^2 in a plot of 0.5 m^2. The maximum stained contact area of 200 cm^2 represents, therefore, only 1% of the potentially available vertical surface of infiltration.

(2) Methylene blue in water and a gypsum slurry were used to trace infiltration patterns in a silt loam soil with vertical worm-channels (Bouma et al., 1982). Excavation of the soil allowed an evaluation of sizes, and shapes of the vertical channels.

2.4 Horizontal cracking

Vertical cracks or worm-channels may result in bypass flow. However, soil shrinkage also causes the formation of horizontal cracks which strongly impede upward flow of water in unsaturated soil (Bouma and De Laat, 1981). A method was devised to stain air-filled horizontal cracks at different moisture contents and corresponding (negative) pressure heads. A cube of soil (30 cm x 30 cm x 30 cm) is carved-out in situ (Figure 2). The cube is encased in gypsum and is turned on its side.

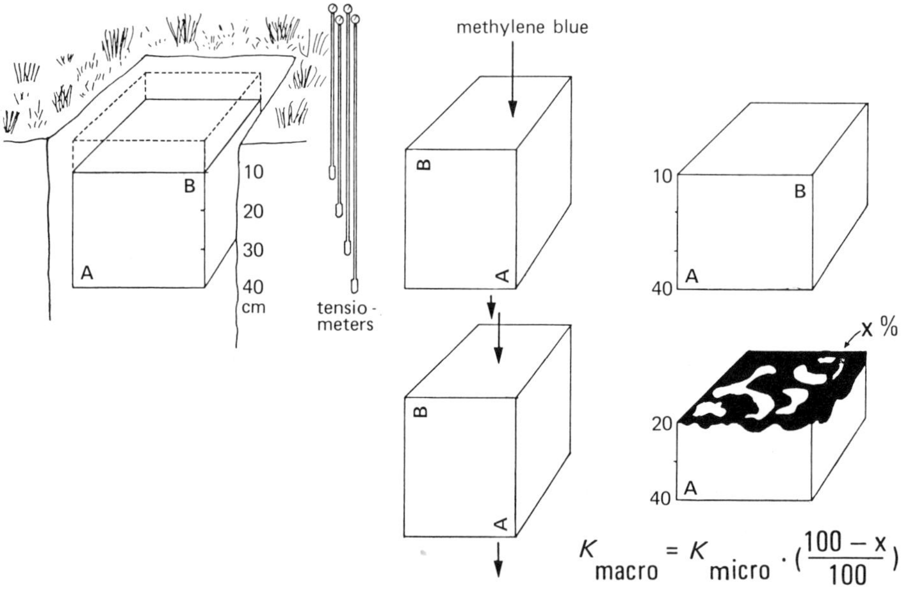

Figure 2. A schematic representation of the method for measuring the area of air-filled horizontal cracks as a function of the pressure head (see text).

The upper and lower surfaces are opened and two sidewalls of the turned cube are closed. Methylene blue in water is poured into the cube and will stain the air-filled cracks. The surface area of these stained cracks is counted after returning the cube to its original position. A separate cube is needed for each (negative) pressure head. The K-curve for the peds (Figure 3) is "reduced" for each pressure head measured in a cube. When, for example, 50% of the horizontal cross sectional area is stained, K_{unsat} for upward flow is 50% of the K_{unsat} at the same pressure head in the peds.

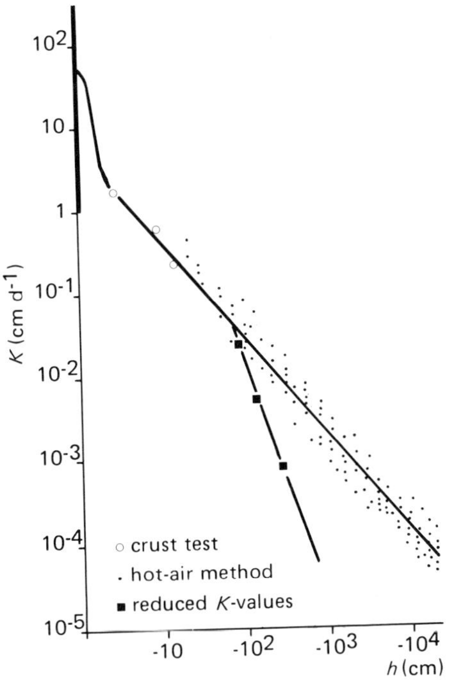

Figure 3. K curve for a heavy clay consisting of the regular curve which defines water movement in the peds and a K_{macro} curve (see Figure 2) which is used to define upward, unsaturated flow from the water-table. Three cubes of soil were used to obtain the three reduced K values that are indicated.

3 Methods

3.1 Introduction

Several methods will be described that have particularly been developed for clay soils with large structural elements ("peds") and large pores. All methods have one feature in common, which is use of large, undisturbed samples. The size of samples is a function of soil structure. Use of standard sample sizes in different soils is incorrect (Bouma, 1983).

3.2 Cube and column method

Measurement of K_{sat} in clayey soils presents the following problems: (i) smearing of the walls of bore-holes may yield unrealistically low K_{sat} values for the auger-hole method, which are in any case an undefined mixture of K_{sat} (hor) and K_{sat} (vert); (ii) small samples give poor results because of unrepresentative large-pore continuity patterns (e.g. Bouma, 1981) and (iii) water movement occurs only along some pores which occupy less than 1% by volume. These pores can be easily disturbed by compaction which may occur when sampling cylinders are pushed into the soil. The cube method (Bouma and Dekker, 1981) avoids these problems and uses a cube of soil (25 cm x 25 cm x 25 cm) which is carved out in situ and encased in gypsum on four vertical walls. First, the K_{sat} (vert) is measured by determining the flux leaving the cube while a shallow head is maintained on top. Next, the cube is turned 90°. The open surfaces are closed with gypsum and the new upper and lower surfaces are exposed. Again, a K_{sat} is measured which now represents the K_{sat} (hor) of the soil in situ. The column method measures only K_{sat} (vert) and uses a cylindrical column of soil with an infiltrometer on top (Bouma, 1977).

3.3 Crust test

Non-steady-state methods, which are widely used to measure K_{unsat}, are not suitable to obtain K values near saturation in all soils, in the range h = 0 cm to, say, h = -15 cm. These values are particularly relevant for describing water flow in clay soils. In these soils there is a strong drop of K upon desaturation owing to emptying of the macropores (Bouma, 1982). Again, large samples are needed to obtain representative results. The cube method can be extended to provide K_{unsat} data near saturation. This procedure represents a version of the crust test (Bouma et al., 1983). Two tensiometers are placed about 2 and 4 cm below the surface of infiltration, which is covered by a series of crusts, composed of mixtures of sand and quick-setting cement. Earlier, the crust test used gypsum but this may dissolve too rapidly. Dry sand and cement are thoroughly mixed. Water is applied and a paste is formed which is applied as a 0.5 cm to 1 cm thick crust on top of the cube. The crust, which has perfect contact with the underlying soil because of the application method, hardens within 15 minutes. Light crusts (5 to 10% of cement by volume) induce pressure heads (h) near saturation and relatively high fluxes. Heavier crusts (20% cement and more) induce lower h values and fluxes. Cement can be added to existing crusts to avoid removal of crusts which could cause damage. Fluxes, when steady, are equal to K_{unsat} at the measured sub-crust h value. Cubes can be placed on a sandbox to create a semi-infinite porous medium. Thus, the range of fluxes can be extended to corresponding pressure heads of approx. -60 cm.

3.4 Bypass flow

Bypass flow can be measured by using large undisturbed cores of surface soil with a height that is equal to rooting depth (Bouma et al., 1981). For Dutch conditions in heavy clay soils, cylinders are used with a height and diameter of 20 cm. Cores include the soil surface with grass, which is closely cropped. The cores are placed in the path of a spraying gun in the field which is commonly used for sprinkling irrigation. In general, sprinkling conditions should correspond to local practices. The mass of the soil-filled cylinder is determined before and after sprinkling and the stove-dry mass is measured at the end, thus allowing calculation of physical constants such as bulk density and moisture contents. Sprinkling intensities and duration should be measured independantly. The volume of water that leaves the column is measured as a function of time, thus allowing an estimate of bypass flow which can be expressed as a percentage of the applied quantity of water. Many measurements can be made in a short time and the effects of using different durations can be easily evaluated. Thus, irrigation efficiencies can be improved because movement of water beyond the root zone often presents a loss of precious irrigation water and surface-applied chemicals (Dekker and Bouma, 1984).

4 Simulation

4.1 Using submodels for characterizing the entire flow system

Complex flow processes in clay soils with macropores become much simpler when submodels are distinguished, which can be defined by using soil morphological and other data and which require specific input data. Three submodels for infiltration are distinguished in Figure 4:

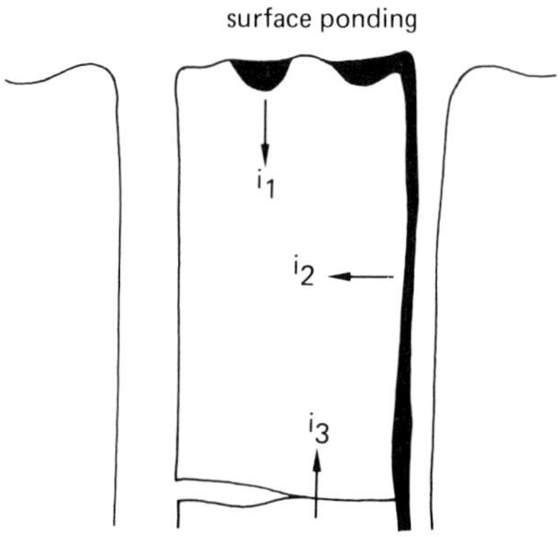

Figure 4. A schematic representation of water infiltration into cracked clay soils, showing surface ponding of water, vertical infiltration (i_1), lateral infiltration from bands along ped faces into the peds (i_2) and upward flow (i_3).

1) vertical infiltration at the upper soil surface between the macropores (i_1),
2) flow of water from the surface into the macropores, after filling of microdepressions at the soil surface,
3) partial or complete filling of the macropores and lateral infiltration into the (unsaturated) soil matrix (i_2).

A separate submodel for upward unsaturated flow (i_3 in Figure 4) is based on the staining-method to characterize the effect of horizontal cracks (Section 2.4). Vertical and lateral infiltration can be characterized by Darcy's equation in combination with the continuity equation. The reader is referred to any current soil physics textbook for specific details. Computer simulation, using CSMP or other user-friendly subroutines, becomes very attractive for the applications being discussed here. This approach involves gross simplifications because volume changes of the soil are not considered. This can be done later without changing the basic concept. Some specific applications will now be discussed.

4.2 Sprinkling irrigation in a clay soil

A model, composed of three submodels as outlined in Section 4.1, was used to predict the infiltration of water during sprinkling irrigation in a dry, cracked soil (Hoogmoed and Bouma, 1980). They used measured K-h, D-Θ and h-Θ functions, as well as independantly measured data for the contact area S at two sprinkling intensities. The extent of surface ponding was estimated. The high percentage of bypass flow (expressed as a percentage of sprinkling rate) was due to relatively low surface ponding but, particularly, to low S values. Simulation of conditions in which all vertical ped faces would be available for lateral infiltration, (S = 20 000 cm^2) resulted in lack of bypass flow: waterwas absorbed within 2 cm below surface.

4.3 Flooding of dry, cracked clay soil

Cracks were not filled with water in the previous example, where water moved as narrow bands on the vertical walls of air-filled cracks. On the contrary, ponding of water results in filling of the cracks. The number of cracks per unit surface area, and their width and depth determine the available volume for storage. Infiltration occurs into the upper soil surface, and laterally from the filled cracks. A field study was made in which these various flow processes were combined (Bouma and Wösten, 1984). The volume of air-filled cracks, available for storage of water, could be reliably estimated by making counts of gypsum filled cracks. Lateral infiltration into the peds was simulated by using a measured D-Θ function in a similation model, which also needed the total length of cracks within a given horizontal cross-sectional area.

4.4 Flooding of soils with worm-channels

Flooding of soils with worm-channels results in deep penetration of water in a short period of time. A field study was made in which vertical infiltration at the soil surface, filling of the worm-channels and lateral infiltration into the soil matrix from the filled channels, was simulated (Bouma et al., 1982). Due to the irregular morphology of the worm-channels, measured infiltration rates into the channels were used as input data for the overall model, rather than calculated infiltration rates assuming channels of perfect cylindrical shape (e.g. Edwards et al., 1979).

4.5 Upward, unsaturated flow in cracked clay soil

Using the technique described in Section 2.4, a "reduced" K-curve (Figure 3) was applied to calculate fluxes from the water-table to the rootzone in the growing season. Good agreement between simulated data and field measurements was found (Bouma and De Laat, 1981) while results were unrealistic when using the K-curve for the peds for calculating these upward fluxes.

5 Conclusions

1) Water and solute movement in structured clays is governed by preferential flow along larger pores, such as cracks and channels. Morphological techniques are needed to characterize those pores in terms of number or length per unit area. Relative volumes are not useful for simulation purposes.
2) Complex flow processes, particularly those associated with bypass flow, can be simulated for field soils by means of a series of coupled submodels which routinely define vertical and horizontal flow using K-h; D-θ and h-θ data. Data obtained by morphological techniques (see point 1) provide essential boundary conditions for the various flow systems.
3) Studies, reported in this paper, characterize physical behaviour at

various distinctly different points in time. The dynamic nature of the processes involved requires a more dynamic definition of points during the year at which critical properties are exceeded. For example, K_{sat} in the clay soils being studied still exceeds 1 cmd^{-1} even after prolonged swelling. This will not be true for other clay soils such as some Vertisols. For example, how much swelling is needed in these soils to arrive at very low K_{sat} values? Another question relates to the drop of K upon horizontal crack formation in various clay soils, which is bound to differ significantly.

4) Several methods, using dyes and gypsum, have been presented which can be used to characterize soil structure. Such characterizations, and others yet to be conceived, are needed to make traditional soil structure descriptions more informative.

References

Beven, K. and P. Germann 1982. Macropores and water flow in soils. Water Resour. Res. 18: 1311-1325.

Bouma, J. 1977. Soil survey and the study of water in unsaturated soil. Soil Survey Papers No. 13, Soil Survey Institute, Wageningen, 107 pp.

Bouma, J. 1981. Soil morphology and preferential flow along macropores. Agric. Water Managem. 3: 235-250.

Bouma, J. 1982. Measuring the hydraulic conductivity of soil horizons with continuous macropores. Soil Sci. Soc. Am. J. 46: 438-441.

Bouma, J. 1983. Use of soil survey data to select measurement techniques for hydraulic conductivity. Agric. Water Managem. 6: 177-190.

Bouma, J. and L.W. Dekker 1978. A case study on infiltration into dry clay soil. I. Morphological observations. Geoderma 20: 27-40.

Bouma, J. and J.H.M. Wösten 1979. Flow patterns during extended saturated flow in two undisturbed swelling clay soils with different macrostructures. Soil Sci. Soc. Am. J. 43: 16-22.

Bouma, J. and L.W. Dekker 1981. A method for measuring the vertical and horizontal K_{sat} of clay soils with macropores. Soil Sci. Soc. Am. J. 45: 662-663.

Bouma, J. and P.J.M. de Laat 1981. Estimation of the moisture supply capacity of some swelling clay soils in the Netherlands. J. Hydrol.

49: 247-259.

Bouma, J. and J.H.M. Wösten 1984. Characterizing ponded infiltration in a dry, cracked clay soil. J. Hydrol. 69: 297-304.

Bouma, J., A. Jongerius, O. Boersma, A. Jager and D. Schoonderbeek 1977. The function of different types of macropores during saturated flow through four swelling soil horizons. Soil Sci. Soc. Am. J. 41: 945-950.

Bouma, J., A. Jongerius and D. Schoonderbeek 1979. Calculation of saturated hydraulic conductivity of some pedal clay soils using micromorphometric data. Soil Sci. Soc. Am. J. 43: 261-264.

Bouma,, J., L.W. Dekker and C.J. Muilwijk 1981. A field method for measuring short-circuiting in clay soils. J. Hydrol. 52: 347-354.

Bouma, J., C.F.M. Belmans and L.W. Dekker 1982. Water infiltration and redistribution in a silt loam subsoil with vertical worm channels. Soil Sci. Soc. Am. J. 46: 917-921.

Bouma, J., C. Belmans, L.W. Dekker and W.J.M. Jeurissen 1983. Assessing the suitability of soils with macropores for subsurface liquid waste disposal. J. Environm. Qual. 12: 305-311.

Dekker, L.W. and J. Bouma 1984. Nitrogen leaching during sprinkler irrigation of a Dutch clay soil. Agric. Water Managem. (in press).

Edwards, W.M.R., R.L. van der Ploeg and W. Ehlers 1979. A numerical study of the effects of noncapillary-sized pores upon infiltration. Soil Sci. Soc. Am. J. 43: 851-856.

Hoogmoed, W.B. and J. Bouma 1980. A simulation model for predicting infiltration into cracked clay soil. Soil Sci. Soc. Am. J. 44: 458-461.

Thomas, G.W. and R.E. Phillips 1979. Consequences of water movement in macropores. J. Environm. Qual. 2: 149-153.

Discussion

L.P. Wilding:

I am concerned about your small percentage of stained pores (macropores) upon rewetting a dry soil. I have observed piping of sediments and 'complete' crack infilling and water flow through cracks caused by high-intensity, short-duration storms of 25 mm/hr. Would you please expand on this point?

Author:

I should emphasize that the described features are associated with Dutch conditions, where rainfall intensities and rates are usually relatively low. Conditions in other climates are quite different, and so are the associated pedological features.

L.P. Wilding:

The limited dye stains on macrovoids indicate that only a fractional percentage of the macropores are effective in saturated transmission. This fact is well established. Wouldn't this situation be quite different in soils that undergo many cycles of desiccation and rewetting during the year. I am implying that a much higher percentage of the ped interfaces would be effective conduits in rewetting a dry soil. Would you comment on this?

Author:

Our staining was done when soils had been naturally saturated for several months. Then, only a small percentage of marcropores was stained. When wetting a dry soil, more ped faces will conduct water, but then there is bypass-flow and not saturated flow with zero or positive pressure heads in the entire soil mass. It is indeed conceivable that a relatively high percentage of ped faces will conduct water after a short-period of saturation and swelling. We have no data on this.

C. Dirksen:

On page 304 you suggest that cement may be added to existing crusts to induce lower h values and fluxes. Do you wash the cement into the existing crust (what does excess cement do to the soil hydraulic properties below?), or do you add a crust with a larger cement percentage? Whatever you do, is this not an undesirable procedure? It will take a long time before the relatively wet soil below the crust will attain the lower water content corresponding to the new, more resistant crust. It should be much faster to start on the dry end (most resistant crust) and let successively wetter fronts overtake each other.

This also prevents the hysteresis problem inherent in the reversed order, increasing the accuracy of the measurements. These arguments would seem to justify the extra work involved in first carefully removing old crusts before new ones are applied.

Author:

The procedure you outline has traditionally been followed when applying the crust test. It is indeed an attractive procedure, except for the required periodical removal of the crust. In the described procedure one starts with a light crust, the resistance of which is increased by adding cement. When this cement hardens before water application, there is no danger of wash-out of cement into the soil. Equilibrium is quickly reached because fluxes are relatively high in the pressure head range to, say,-0.4 m. Hysteresis is not very pronounced either in that range.

D.H. Yaalon:

Which morphological or field methods would you suggest would best document lateral moisture flow, e.g. at the top of the argillic horizon?

Author:

Mottling patterns give an indication of saturation. Lateral flow itself can be documented by collecting the water downslope at the upper boundary of the argillic horizon. A horizontal (lateral) K_{sat} can well be measured with the cube method (Bouma and Dekker, 1981).

D.H. Yaalon:

In semi-arid regions and in irrigation in general we are interested in infiltration and depth of wetting. Since bypass flow will increase depth of wetting, it is beneficial.

Are there simple functional relationships between macropores (e), depth of wetting (z), and rate of application (P/t)? How would you measure bypass flow for such conditions?

Author:

Bypass flow is not necessararily beneficial when water runs down a few cracks without wetting surface soil. Functional relationships as suggested have been defined by Hoogmoed and Bouma (1980), as referred to in my paper. However, their data refer to two types of showers only. It would be very attractive to develop the suggested relationship that will be very different for different soils as a

function of the microrelief (surface storage), soil texture (infiltration rate in matrix), and macropore quantity (vertical bypass flow). Direct measurement of macropore flow is well possible with the field technique published by Bouma et al. (1981).
Your suggestion, and a similar one from Dr. Bolt, to use bypass flow rather than short-circuiting has been followed.

M. Kutilek:

1. Ch.2.2. What was the percentage of narrower necks (e.g. 22 μm) resulting in the given Ks reduction (e.g. K_s = 50 mm day $^{-1}$).
2. Ch.2.4. When mapping the horizontal cracks on cubes 0.3 × 0.3 × 0.3 m, you have to assume that the representative elementary volume is equal to or smaller than this size. Do you have experimental proof for this? Similary for 0.25 × 0.25 × 0.25 m in Ch.3.2.

Author:

1. The percentage by volume of the necks themselves was only 0.2%.
2. The sizes contain about ten elementary units of structure. Theoretically we would prefer to have some more. However, handling such larger samples creates many practical problems, so we have settled for the sizes indicated, which still represent very large samples as such.

A.M. Silot:

In some Sahelian soils with heterogenous, large cracks (20-30 mm wide), it is possible to infiltrate more than 1000 l of water through a 1 m^2 area in 10 minutes, while in other spots of the same area one cannot enter 200 l of water during 3 months. My question is: Are there any limits to the use of your different measurement methods, in particular the bypass-flow methods? Is it possible to use them in such soils to calculate hydraulic conductivities?

Author:

First, we should not confuse short-circuiting (bypass flow) with hydraulic conductivity. Bypass flow is vertical movement of free water through an unsaturated soil horizon. Hydraulic conductivity is a flux at unit hydraulic gradient in a soil horizon with a homogeneous, defined moisture content. Be that as it may, your question refers to the very important problem of spatial variability. Having the large differences you describe, one would have to start with a survey of the area to distinguish sub-populations. If this is not done, quite variable data on different populations are obtained, which are impossible to interpret. The survey could consist of soil structure observations, perhaps in combination with penetrometer readings.

J.J.B. Bronswijk:

1. You determine the horizontal cracks in a soil profile by infiltration of a methylene blue solution in a soil cube that is turned on its side. Are you not afraid that this procedure underestimates the real area of horizontal cracks because of the presence of non-continuous cracks that do not reach the surface of your turned soil cube, and thus will not be stained?
2. In your model for capillary rise, do you take into account some effect of the arrangement of horizontal cracks at different depths with respect to each other?
3. At this symposium, field workers have reported the great importance of opening and closing of shrinkage cracks on heavy clay soil behaviour. Most of the presented simulation models for water and solute movement in macropores and shrinkage cracks, however, are still treating the soil as a fixed matrix with fixed macropores.

To obtain a simulation model for the moisture behaviour of heavy clay soils, for longer periods, say, one growing season, I think one should combine these 'fixed-pore models' with models describing the swelling/shrinking process, but surely that would be rather difficult. What is your opinion about the possibilities and problems of such a combination?

Author:

1. The real area of horizontal cracks will probably be underestimated even though dyed water is infiltrated from two sides into the cube. Underestimation is due to 'shadow' effects. Besides, it is unlikely that all very small cracks that are effective in blocking flow are being stained. Even though the overall effect is probably underestimated, the effect as measured is still major.

2. As is, we do not take this effect into account because horizontal cracks become less developed at greater depth, and peds are usually rather high. A simple statistical model could be used, however, to express the cumulative effects of several horizontal cracks at different levels.
3. The swelling and shrinkage effects are incorporated in our model by including bypass flow as a function of the moisture content of surface soil (which varies during the year), and by including horizontal cracking as a function of the pressure head (which also varies during the year). Important for our soils is the fact that the cracks do not completely close during winter and early spring. In fact, it does not make much of a difference whether a crack is 20 mm or 20 μm wide. Conditions in some other soils, where cracks do close completely, are significantly different. Then, it is very important to be able to predict the moment of closure.

K.H. Hartge:
You stated, 'Measurement of fluxes provides no clue to flow patterns'. So you take morphological evidence: distribution and frequency of stained slits.

1. Can you think of measuring permeability by flux methods on a set of ten samples placed at regular distances (same depth) and evaluating distribution of κ-values as a characteristic of the soil?
2. Do you think it would be possible to work out a κ-value from such a distribution in a way roughly analogous to that which you used for equation (1)?

Author:
1. This procedure would be very suitable to obtain a measure for spatial variability patterns. However, the size of each sample should be large enough to be representative for the structure to be characterized.
2. Data obtained could be used to calculate average and median κ-values on the basis of measurements. Data would not be suitable to calculate κ with equation (1), which is based on observation of water-conducting voids.

NMR Measurement of water in clay

R. F. Paetzold
U.S. Dept. of Agriculture, Soil Conservation Service
BARC-West, Rm. 139, Bldg. 007
Beltsville, Maryland 20705 USA
G. A. Matzkanin
Southwest Research Institute, P. O. Drawer 2851
San Antonio, Texas 78510 USA

Recent advances in nuclear magnetic resonance (NMR) technology, including improved electronics, emergence of pulsed NMR techniques, improvements in magnet materials and new designs of magnet configurations, have greatly expanded the usefulness of NMR. Once used primarily by chemists to learn about molecular structure, NMR now is receiving considerable attention as a tool for many practical applications, including detection of explosives and illegal drugs, quality control, medical diagnostics and water content measurement in a variety of materials. For practical problems, NMR is very attractive since it is both non-destructive and non-invasive.

A variety of measurements can be made with NMR. A spectrum of resonance frequencies can be obtained by sweeping through a range of magnetic field strengths while holding the radio-frequency input constant (or vice versa) to give a signal pattern based on the chemical shift caused by local magnetic field variations due to the effects of neighboring atoms. The same information can be obtained through a Fourier transform of pulsed NMR generated free induction decay signals. These techniques are most useful in identifying molecular structure and in detecting the presence of specific molecules or groups in a sample.

Spin-echo and free induction decay signals are produced by applying sequences of radio-frequency (RF) pulses of specific durations and intervals. These signals can be used to measure the quantity of a specific atomic nuclei, e.g., hydrogen. Application of certain pulse sequences can provide information about the spin-spin and spin-lattice relaxation constants of the atomic nuclei under investigation. These relaxation constants are functions of the local molecular environment and the physical state of the sample.

One area where NMR is particularly useful is the measurement and characterization of water in materials. For this application, free induction decay or spin-echo NMR signals from hydrogen nuclei in the water molecules are sensed and measured. Not only can the amount of water present be determined, but through the use of relaxation characteristics the physical state and local environment of the water molecules can be investigated. The NMR signal from liquid water can be distinguished from hydrogen NMR signals from other materials, such as organic matter, oil, coal, and even ice. NMR can be used to study the orientation of adsorbed water molecules on clays, molecular diffusion of water within soil pores and bulk flow of water in soil. Determination of molecular diffusion can be made from spin-echo measurements of the spin-spin relaxation times with a well defined magnetic gradient superimposed on the static magnetic field.

The objective of the research reported here was to develop an instrument capable of measuring volumetric soil water content using NMR techniques under field conditions. An NMR instrument was designed to continuously measure volumetric water content in the soil surface at ground speeds of up to 15 km/h. The instrument will to be used to gather ground truth data for remote sensing studies. This application requires a method that is relatively accurate, rapid and independent of soil type, organic matter content, texture and clay mineralogy.

Measurements of the relaxation constants of water in various soils were required for system design. These measurements were made at various water contents on small soil samples using a laboratory NMR instrument. The spin-lattice and spin-spin relaxation constants, T_1 and T_2, respectively, are the slope of the log of NMR signal intensity (free induction decay for T_1 and spin-echo for T_2) plotted against time between pulses (the pulse for T_2 is twice the duration of the pulse for T_1). For soil water, both the spin-lattice and spin-spin relaxation curves are multi-exponential. Figure 1 is a typical example. The multi-exponential nature of these relaxation curves are the result of the different structural orientation and the degree of restricted mobility of water molecules as a function of distance from the soil particle surfaces. Further, the curves indicate that there is little or no exchange among the water states.

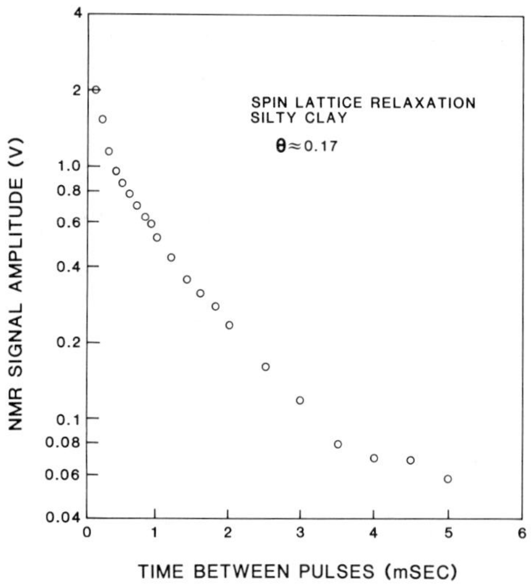

Figure 1. Typical spin-lattice relaxation for water in soil

Relaxation times for hydrogen associated with various forms of water are given in Table 1. The short spin-spin relaxation time for hydrogen in solids is a manifestation of tight molecular coupling and restricted motion. The long relaxation time constant for free water is due to the rapid molecular motion in the water.

Table 1. Spin-Lattice and Spin-Spin Relaxation Time Ranges for Various Samples

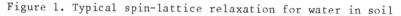

Sample	T_1		T_2
Loamy Fine Sand	400 μs - 40 ms	depending on water content	200 μs - 1 ms
Silty Clay	200 μs - 5 ms		130 - 700 μs
Tap Water	200 - 300 ms		200 - 300 ms
Pure Water	2 - 3 s		2 - 3 s
Hydrogen in Solids	1 ms - 10 s		30 - 40 μs

The relaxation information was used to design a pulse sequence for the field instrument to minimize the effects of hydrogen in solid forms, such as organic matter, and to allow the measurement of water in various soil types. The information was also used in the physical design of the magnet used in the instrument. The instrument was designed to measure soil water content at 3 depths, 38, 51, and 64 mm, using 2 different sensor magnet configurations.

Calibration measurements were made on large soil samples of various soil water contents. In addition, field measurements were made at various ground speeds from 0 to 17 km/h. A sandy loam soil with 5 levels of organic matter content and three pure clays were studied. A typical calibration curve is shown in Figure 2. These preliminary calibration curves appear to be independent of organic matter content, texture and clay mineralogy. Field tests show the measurements to be independent of ground speed up to 17 km/h. Table 2 gives a summary of the calibration results. Further tests are scheduled to include the results from other soil types and investigate salinity effects, if any, on sensor output.

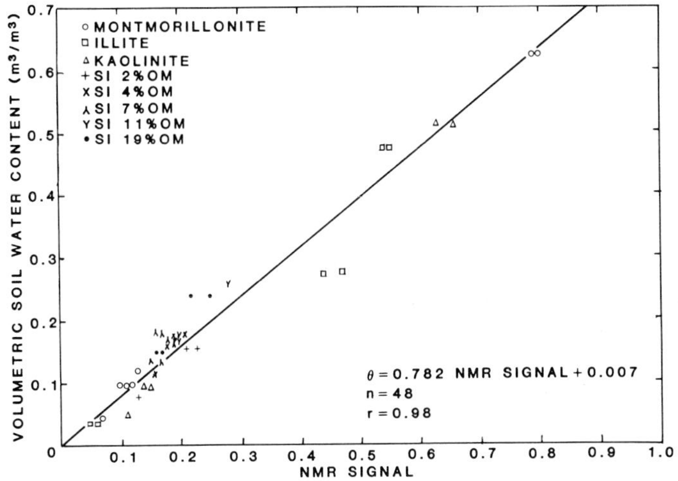

Figure 2. Calibration curve for NMR measurement of soil water content using the field instrument. This curve is for the 38 mm depth and flat sensor configuration

Table 2. NMR Calibration Summary[1]

Configuration	Depth (mm)	# Samples	Slope	Intercept	r
Extended	38	46	0.115	0.011	0.980
Extended	51	46	0.278	0.007	0.987
Extended	64	30	0.980	-0.018	0.982
Flat	38	48	0.782	0.007	0.978
Flat	51	49	1.731	-0.051	0.972
Flat	64	49	4.153	-0.242	0.946

1 Volumetric soil water content = slope x NMR signal + Intercept

The time response characteristics of tensiometers in heavy clay soils

G.D. Towner
Rothamsted Experimental Station,
Harpenden, England.

A tensiometer must necessarily take a finite time to register a change in pore-water pressure. The tensiometer gauge requires an exchange of water with the soil, and the rate of transfer of this water is controlled not only by the conductance of the tensiometer cup but also by the hydraulic conductivity and water capacity of the soil. Thus the oft-quoted time constant of pressure transducer tensiometers, obtained by measuring the response with the cup in free water, may be quite meaningless in heavy clays which characteristically have very low hydraulic conductivities. To avoid confusion, it is useful to refer to this latter value as the 'instrument time constant', and that effective in the soil as the '*in situ* time constant'.

The basic equations governing the response of a tensiometer in the soil were given by Towner (1980), but it is not possible in general to derive analytic solutions of them without making simplifying physical or mathematical approximations appropriate to particular cases. Thus an estimate of the time constant of a tensiometer embedded in a clay ped can be obtained by making the following simplifications. The cylindrical cup and the irregular shaped ped are replaced by a spherical cup at the centre of a spherical ped. The soil matrix is assumed to remain at constant water content and remain rigid during the passage of water from the surrounding fissures to the tensiometer. The influence of the gravity gradient is ignored. Hence the *in situ* time constant, τ', is given by (Towner, 1981),

$$\tau' = 1/CS + 1/(4\pi aSK) = \tau_{tensiometer} + \tau_{soil}$$

where

a = radius of the cup (assumed small compared with that of the ped).

C = conductance of the cup

S = sensitivity of the pressure gauge

K = hydraulic conductivity of the saturated soil.

$\tau_{tensiometer} = 1/CS$ (Klute and Gardner, 1962)

$\tau_{soil} = 1/(4\pi aSK)$.

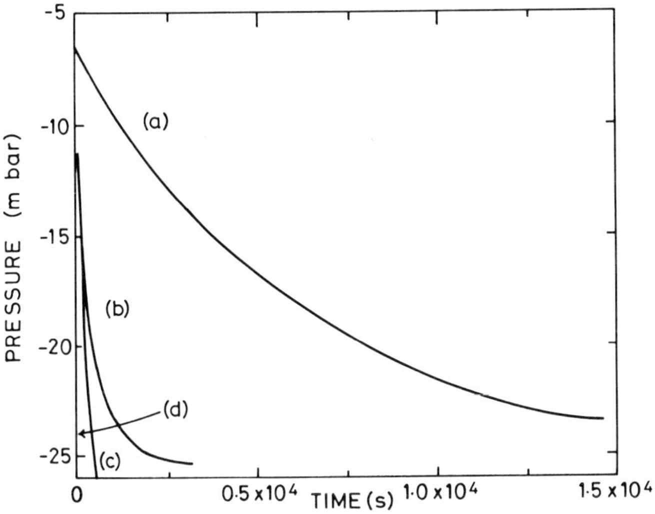

Figure 1. Time response curves: (a) In situ 'slow' tensiometer response to transducer pressure change; (b) As (a) for 'fast' tensiometer; (c) 'Slow' tensiometer response in water; (d) 'Fast' tensiometer response in water (indistinguishable from the vertical axis).

An experimental demonstration of the influence of the soil on the tensiometer response is illustrated in Figure 1. A small laboratory tensiometer cup was situated at the centre of a saturated clay paste contained in a cell, and could be switched to either of two pressure transducers of different sensitivities. The return of each transducer to equilibrium after perturbing the fluid pressure within it (Figure 1, curves (a) and (b)), is compared with the corresponding behaviour obtained with the cup in free water (curves (c) and (d) respectively).

Table I summarises the experimental details of a number of laboratory

experiments performed with several different types of small laboratory tensiometers embedded in different clays, together with the corresponding measured and calculated time constants. The theory tends to underestimate the measured values.
The tensiometer cups used in these experiments were small laboratory

Table I. Measured and calculated <u>in situ</u> time constants, τ'

Experiment No.	1	2	3	4	5
Soil*	a	a	b	c	d
K, mm/s	$4x10^{-6}$	$4x10^{-6}$	$8x10^{-2}$	-	$3x10^{-5}$
Tensiometer**	e_1	e_2	f_1	f_2	f_1
Radius a, mm	2.0	2.0	12.0	12.0	12.0
Sensitivity S, $mbar/mm^3$	$3x10^{-1}$	$4x10^0$	$2x10^{-2}$	$2x10^{-2}$	$2x10^{-2}$
τ, s (instrument)	14	180	270	540	270
τ', s (calculated)	$3x10^2$	$4x10^3$	$3x10^2$	-	$2x10^3$
τ', (measured)	$4x10^2$	$8x10^3$	$3x10^2$	$1x10^3$	$6x10^3$

* a-Kaolinite; b-loosely packed Kimmeridge clay; c-densely packed Kimmeridge clay; d-re-packed Kimmeridge clay.

** e_1-disc 'cup'/pressure transducer A; e_2-disc 'cup'/pressure transducer; B; f_1-cylindrical cup 1/water manometer; f_2-cylindrical cup 2/water manometer.

Table II. Theoretical values of time constants for a representative tensiometer in soils of various hydraulic conductivities

K, mm/s	τ', s
10^{-5}	$6x10^0$
10^{-6}	$6x10^1$
10^{-7}	$6x10^2$

types, not typical of those used in the field. The following set of data may be considered representative of rapid recording field systems, and are used in the calculations following: $C = 0.2\ mm^3\ s^{-1}\ mbar^{-1}$; $S = 10\ mbar/mm^3$; $a = 13$ mm (equivalent sphere). The instrument time constant is thus 0.5 seconds. Table II gives the *in situ* values calculated from equation (1) for a range of hydraulic conductivities appropriate to clay soils. The relevant value for K is that measured in the ped. Since equilibrium is only 99% complete by $t = 4.6\tau'$, the specified tensiometer, not atypical, would be virtually useless for following changes in the heaviest soils, and may not respond rapidly enough in field scanning systems that allow only a fraction of a minute for scanning each tensiometer cup.

To decrease the response time in soils of low conductivity such as heavy clays, it is necessary to increase the gauge sensitivity. Some attempts have been made by the author to develop a null-point pressure transducer tensiometer. The movement of the diaphragm in response to the changing soil-water pressure is prevented by applying a counter pressure to the other side of the diaphragm, so that the effective sensitivity tends to become infinite.

The analysis indicated that the soil properties must be taken into account in selecting suitable devices for use in heavy clay soils. Under natural conditions, the soil will swell and shrink and may become unsaturated, so that the above simple theory cannot be applied, except as an approximation. In so far as a time constant can be defined for such conditions, it will be larger than that given by the equation. In practice, the available choice of design parameters is limited. Thus one is probably only able to assess the performance of a possible system for a given purpose or interpret the value of observations. It is therefore recommended that the time constant should always be measured *in situ*, preferably over the expected water-content range.

References

Klute, A. and W.R. Gardner 1962. Tensiometer response time. Soil Sci., 93: 204-207.

Towner, G.D. 1980. Theory of time response of tensiometers. J. Soil Sci., 31: 607-621.

Towner, G.D. 1981. The response of tensiometers embedded in saturated soil peds of low hydraulic conductivity. J. agric. Engng Res., 26: 541-549.

The moisture characteristic of heavy clay soils

L. Stroosnijder
G.H. Bolt

In dealing with problems of flow of water in unsaturated soils, there is need for a relationship between the volume fraction of water and an appropriate component of the water potential. In rigid soils, the relation between the matric potential, p_m (Pa), and the volume fraction of water θ (m^3 water per m^3 bulk soil), is named the moisture characteristic (MC) or retentivity curve of the soil. If the soil air pressure is atmospheric (which usually is the case) p_m can be replaced by p_t, i.e. the pressure as can be determined experimentally with a tensiometer. Apart from hysteresis, the p_m-θ relation is a unique one. It shows the soil's capacity for water uptake or release, $d\theta/dp_m$ and also the p_m range for which the soil remains saturated, i.e. $d\theta/dp_m = 0$.

For heavy clay soils the concept of the MC is more complicated. The tensiometer pressure (of which the gradient together with the gradient of the gravitational potential forms the driving force for water flow as in rigid soils) now consist of two component pressures, i.e. the wetness pressure p_w and the envelop pressure p_e. The latter is the result of a load P on the soil. Because of the absence of sufficient contact points between solid phase particles, there will be insufficient reaction forces between those solid particles which implies that a load P on the clay will influence the pressure potential (i.e. p_t) of the water.

If one measures the MC of a heavy clay soil with the equipment normally used for rigid soils, one obtains a p_t-ν curve for load P=0 as schematically given in fig. 1.

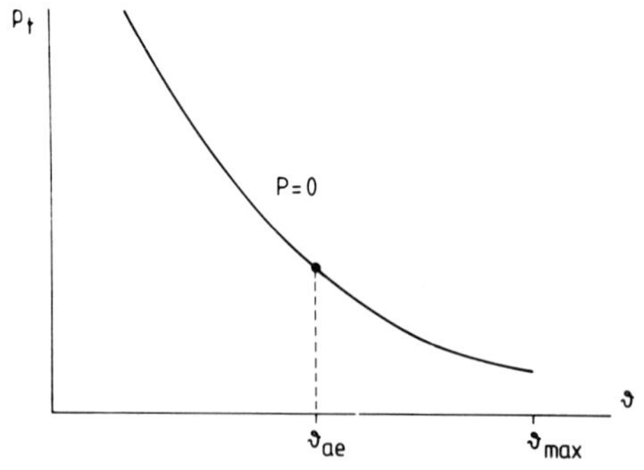

Fig.1 The Moisture Characteristic of a heavy clay soil. ν is the moisture ratio (m^3 water per m^3 solid phase), p_t is the tensiometer pressure (Pa).

The above MC is named the unloaded MC. There is a maximum ν value at which p_t is still < 0. This indicates that there is a remainder of the swelling pressure π which is not able to let the clay swell further due to crosslinking forces between the clay plates.

Another typical aspect of this MC is the shrinking of the soil matrix. In 'drying' from ν_{max} to the air entry value, ν_{ae}, the soil remains saturated. Note that in this range $d\nu/dp_t$ is not equal to zero as is the case in rigid soils. This type of shrinkage is named proportional or normal shrinkage. Shrinkage continues (i.e. the void ratio, e, (m^3 voids per m^3 solid phase) decreases) upon further drying below ν_{ae}. This is named unsaturated shrinkage. Obviously, this MC applies only for P=0, e.g. at the soil surface.

As to the situation below the soil surface where P≠0, the relevant information may be obtained with two different experimental methods, viz.

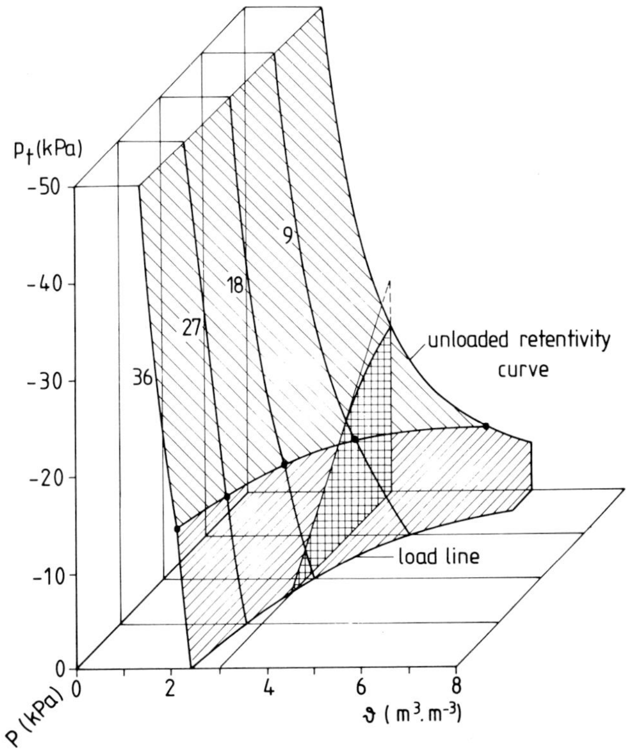

Fig. 2. Three-dimensional representation of a hypothetical family of retentivity curves.

a) One measures (p_t- ν) curves for many P values. The complete MC then consists of a $\nu(p_t$-P) plane as shown in fig. 2.

A special feature of the above MC is the $(\nu -P)_{p_t=0}$ curve which is named the load line, i.e. the reaction of ν on load under the outflow of free water (p_t=0). This is a curve commonly determined in civil engineering. As with 'drying' the soil, loading the soil also causes shrinkage of the soil matrix. In the extreme case of the load line (p_t=0), this shrinkage is named consolidation. If the full $\nu(p_t,P)$ plane is measured, the contribution of the load P to the tensiometer pressure p_t is automatically taken into account. This is shown in fig. 2 where a cross-section for =3 illustrates the phenomenological expression for p_e, i.e. $p_e = \int_0^P \left(\frac{\partial p}{\partial P}\right)_\nu d\alpha$. The above implies that the MC plane is also fully determined if besides measurement of the unloaded MC, $p_e(\nu,P)$ is obtained from a separate measurement. This leads to the second experimental technique for the determination of the MC plane.

b) It was proven thermodynamically by Groenevelt and Bolt (1972) that p_e can also be written as: $p_e = \int_0^P \left(\frac{\partial e}{\partial \nu}\right)_\nu d\alpha$ sothat the e(ν,P) plane which is named the Shrinkage Characteristic (SC) provides the same information as the extension of the $(p_t - \nu)_{P=0}$ curve into a $\nu(p_t,P)$ plane. A hypothetical e(ν,P) plane is given in fig. 3 (as a projection on the e(ν) plane, showing the result for different values of the overburden P).

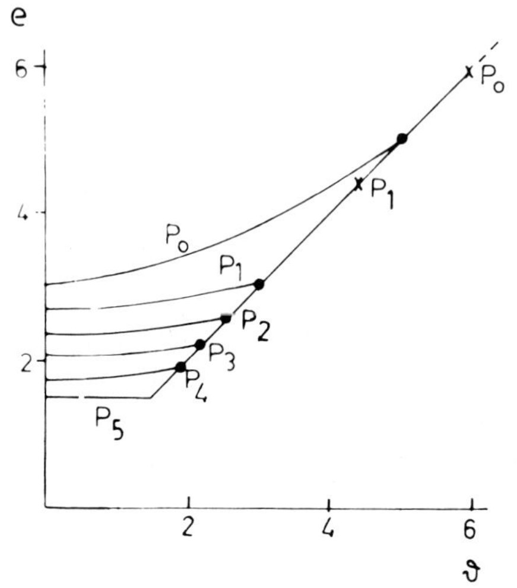

Fig. 3. Hypothetical shrinkage characteristic. •——x: Part of curve, for constant P, where normal shrinkage prevails. P=0 is indicated as P_0.

The above has made clear that there are 4 variables which describe shrinkage and drying of heavy clay soil: ν, p_t, P and e. The plane of the MC includes the effect of P on p_t. However, if the unloaded MC is known, the complete MC can also be obtained from a measured SC. The latter measurement has the advantage that also information about e is obtained sothat then all variables are sufficiently known.
It was tried to measure both the MC and the SC of a margalite clay (Koenigs, 1961) in order to verify that both give the same effect of P on p_t, i.e. p_e. The MC was measured on small samples 4.5 cm in diameter and 1 cm high. These were covered with a course porous plate on which a weight could be placed. The sample was then placed on a fine porous plate. The matric potential of the water in the pores of that plate could be decreased by means of a hanging water column. Some preliminary results are given in fig. 4.

The SC of the margalite clay was determined on samples of 12 cm in diameter and 3 cm high in an experimental set-up shown in fig. 5. The samples in this set-up were dried in an oven at 50°C during 12 days. On 8 days non-destructive measurements of ν and e were made using a double-beam gamma scanner (Stroosnijder and De Swart, 1974). Results of an unloaded and a loaded sample are given in fig. 6 and 7 respectively.

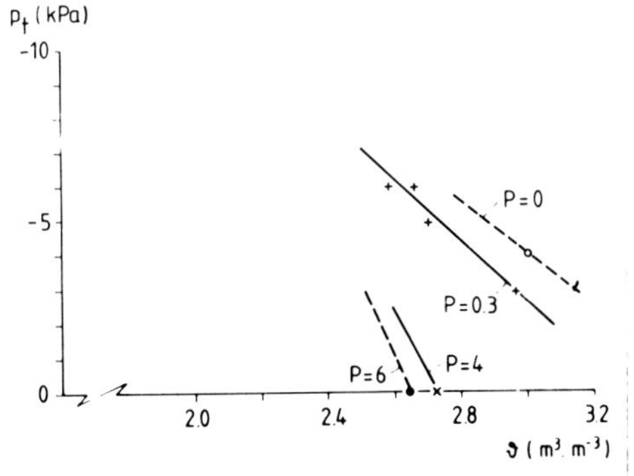

Fig. 4. Some points from the retentivity curve of margalite clay.

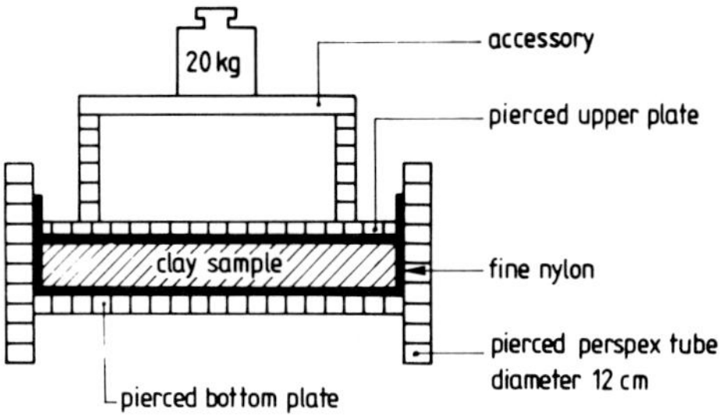

Fig. 5. Cross section of a ring used to estimate the shrinkage characteristics.

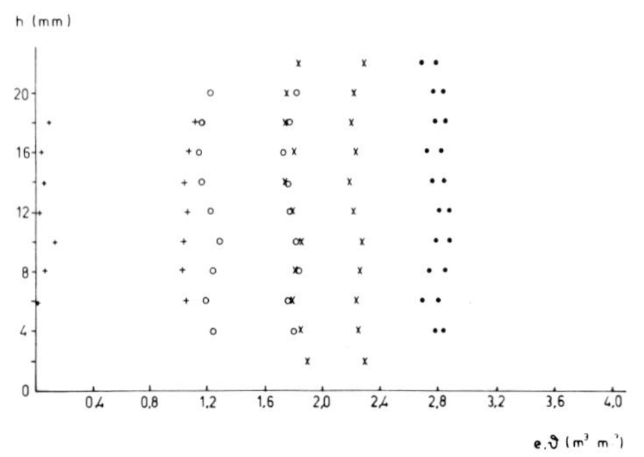

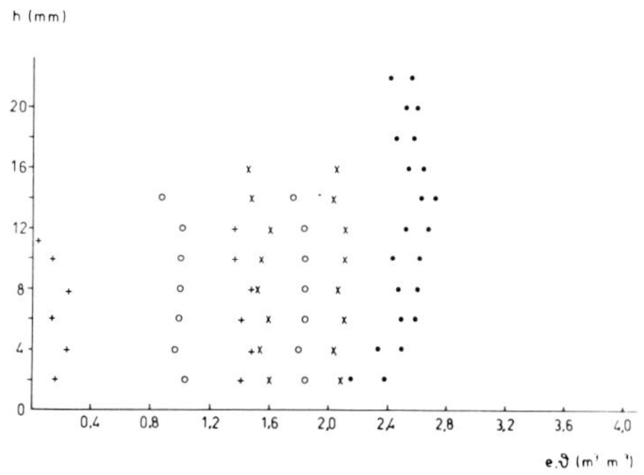

Fig. 6. Drying of an unloaded margalite clay. Points for e are on the right of those for ν.
●: Initial situation. x: After drying for 1 d at 50°C.
o: After drying for 2 d. +: After drying for 7 d.

Fig. 7. Drying of a margalite sample under load, P = 17.8 kPa. Key as in Fig. 6.

The SC which could be calculated from the results as given in fig. 6 and 7 is presented in fig. 8. At ν = 1.8 the curves for P = 0 and for P = 17.8 kPa cross. This is not conform theoretical expectation. The latter is due to the fact that the measured e consisted of the total of small voids between the clay plates and which probably are still saturated and a circular crack that circumferenced the shrunken sample. If this crack is not included in e, the curve for P = 17.8 drops and will lie below the curve for P = 0 as theoretically is expected. The need for such a correction clearly shows the 'problem' that cracks give in theoretical considerations.

The above measurements are only preliminary and do not suffice for a true verification of the theory as presented above. Further experimental data are planned with an instrument as shown in fig. 9 where the MC and the SC can be measured simultaneously. With piston 2 a load P1 can be apllied on the soil sample 1 which is on a water saturated porous plate 3 sothat with a gas pressure P2 the tensiometer pressure p_t can be regulated. ν and e can be measured simultaneously and non-destructive with dual gamma transmission. If water can flow both out of and into the sample also hysteresis can be measured with the above instrument.

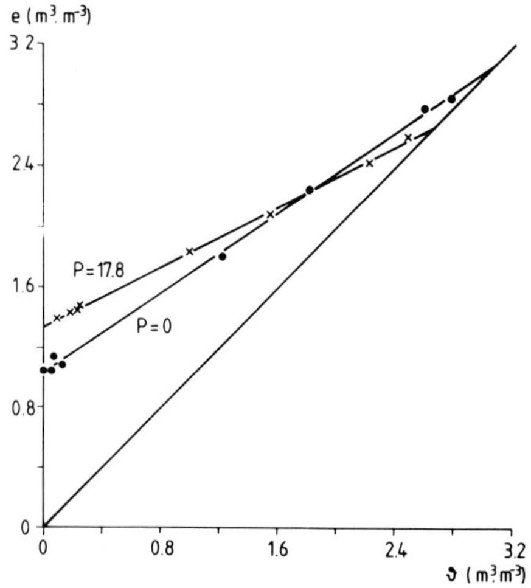

Fig. 8. The shrinkage characteristics of margalite clay as calculated from Fig. 6 and 7.

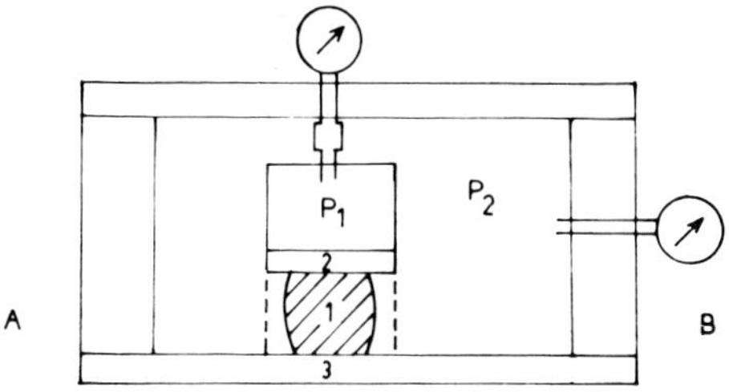

Fig. 9. Scheme of an instrument for estimating retentivity curves of swelling soils. 1: Soil sample. 2: Air-tight piston 3: Pourous membrane. P1: Air pressure estimating envelope pressure. P2: Air pressure estimating tensiometer-pressure. AB: Direction of gamma transmission estimating e and ν continuously.

References.

Groenevelt, P.H. and G.H. Bolt, 1972. Water retention in soil. Soil Sci. 113: 238-245.

Koenigs, F.F.R., 1961. The mechanical stability of clay soils as influenced by the moisture conditions and some other factors. Versl. Landbouwk. Onderz. nr. 67 Wageningen.

Stroosnijder, L. and J.G. de Swart, 1974. Column scanning with simultaneous use of ^{241}Am and ^{137}Cs gamma radiation. Soil Sci. 118: 61-69.

Computer optimization of a heavy soil drainage system by two-dimensional saturated-unsaturated water flow modeling

N.Shopsky, I.Nickolov, E.Doneva
Department of Soil Physics, N.Poushkarov Institute of Soil Science and Yield Prediction, Sofia 1080, P.O.Box 1369,Bulgaria

In Bulgaria more than 10% of the area under cultivation is occupied by surface flooded soils. These areas have low permeabilities in the upper layer. For high and stable yields to be obtained scientific and engineer problems have to be solved.

The objective of the present investigation is the optimization of an amelioration system covering an area of 2 000 ha of surface flooded area with heavy clay soils in the Mikhailovgrad district of North Bulgaria.

In this optimization a method of mathematical modelling of soil water transfer was used. It allowed a selection among all possible variations of the grid system to attain the best technical characteristics. Then the optimal variant was singled out combining good hydraulic characteristics with suitable economic indexes.

Moisture transfer was investigated by solving a two-dimensional equation /1,2,3/ for water movement in a saturated-unsaturated water medium:

$$\frac{\partial W}{\partial P}\,\frac{\partial H}{\partial t} = \frac{\partial}{\partial x}\left(K\,\frac{\partial H}{\partial x}\right) + \frac{\partial}{\partial y}\left(K\,\frac{\partial H}{\partial y}\right) \qquad (1)$$

where $H = P + y$ = total hydraulic potential (m); P = pressure head (m); W = water content (m^3/m^3); K = hydraulic conductivity (md^{-1}); x, y = axes of co-ordinate.

The soil moisture characteristic explains the relationship between water content (W) and pressure head (P). The following relationship /4/ was used in this study:

$$W = \frac{W_S}{1 + \left(\frac{P}{A}\right)^B} \qquad (2)$$

where W_S = saturated water content; A and B are empirical coefficients.

The value of hydraulic conductivity K depends on the water content W, hence on the pressure head P. The experimental data could be approximated with sufficient accuracy by /5/:

$$K = \frac{K_S}{1 + \left(\frac{P}{M}\right)^N} \qquad (3)$$

where K_S = hydraulic conductivity at saturation (K_{sat}); M and N = empirical coefficients.

Equation (1), modelling water movement in the drained area, was solved for specific boundary conditions taking into account rainfall or irrigation recharge, discharge to the drainage collector, the impermeability of the lower layers and the symmetry of the system.

The initial condition for nonstationary problems assumed a condition corresponding with long drainage in the absence of moisture flow across the soil surface.

Stationary problems were solved using relationships defined in equations (2) and (3) and equation (1) in which the time derivative was equal to zero. The nonlinear partial differential equation (1) describing water movement was solved numerically by the finite element method /6/ using iterative techniques.

For collection of sufficient and representative information physical investigations of two different types of soils were made together with parametrical relationships in equation (1).

The moisture characteristic was determined by means of a tension plate assembly /7/ for pressure heads higher than -1 m and with a pressure membrane apparatus /7,8/ for lower pressure heads. From the experimental data for the upper layer coefficients in equation (2) were determined by the least square analysis, yielding:

A = -88 /m/ B = 0.369 W_S = 0.464 /m^3/m^3/

The Gamma apparatus /8/ was used to obtain the relationship between hydraulic conductivity (K) and the pressure head or water content. Experimental data for the upper layer yielded the following values for equation (3) coefficients:

K_S = 0.382 /m/day/ M = -0.426 /m/ N = 2.1585

The lower layer below 40 cm had very low K_{sat} values (0.016 - 0.024 m/day).

Calculations were made for three types of drain grid system: (1) tube drainage, (2) tube drainage combined with deep loosening at 60, 80 or 100 cm below surface and (3) mole drainage combined with relatively widely spaced tube drainage. Many possible drain spacings and depths were taken into account together with changes in K_{sat} following deep loosening /9/. The standard criterion for drainage efficiency of

Table 1. Basic physical and chemical characteristics of strongly leached chernozem of Northwestern Bulgaria

Horizon	A	B
Depth (cm)	0 - 40	40 - 120
Bulk density (g/cm^3)	1.34	1.42
Total porosity, %	48.49	46.54
Particle size distribution, %		
Loss in HCl-treatment	1.4	1.8
1.000 - 0.250 mm	6.2	5.2
0.250 - 0.050 mm	12.7	11.0
0.050 - 0.010 mm	23.3	20.6
0.010 - 0.005 mm	7.7	7.1
0.005 - 0.001 mm	9.5	8.1
less than 0.001 mm	39.2	46.2
Saturated water content, %	46.40	46.10
Field capacity, %	37.61	41.07
Wilting percentage, %	22.32	29.48
Hydraulic conductivity at saturation (m/day)	0.382	0.0256
pH(KCl)	5.0	5.6
$CaCO_3$, %	0.000	0.000
Total N, %	0.132	0.064
Total P, %	0.181	0.099
Organic matter, %	2.30	1.07

Table 2. Summary of upper and lower limits for optimization runs

Type	Spacing (m)	Depth below surface (m)	Deep loosening	
			Depth below surface (cm)	Period (years)
1. Tube drainage	10-16	110	-	-
2. Tube drainage combined with deep loosening	16-30	90-130	60-100	2-5
3. Mole drainage combined with tube collectors at 80 (m)	4-10	45-70	-	-

the system is the position of the water table under stationary conditions. The efficiency of selected variants of the system during intensive rainfall and during dry periods was tested for nonstationary conditions. The final index for the system was the annual per ha expenditure for installation and maintenance.

Mole drainage was defined as a basic variant with 4-5 m drain spacing and collectors at 80 m. The gradients were more than 2-3°/oo. For the rest of the area without marked sloping of the ground the gradients were less than 2-3°/oo, and tube drainage was envisaged combined with deep loosening. The results of the study show that the best drainage system consists of drains at a depth of 1.3 m below surface with spacings of 25 m. Drains are placed in sand and gravel filters. The soil should be loosened every four years. Thus, a management system is obtained that optimizes hydraulic and economic aspects.

REFERENCES

1. Bouwer H.,Little W.C. A unifying numerical solution for two-dimensional steady flow problems in porous media with an electrical resistance network. Soil Sci.Soc.Am.Proc.,vol.23,pp.91-96 (1959).

2. Rubin J.,Theoretical analysis of two dimensional, transient flow of water in unsaturated and partly saturated soils. Soil Sci.Soc. Am.Proc., vol.32, pp.607-615 (1968).

3. Hornberger G.M.,Remson I.,Fungaroli A.A. Numeric Studies of a Composite Soil Moisture Ground Water System, Water Resour.Res.,vol.5,pp.797-802,1969.

4. Vauclin M.,Khanji J.,Vachaud G. Etude experimentale et numerique du drainage et de la recharge des nappes à surface libre,avec prise en compte de la zone non saturée. Journ.de Mecanique, vol.15,pp.307-348,1976.

5. Gardner W.R. Some steady state solutions of the unsaturated moisture flow equation with application of evaporation from soil and water table. Soil Sci.,vol.85,pp.228-232, 1958.

6. Zienkiewicz O.C. The finite element method in engineering science. McGraw Hill, London, 1971.

7. Globus A.M. Experimental soil hydrophysics, Gidrometeoizdat, Leningrad, 1969 (in Russian).

8. Nerpin S.,Khlopotenkov E. Methods and devices to determine the main parameters of equations, modelling the process of moisture transport to zone of incomplete saturation. Agrophysical Studies, v.1, pp.161-167, Sofia, 1973 (in Russian).

9. Maslov B.S. Deep loosening of soils, experience and problems of science. Hydrotechniques and melioration. No.7,pp.28-33, 1979 (in Russian).

Use of the neutron probe and tensiometers to monitor gravity irrigation in soils of low permeability

J.M. Allard and O. Auriol
Engineers in Sogreah's
Agronomy Department
Grenoble - France

1 Introduction

Irrigation tests conducted in the Kirkuk region (North Iraq) were aimed at determining strip irrigation characteristics in an extensive rain-fed cultivated area where practically no local references exist in the field of irrigation. The tests were carried out throughout the 1982 irrigation season.
An attempt was made to obtain homogeneous infiltration and distribution of the water over the entire plot, while at the same time limiting losses by percolation and surface runoff. The soil moisture content was monitored with a neutron-probe and tensiometers.

2 Natural environment

The soils of the Kirkuk plain are generally fairly homogeneous silty wind-blown deposits with massive structure when dry; under irrigation, they produce a hard surface crust with the formation of contraction cracks after drying out.
The structural instability observed results mainly from the physical properties of the deposits (the 0-50μ volume fraction accounts for more than 80 % of the total in most profiles observed). The soils on which trials were made are not saline (conductivity of saturation extract < 1 mmhos).

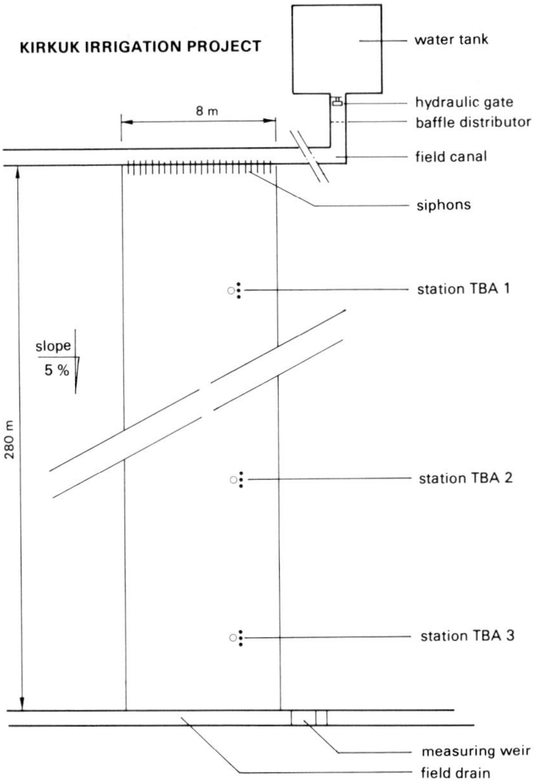

Fig. 1 Location of the experimental plot

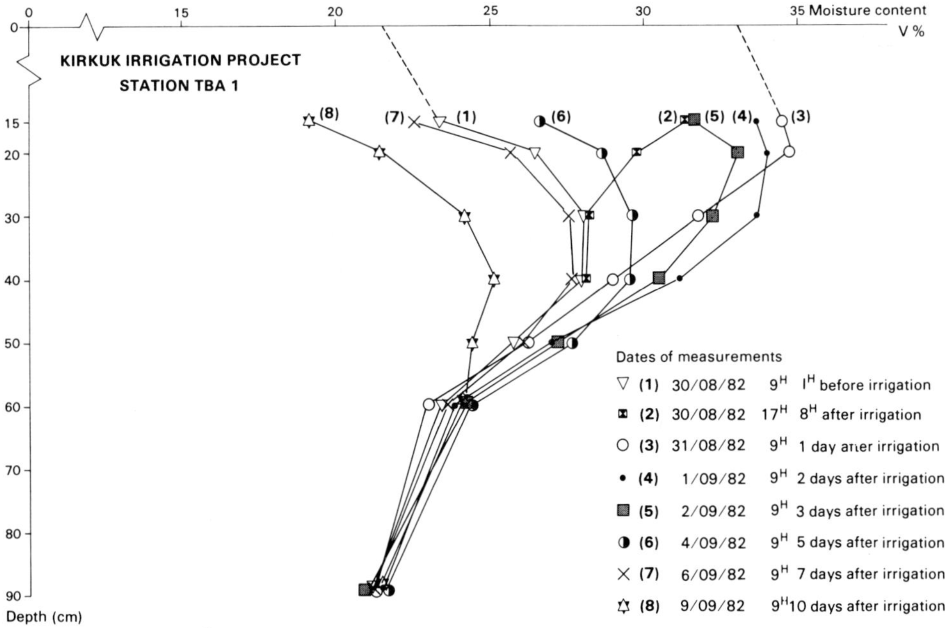

Fig. 2 Variation of moisture front between two irrigations

The climate is arid with considerable water deficiencies. Winter rains (350 mm) are not sufficient to sustain anything other than extensive cereal crop production. The very high evaporation rate from June to October (15 to 20 mm/day) effectively precludes any summer crop and compromises long-cycle winter crops without irrigation. Evapotranspiration is higher than average rainfall nine months out of twelve (annual Penman ETP = 2260 mm).

3 Experimental conditions (Figure 1)

The variation in soil water content was observed in an 8 m wide, 280 m long strip with a slope of 5‰ planted with beetroot. The following measurements were taken in the experimental plot:

- intake flow, controlled by baffle distributors,
- surplus runoff flow by a notched weir,
- advance of the water front in the field,
- infiltrated volume of water and evapotranspiration between two irrigations assessed from neutron-probe humidity profiles. The measurements were taken at ten depths in each profile (15, 20, 30, 40, 50, 60, 70, 80, 90, 120 cm).

Three measurement stations were located in the observation field. The flow direction was assessed from tensiometers placed at different depths (15, 30, 60, 90 and 120 cm). The simultaneous neutron-probe and tensiometer measurements were performed at regular intervals: one hour and eight hours after irrigation on the first day and, on the following days, at the same time as the soil was irrigated on the first day.

The field was irrigated in two stages: the first stage involved a high initial discharge to ensure sufficiently rapid advance of the sheet of water over the soil surface, followed by a second stage with a small discharge and of variable application time in order to obtain a higher volume of infiltrated water.

From subsequent water content readings the optimum irrigation frequencies were determined: in summer one application every five days. However, the last two irrigations in September were made with 10-day intervals as a result of pumping station difficulties. The effects of this change are very noticeable on the graphs of time-dependent water content at three depths shown in Fig. 4.

4 Assessment of water profiles (Figure 2 and Figure 3)

4.1 During an irrigation period, two stages can be observed:

- An infiltration stage over the first 24 hours following irrigation. The water infiltrates rapidly first of all to a depth of 20 to 30 cm (2 mm/h) owing to the wide and deep contraction cracks, and then more slowly (< 1 mm/h) once these cracks have closed. Eight hours after the start of irrigation, the soil condition ranges from saturated at ground level (40% humidity by volume), to 28-30% at 30 cm depth. Twenty-four hours later, the humidity front reaches the maximum depth of 40-50 cm.
- An evaporation stage, starting on the second day, accentuates with the formation of contraction cracks. The initial water profile returns after six days: all the infiltrated water has been extracted by evapotranspiration.

The water balance indicates a small amount of infiltrated water (about 30 mm) and corresponding high losses by surface runoff.
At depths in excess of 50-60 cm, there is practically no variation in water content.

There may be two combined reasons for stoppage of water infiltration at 50 to 60 cm:
- presence of a more compact and less permeable horizon at this depth, created by slight leaching of clayey colloids by winter rainfall,

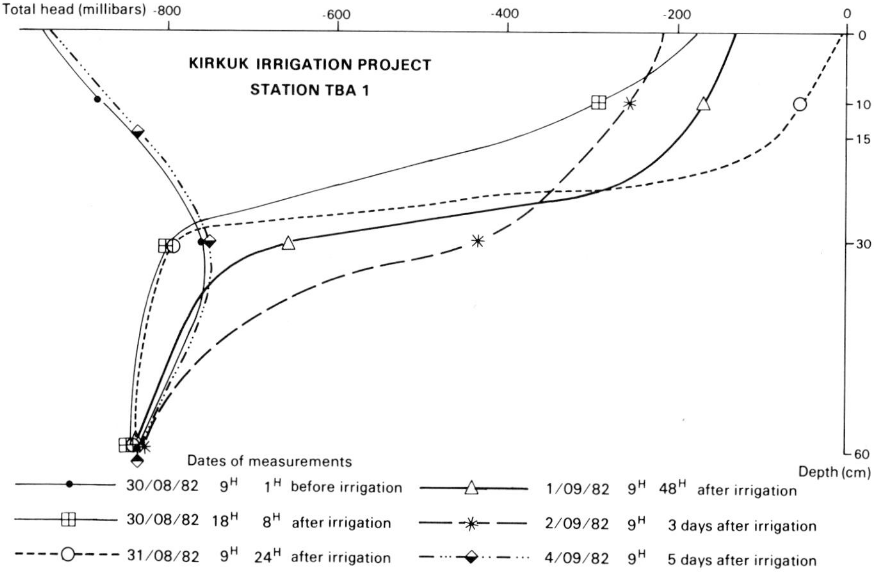

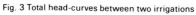

Fig. 3 Total head-curves between two irrigations

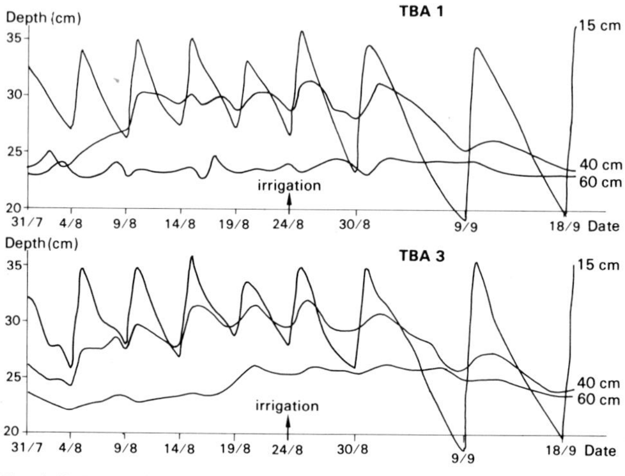

Fig. 4 Variation of moisture at several depths

- reforming of cracks by the second day after irrigation under the effect of evaporation; this is why evaporation is preponderant in relation to infiltration and thus prevents water redistribution at depth.

4.2

During the irrigation season, very sudden changes in surface humidity related to the frequency of inflows may be observed together with a sustained and regular increase as far as 40 cm depth, and then complete absence of change to 60 cm depth, thereby confirming the small thickness of the soil profiles affected by irrigation (Figure 4).

4.3

Water balances show that the total infiltrated depth, essentially in the first 40 cm of soil, is not very high (25 to 40 mm) and is not dependent to any great extent on the total irrigation duration.

Water infiltrates above all at the start of irrigation when the soil is dry and cracked on the surface.
As soon as the first 25 to 30 cm are saturated, the contraction cracks close and the infiltration rate slows down dramatically.
During the second stage, losses are extremely high: 60 to 70% of the water is lost through surface runoff. The fact of increasing submersion time in order to achieve better infiltration therefore leads to enormous losses for a poor result (a few extra mm of infiltration).
Irrigation efficiency at field level is low: 0.4 to 0.45. However, this value improves when the hydraulic conductivity increases (case of the two irrigation applications in September when efficiencies of 0.58 and 0.72 were obtained).

5 Conclusions

Using a neutron probe, it has been possible to determine the behaviour of a soil of low permeability under irrigation and to assess its change during the irrigation season thanks to regular in-situ soil humidity measurements.

The conclusions that can be drawn from these experiments concern:

- the irrigation applications which, under present conditions, should be more frequent and of shorter duration because of the low soil permeability,
- working of the soil, subsoiling in particular, to a depth of 60 to 80 cm before planting crops in the fields, which improves water infiltration at depth,
- the volumes of irrigation water to be provided in the future irrigation area which will have to be revised upwards because of the low efficiency of field irrigation.

References

Vachaud, G. 1968. Contribution à l'étude des problèmes d'écoulement en milieux poreux non saturés. Thèse doctorat d'Etat, Univers. Grenoble, Institut de Mécanique, 163 pp.

Vachaud, G. and J.M. Royer 1974. Détermination directe de l'évapotranspiration et de l'infiltration par la mesure des teneurs en eau et des succions. Bull, Sci-Hydrol, AISH 17(3), 319-336 pp.

Raats, P.A.C. and W.R. Gardner 1974. Mouvements of water in the unsaturated zone near water table. American Society of Agronomy, Reprint from Agronomy n° 17, Drainage for Agriculture, 311-405 pp.

Daudet, F.A. and G. Vachaud 1977. Mesure neutronique du stock d'eau et de ses variations. Application à la détermination du bilan hydrique. Ann. Agro, 28(5), 503-519 pp.

Mechergui, M. and J. Vieillefon 1981. Utilisation des méthodes neutroniques et tensiométriques pour l'étude du régime hydrique en sol non saturé en présence d'une nappe peu profonde. Bull. du Groupe Français d'Humidité Neutronique 10, 29-52 pp.

APPENDIX

WATER CONTENT BALANCE

Date of irrigation	Initial water content (%) at 25 cm	(1) Dose supplied	(2) Runoff losses	(3) Evaporation during irrigation (mm)	(4) Infiltration depth 1-(2+3) (mm)	(5) Infiltration depth estimated from water content balances* (mm)	(5)/(1) Efficiency
30/7/82	26	122 m^3 (53 mm)	46 m^3 (20 mm)	10	23	26	0.49
4/8/82	28	130 m^3 (58 mm)	36 m^3 (16 mm)	10	32	24	0.41
9/8/82	29	133 m^3 (59 mm)	25 m^3 (11 mm)	10	38	23	0.40
14/8/82	29	144 m^3 (64 mm)	39 m^3 (17 mm)	10	37	22	0.34
19/8/82	29.5	169 m^3 (75 mm)	58 m^3 (26 mm)	10	39	22	0.29
24/8/82	29.5	134 m^3 (60 mm)	46 m^3 (20 mm)	10	30	24	0.40
30/8/82	28	173 m^3 (77 mm)	54 m^3 (24 mm)	10	43	25	0.32
9/9/82	24	119 m^3 (53 mm)	25 m^3 (11 mm)	10	32	38	0.72
19/9/82	24	143 m^3 (64 mm)	49 m^3 (22 mm)	10	32	37	0.58

* Average value of the three field measurements

The role of structure for the compressibility and trafficability of heavy clay soils

R. Horn
Lehrstuhl für Bodenkunde und Bodengeographie, Abteilung Bodenphysik, Postfach 3008, 8580 Bayreuth, West Germany

1 Introduction

Cultivation methods are being evaluated critically because increasing loads of agricultural machinery often induce soil compaction or destruction of the soil structure by kneading under wet conditions and thus worsening growing conditions for plants. Such problems are frequent in clayey soils which, because of their high clay content, may be stabilized by swelling and shrinkage (Kuntze (1965), Horn (1981)). However clay soils may also be sensitive to changes in aggregate stability (Satyavaniya et al. (1971), Horn (1978)). This paper presents data on the quantification of the mechanical compressibility of clay soil, because in the literature such data are rare.

2 Material and methods

Investigations were made in a vertisol, formed out of mesozoic parent material (pH 7.5, CEC: 33 mval/100 g). Soil cores (250 cm^3) and soil monoliths (0.0013 m^3) were taken under natural conditions. Some samples were homogenized and cylinders were refilled. All soil cores were irrigated and desorbed at suctions of 60 and 300 mbar.

In addition a few mechanically homogenized soil samples were frozen for elucidating the freezing effect on the aggregation process.

All differently treated soil samples were investigated as follows:

(1) The compression-settlement behaviour (load range: 0 - 15 N/cm^2) was measured whereby the changes in water pressure in dependence of time and load were registered. Thus the preconsolidation load could be computed and the saturated hydraulic conductivity in dependence of the compression could be measured. The value of the preconsolidation load indicates that load range up to which the soil sample is overconsolidated, $\pm$ incompressible and thus stable while exceeding that value primary consolidation starts.

(2) The spatial pressure transmission was measured at different surcharges by strain gages (see Horn 1980) and the concentration factors V_K were determined by

$$V_K = \frac{\log \left(\frac{\sigma_o}{\sigma_o - \sigma_z}\right)^2}{\log \left[\left(\frac{R}{z}\right)^2 + 1\right]} \qquad \text{(eq. 1)}$$

where σ_o is the vertical load x 10^4 (Pa) at the top of the soil, σ_z is the normal stress x 10^4 (Pa) in the depth z (cm) below the surface, R is radius (cm) of the compressed surface area. This factor describes the form of isobars in the soil, and it increases as soils become softer (for further information see Newmark (1942) and Horn (1981)).

(3) The depth function of the saturated hydraulic conductivity in dependence of the vertical load at the top of the soil could be calculated by combining the transformed equation 1.

$$\sigma_z = \sigma_o \left[1 - \left(\frac{z^2}{R^2 + z} \right)^{V_{K/2}} \right] \quad \text{(eq. 2)}$$

(abbreviations see equation 1)

with the load dependent values of the saturated hydraulic conductivity.

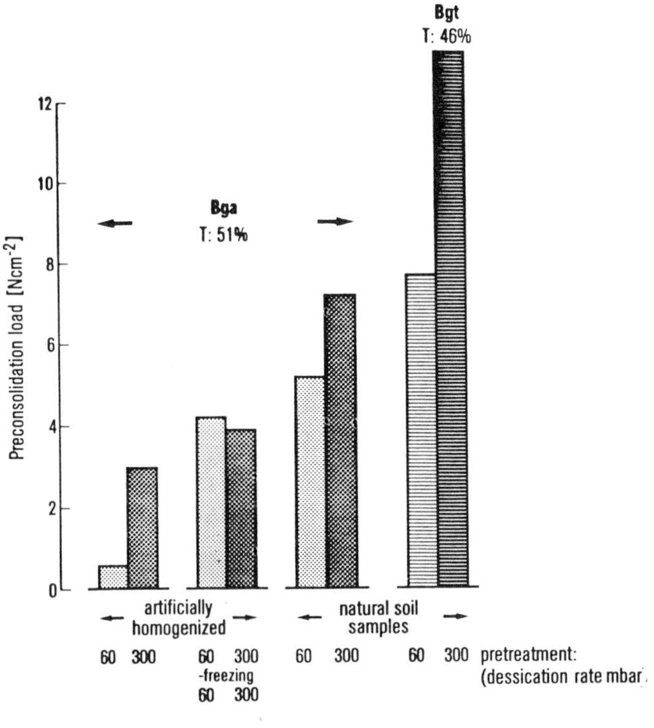

Figure 1: Variation of the preconsolidation load (N cm^{-2} ≙ 10^4 Pa of two differently pretreated clayey soil horizons (B_{ga} identical to B_{gw}, and B_{gt} (according to FAO nomenclature), T = clay content (%))

While the values of artificially homogenized soil samples are identical with the degree of desiccation, freezing as well as repeated drying and wetting increases the stability of the soils up to values measured in the natural soil samples.

Beyond it that change of the stability can be derived from the values, summarized in the general stress equation of Terzaghi and partly shown in figure 2.

$$\sigma = \sigma' + u \qquad \text{(eq. 3)}$$

where σ is the total stress in the soil

σ' is the effective stress, acting between the solid phase and

u is the pore water pressure.

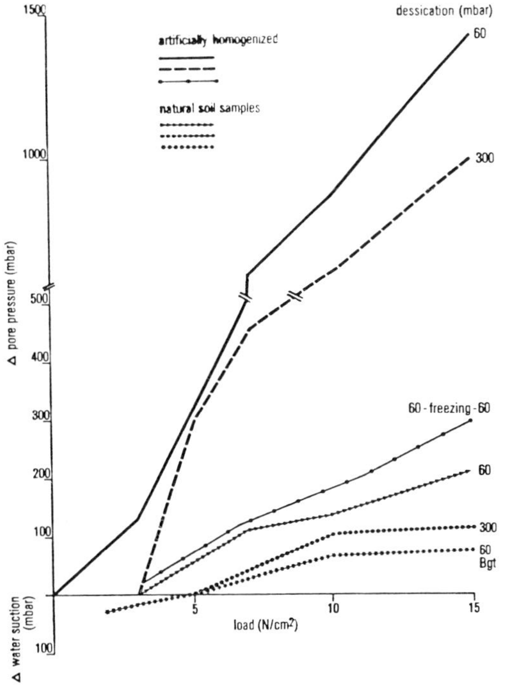

Figure 2: Change of the water suction and/or pore pressure (mbar) with the load ($N/cm^2 \mathrel{\hat{=}} 10^4$ Pa) of differently pretreated clayey soil samples

The higher the value of the maximum pore water pressure which arises during the compression test is, the smaller is the effective stress and therefore the shearing resistance at constant total stress and vice versa.

That effect is also confirmed by the measured and calculated dependence of the concentration factors upon the vertical load shown in table 1. The reason why the concentration factor depends on the load is discussed by Horn (1983).

While the values of the concentration factor of the mechanically homogenized soil samples were very high even after desiccation at 300 mbar and under small loads, corresponding values for the natural soil samples were much lower.

desiccation rate (mbar)		vertical load (N/cm^2)					
		2	5	7	10	15	20
60	natural	12.9	12.9	14.1	11.5	8.9	7.5
	artificially homogenized	18	18	i	i	i	i
300	natural	1.1	1.1	1.1	3.0	2.5	2.0
i = indeterminable							

Table 1: Load dependent values of the concentration factor for clay soils (desiccation rate: 60, 300 mbar)

The natural aggregation effect is thus demonstrated as well as the stabilization effect in relation to the desiccation rate.

The consequence of these stabilisation effects due to aggregation for the trafficability and compressibility of clayey soils will be demonstrated by the relation between the normal stress at different depths and the values of the saturated hydraulic conductivity. The latter is shown in figure 3 in dependence of the vertical load at the top of the non plowed soil horizon and the concentration factors.

While the saturated hydraulic conductivity of the mechanically homogenized soil is intensively reduced not only directly below the surface of the compressed and untilled soil horizon (= plow pan) but also in the following 20 to 30 cm, increasing structural stability leads to higher values in the same depth and to a more intensive pressure compensation. That can be derived by the depths where the vertical K-lines begin.

Higher load at the top of the soil horizon (right diagram in fig. 3) induces a further reduction of the saturated hydraulic conductivity in all depths.

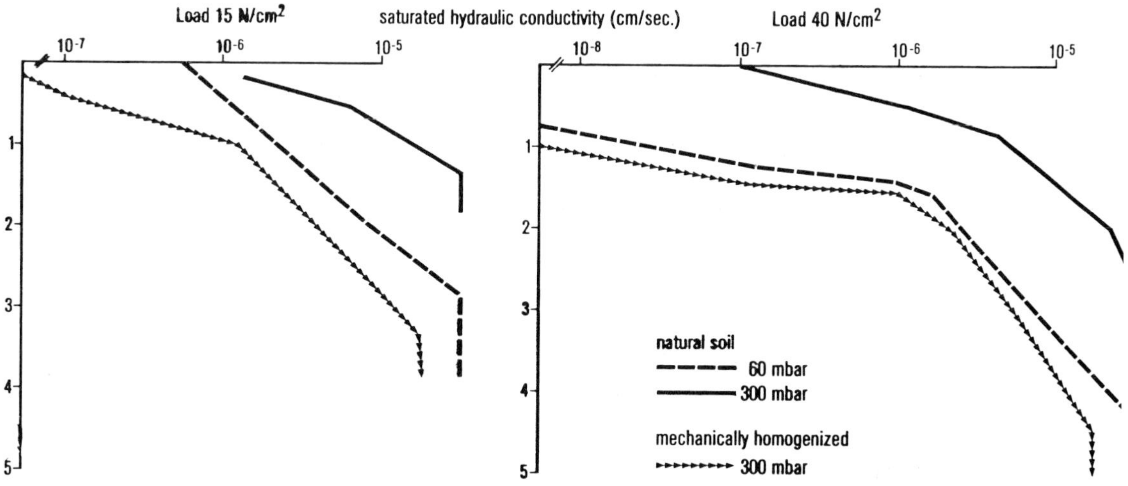

Figure 3: Change of the saturated hydraulic conductivity of clayey soil samples with depth in dependence of the load ($N/cm^2 \hat{=} 10^4$ Pa) and the pretreatment

Because up to now there exist not a single applicable equation for the prediction of the compressibility or trafficability of differently aggregated clay soils a first approach can be made by use of multiple regression analysis. The coefficient of volume compressibility (Sm) $\left[\frac{1}{10^5 \text{ Pa}}\right]$ as the dependent variable, which describes the settlement of the soil under the defined load of 10^5 Pa, can be estimated highly significant for a number of clay soils by the following equation

$$Sm = (72{,}1\ c + 6.83\ U - 3.16\ LK - 997.28\ d_B - 1.85\ TW + 1211.26)^{-1} \qquad \text{(eq.4)}$$

where c = cohesion x 10^4 (Pa), U = silt (%),
LK = air capacity (%), d_B = bulk density (g/cm^3) and
TW = non available water (%) (pF > 4,2).

Adding values for the concentration factors as well as equations for the load dependent change of soil physical parameters, it should be possible to predict the depth function of the mechanical compressibility more accurately.

Literature

Horn, R., B. Mattiat and E. Knickrehm, 1978: Die Ermittlung der Porengrößenverteilung eines Tonbodens über Entwässerung und auf rasterelektronischem Wege - Ein Vergleich. Catena 5, 9 - 18

Horn, R., 1980: Die Ermittlung der vertikalen Druckfortpflanzung im Boden mit Hilfe von Dehnungsmeßstreifen. Z. f. Kulturtechnik und Flurbereinigung 21, 343-349

Horn, R., 1981: Die Bedeutung der Aggregierung von Böden für die mechanische Belastbarkeit in dem für Tritt relevanten Auflastbereich und deren Auswirkungen auf physikalische Bodenkenngrößen. Schriftenreihe TU Berlin, H. 10, ISBN 3 7983 0792 x

Horn, R., 1983: Die Bedeutung der Aggregierung für die Druckfortpflanzung im Boden. Z. f. Kulturtechnik und Flurbereinigung 24, 238-243

Kuntze, H., 1965: Die Marschen - schwere Böden in der landwirtschaftlichen Evolution. Verlag P. Parey

Newmark, N. M., 1942: Influence charts for computation of stresses in elastic foundations. Uni Illinois Bull. No. 338

Satyavaniya, P. and J. D. Nelson, 1971: Shear strength of clays subjected to vibratory loading Proc. 4. Asian Reg. Conf. Soil Mechanics Foundation Eng . Bangkok 1, 215-220.

The measurement of soil structural parameters by image analysis

A. Ringrose-Voase and P. Bullock
Soil Survey of England and Wales,
Rothamsted Experimental Station,
Harpenden, Hertfordshire, AL5 2JQ, U.K.

INTRODUCTION

Soil structure is the most important property affecting the storage and movement of water in soils. Modelling water retention and flow in soils requires measurement of pore sizes and arrangements, but until recently neither of these requirements has been obtainable directly. Instead there has been a dependence on indirect measurements e.g. water retention, hydraulic conductivity, infiltration rate.
The currently most satisfactory method of directly measuring pore space seems to be through resin impregnation of undisturbed samples and the measurement of structural properties in the images derived. All stages in this approach have been improved recently: (i) larger samples can be impregnated, giving a more representative volume; (ii) drying by solvent extraction of water reduces the likelihood of artifacts; (iii) more sophisticated photographic techniques enable high quality images to be derived (Bullock and Murphy, 1980).

IMAGE ANALYSIS

The introduction of the image analysing computer, e.g. Quantimet 720, has revolutionised the range of structural measurements possible and the rate at which they can be made. Basic measurements of pore space include: area, number, perimeter, convex perimeter, Feret's diameter and intercept. From these, a variety of derived measurements are obtained

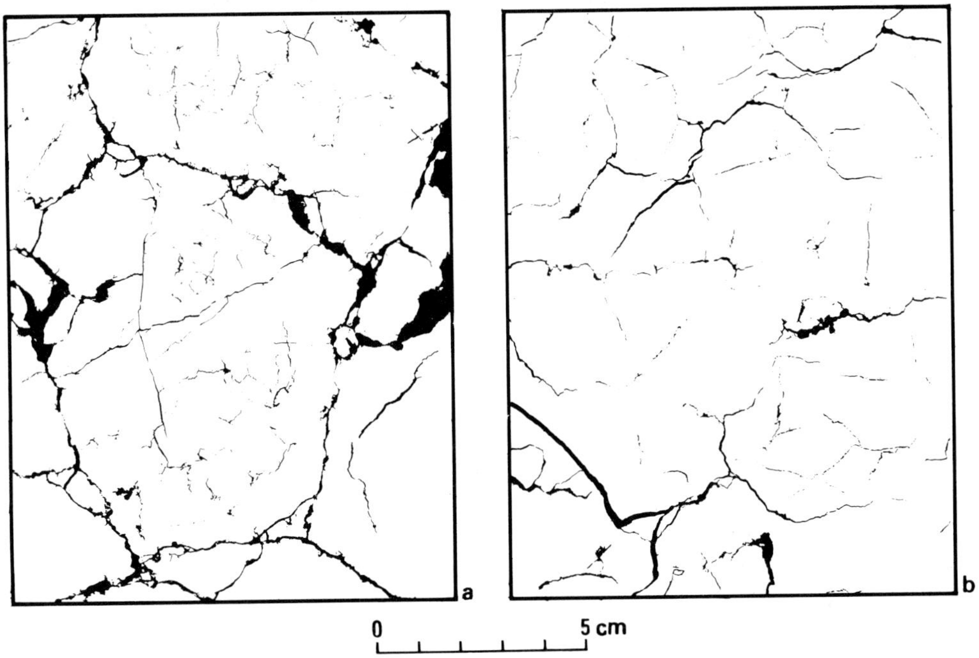

Figure 1. Horizontal sections of subsoil horizon (55 cm) of Denchworth soil, treated to show only fissures. a) Dry soil b) Wet soil

including size, shape, orientation and irregularity (Murphy et al. 1977). The data are from 2-D sections but to increase the value of this approach for deriving accurate models of pore space, the 2-D measurements can be translated into 3-D volumes using the mathematical tools of stereology (Weibel, 1979). Recently, a system has been developed for the automatic recognition and measurement of different pore types, each with a distinct shape and function with respect to water flow (Ringrose-Voase and Bullock, in press). This allows more detailed interpretation for each pore type. This paper illustrates some of these developments with respect to one pore type, namely fissures.

STEREOLOGICAL CONCEPTS

The volume of a structural component such as a pore is expressed as area in a 2-D image. The volume of pores (V_V) per unit volume of soil can be

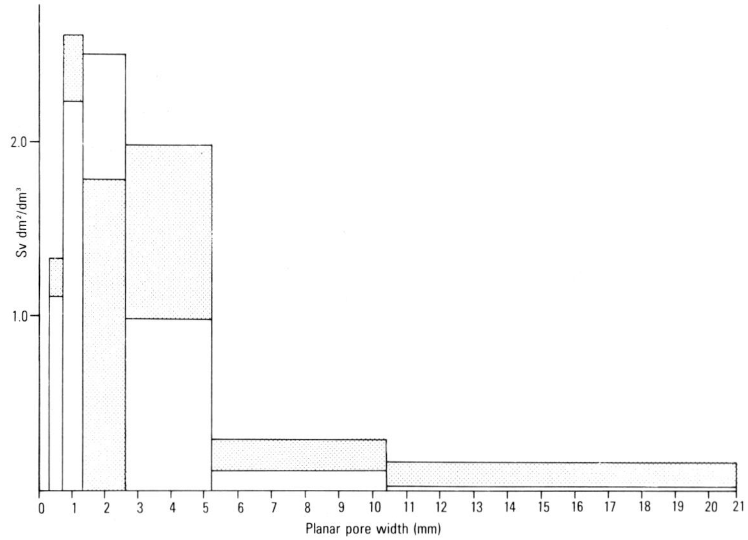

Figure 2. Histogram of fissure width distribution. Stipple = dry soil, blank = wet soil

estimated from the area of pores (A_A) per unit area of cross section by the relationship $V_V = A_A$ (Delesse, 1847).
The surface area of a pore is reflected in 2-D section by its perimeter. The surface area (S_V) per unit volume can be determined from the length of the perimeter (B_A) per unit cross-sectional area. The length of the perimeter can either be measured directly or determined by counting the number of intercepts (I_L) per unit length of scan line, the scan lines being taken at several angles across the image. S_V, B_A and I_L are related by

$$S_V = 4/\pi \cdot B_A = 2I_L \qquad \text{(Weibel, 1979, pp 30-37)}$$

A common pore type in clay soils is the fissure. This can be modelled as a sheet of varying width. The areal extent of this sheet per unit volume ($S_V(s)$) can be considered as the area of a plane running along its centre. The mean width of the sheet can be estimated from:

$$\bar{t} = V_V/S_V(s)$$

The distribution of sheet widths can also be obtained stereologically. Consider a sheet of constant width in 3-D and isotropic orientation. When sectioned, this sheet appears in 2-D as a line of varying thickness greater than or equal to the 3-D width. For a given width in 3-D there is a predictable distribution of 2-D widths. Therefore, for a sheet of varying width, a 2-D width distribution can be predicted for each 3-D width class. This allows the 3-D width distribution to be 'unfolded' from the measured distribution of 2-D widths according to the method of Cruz-Orive (1979), as in Figure 2.

MATERIALS AND METHODS

Samples were taken of a Denchworth soil, a pelostagnogley with 60% clay and a clay mineral suite dominated by interstratified mica-smectite. Samples were taken at 35% and 45% moisture contents as measured by neutron probe. The samples were dried by acetone replacement of the water, impregnated with a Crystic resin containing a fluorescent dye (Bascomb and Bullock,1974) and images taken by photographing a flat face of the impregnated block in UV light (Figure 1).

RESULTS AND CONCLUSIONS

The fissure patterns in the soil at the two moisture states were analysed on Quantimet. The data from horizontally and vertically oriented sections were averaged for each moisture state to allow for anisotropy of the fissure pattern. The Quantimet measurements were used to derive V_V, S_V, $S_V(s)$ and $\bar{t}$.

The results clearly indicate a difference in structure between wet and dry conditions. Porosity associated with fissures is reduced by 55% in the wet state compared with the dry one (Table 1).

The areal extent ($S_V(s)$) of the fissures is reduced by 22% in the wet soil and there is a reduction of mean fissure width of 42% from 1.08 mm in the dry soil to 0.63 mm in the wet soil.

The fissure width distribution is skewed at both sampling dates (Figure 2) but there is a shift to smaller widths in the wet soil. The areal

Table 1. Fissure measurement

	DRY SOIL	WET SOIL
Porosity (V_V ·100) (%)	9.6	4.3
Surface area per unit volume (S_V) dm^2/dm^3	16.4	13.8
Areal extent of fissures ($S_V(s)$) dm^2/dm^3	8.6	6.7
Mean fissure width ($\bar{t}$) mm	1.08	0.63

extent contributed by larger fissures is small compared to smaller ones but is significant in view of the fact that water flow through fissures is proportional to the cube of the width. Thus the decrease by 92% and 63% in areal extent of the two largest classes on wetting is important in relation to hydraulic flow.

Using image analysis, the proportion of conducting pores, the areal extent of fissures and their mean width in a clay soil at different moisture contents can be obtained. As well as being valuable in monitoring seasonal changes in structure, it can also provide a stronger numerical input into modelling for water movement and retention.

REFERENCES

Bascomb, C.L. and Bullock, P. 1974. Sample preparation and stone content. In: Avery, B.W. and Bascomb, C.L. (eds.). Soil Survey Laboratory Methods. Soil Surv. Tech. Monogr. 6, Harpenden, 5-13.

Bullock, P. and Murphy, C.P. 1980. Towards the quantification of soil structure. J. Micros. 120, 317-328.

Cruz-Orive, L.M. 1979. Estimation of sheet thickness distribution from linear and plane sections. Biometrical J. 21, 717-730.

Delesse, M.A. 1847. Procédé mécanique pour déterminer la composition des roches. C.R. Acad. Sci. (Paris) 25, 544.

Murphy, C.P., Bullock, P. and Turner, R.H. 1977. The measurement and characterisation of voids in soil thin sections by image analysis. Part I. Principles and techniques. J. Soil Sci. 28, 498-508.

Ringrose-Voase, A.J. and Bullock, P. The automatic recognition and measurement of soil pore types by image analysis. In press.

Weibel, E.R. 1979. Stereological Methods. Vol. 1. Practical Methods for Biological Morphometry. Academic Press, London.

Discussion

A.C. Armstrong:

In using the Bayesian pattern recognition algorithm, you can assign differing prior probabilities to each of the classes. Do you use this facility or do you assume equal prior probabilities? What sort of clarification accuracies do you achieve?

Author:

The procedure which allocates pore outlines to pore classes does indeed use Bayesian pattern recognition (see Ringrose-Voase and Bullock, Journal of Soil Science, 1984, Volume 35, No. 4). We do not assign a priori probability to each class (other than equal ones), because it is the frequency of each class in a given sample that we are trying to measure. The classification accuracy depends on two factors:

a) The Quantimet occasionally makes wrong measurements of the Calliper diameters. This leads to incorrect shape factors being calculated and to misallocation. However, the Calliper diameters are used only in the calculation of the shape factor, and not as one of the final measurements.

b) The Quantimet has to recognize each pore outline in isolation, whereas we can recognize pores in the context of the whole image. This leads to some misclassification.

These errors can be corrected within the allocation program by overriding individual allocation decisions. We would expect no more than 20 pore outlines to be incorrectly classified in an image containing up to 500 outlines.

G.H. Bolt:

Imagine analysis appears, at first sight, to be the answer to analyse the relation between physical behaviour of a soil and its structure. There is no doubt a causal relationship but, and that is the second thought, is there much hope that one could eventually derive field properties from the extreme precision of a point analysis of the geometry?

Author:

In the near future, at least, it is not conceivable that pore geometry alone could be used to predict the physical behaviour of a soil. By studying the measurements of physical behaviour in relation to geometrical measurements it should become clearer what the functions of the different pore types are, and how they interact. Thus, these types of measurements are seen more as a means of extrapolating the measurements of physical behaviour made on one soil to another. This would allow improvements to be made on many of the models of water movements we have heard about this week, and would make them applicable to a wider range of soils than those on which they were developed. There is also the problem of predicting behaviour on a field scale from exact measurements made at individual points. The field variability of structure can be large, and we are currently investigating this. However, these sorts of measurements could have a use in interpreting field descriptions of structure quantitatively, so that survey descriptions could be used in models of physical behaviour.

List of participants

Arabia

G.S. Gumaa
The Arab Center for Studies of arid zones and dry lands
P.O. Box 2440
Damascus - S.A.R.

Australia

Warren J. Bond
CSIRO Division of Soils
GPO Box 639
Canberra City, A.C.T. 2601

J. Philip
CSIRO Division of Environmental Mechanics
GPO Box 821
Canberra City, A.C.T. 2601

G. Smith
Queensland Wheat Research Institute
Department of primary industries
P.O. Box 5282
Toowoomba Queensland 4350

Belgium

N. Kihupi
K.U. Leuven
Fac. Landbouwwetenschappen
Kard. Mercierlaan 92
3030 Heverlee

Bulgaria

N. Shopsky
Institute for Soil Science and Yield Protection "N. Poushkarov"
Shosse Namkia str. no. 5
Sofia 1080

Canada

J.C.W. Keng
Canada Department of Agriculture
Research Station
P.O. Box 1000
Agassiz, BC VOM 1AO

Czechoslovakia

M. Kutilek
Soil Science Laboratory
Technical University Thakurova
16000 Prague 6

F. Zrubec
Research Institute of Soil Science
Roznavska 23
Bratislava

France

Y.G. Amiet
Service drainage - étude des sols
4 Promenade Madame de Sevigne
14039 Caen Cedex

J.C. Chossat
CEMAGREF
50 Avenue de Verdun
33160 Cestas Principal

Ph. Collas
CEMAGREF
50 Avenue de Verdun
33160 Cestas Principal

P. Curmi
Institut national de la Recherche Agronomique
Lab. du Science du Sol
65 Route de Saint Brieux
35042 Rennes Cedexe

L. Florentin
INPL-ENSAIA
38, Rue St. Catherine 54000
Nancy

France
V. Hallaire
INRA Science du Sol
Domaine St. Paul
84140 Montfavet
Avignon

B. Lesaffre
CEMAGREF
B.P. 121
92164 Antony

M. Normand
CEMAGREF
B.P. 121
92164 Antony

Y. Pons
INRA
Douanie Experimental de
St. Laurent
17450 Fouras

P. Stengel
INRA
Station de Science du Sol
B.P. 91
84140 Montfavet
Avignon

Hungaria
G. Varalyay
Research Institute for Soil
Science and Agricultural Chemistry
Hungarian Academy of Sciences
Budapest 11
Herman O.u. 15

A. Muranyi
Research Institute for Soil
Science and Agricultural Chemistry
Hungarian Academy of Sciences
Budapest 11
Herman O.u. 15

Iran
S.A. Rouhi
Y kom Consultant Engineers
Np. 77, North Saba Ave
Teheran 14

Israël
D.H. Yaalon
Department of Pedology
The Hebrew University
Jerusalem

Italy
A. Carmine Dimase
Instituto di Geopedologia
Piazzale delle Cascine, 15
50144 Firenze

N. Rossi
Instituto di Chimica Agraria
Via S. Giacomo 7
40126 Bologna

Guido Sanesi
Instituto di Geopedologia
Facoltà di Agraria Piazalla
Cascine
15-50143 Firenze

Jamaica
V.A. Campbell
Soil Survey
Ministry of Agriculture
Hope Gardens
P.O. Box 480 Kingston 6

H. de Wit
Soil Survey Project, Jamaica
c/o Koeriersdienst Min. BuZa
Casuariestraat 16
2511 VB Den Haag, The Netherlands

The Netherlands
H. ten Berge
University of Agriculture
6703 BC Wageningen

W.A. Blokhuis
University of Agriculture
6700 AA Wageningen

D. de Boer
University of Agriculture
Amsterdam

The Netherlands

J.J. Boesten
Institute for Pesticide Research
Marijkeweg 22,
6709 PG Wageningen

J. Bouma
Soil Survey Institute
6700 AB Wageningen

W. Bouten
Lab. of Physical Geography and
Soil Science
University of Amsterdam
Dapperstraat 115
1093 BS Amsterdam

H. Bronswijk
Institute for Water Management
Research (ICW)
Postbus 35
6700 AA Wageningen

J. Brouwer
Abstederdijk 200
3582 BW Utrecht

M.G.M. Bruggenwert
University of Agriculture
6703 BC Wageningen

J. Brugmeijer
University of Agriculture
Wageningen

L.W. Dekker
Soil Survey Institute
P.O. Box 98
6700 AB Wageningen

G. van Dregt
National Institute of Public
Health and environmental Hygiene
P.O. Box 150
2260 AD Leidschendam

J.J.P. Gerits
Student Lab. of Physical Geogr.
and Soil Science
Dapperstraat 115
1093 BS Amsterdam

H. van Grinsven
University of Agriculture
6703 BC Wageningen

J. Halbertsma
Institute for Water Management
Research (ICW)
Postbus 35
6700 AA Wageningen

J.W. van Hoorn
University of Agriculture
Nieuwe Kanaal 11
6709 PA Wageningen

F.F.R. Koenings
Hullemburgweg 5
6721 AN Bennekom

M. Kooistra
Soil Survey Institute
P.O. Box 98
6700 AB Wageningen

M. Loxham
Laboratory for Soil Mechanics
Postbus 69
2600 AB Delft

J. van der Molen
University of Agriculture
6701 DM Wageningen

B. Overmars
University of Agriculture
6703 PA Wageningen

L.J. Pons
University of Agriculture
6700 AA Wageningen

The Netherlands
B.J.A. van der Pouw
Soil Survey Institute
P.O. Box 98
6700 AB Wageningen

P.A.C. Raats
Institute for Soil Fertility
P.O. Box 30003
9750 RA Haren (Gr.)

J. Reijerink
University of Agriculture
Haarweg 139
Wageningen

H. Rogaar
University of Agriculture
Postbus 37
6700 AA Wageningen

E. Romijn
Provincie Gelderland
Postbus 9090
6800 GX Arnhem

K. Rijniersce
IJsselmeerpolders Authority
Postbus 600
8200 AP Lelystad

L. Stroosnijder
University of Agriculture
6703 BC Wageningen

Van den Tellaart
University of Agriculture
Wageningen

C. de Valk
University of Agriculture
Wageningen

T.J.F. Vollebergh
University of Agriculture
6708 DX Wageningen

W. de Vries
Soil Survey Institute
P.O. Box 98
6700 AB Wageningen

P. de Willigen
Institute for Soil Fertility
P.O. Box 30003
9750 RA Haren (Gr.)

G.P. Wind
Institute for Water Management
Research (ICW)
Postbus 35
6700 AA Wageningen

J.H.M. Wösten
Soil Survey Institute
P.O. Box 98
6700 AB Wageningen

S. van der Zee
University of Agriculture
6703 BC Wageningen

South Africa
D.J. Nel
University of Potchefstrom
Dept. of Soil Science, P.U.
Potchefstrom

Sovjet-Union
V. Kashyap
Department of Soil Fertility
Dokuchaev Soil Institute
Pyjevski 7
109017 Moscow

Spain
J.V. Giraldez Cervera
INIA Centro Nacional de Plantas
Oleagionosas
Apdo de Corresos 240
Cordoba

J. Martinez-Beltran
Calle de Velazquez 147
Madrid-Z

Sudan
O.A. Fadl
University of Gezira
P.O. Box 20
Wad Medani

M. el Hassan A/Karim Dirar
Soil Survey Administration
P.O. Box 388
Wad Medani

Sweden
L. Henriksson
Department of Soil Sciences
75007 Uppsala

I. Messing
Södra Parkvägen 20a
75245 Uppsala

Switzerland
R. Schulin
Swiss Federal Institute of
Technology
Dept. of Soil Physics
HG G 24.2, ETH-Zentrum
8092 Zürich

United Kingdom
T.M. Addiscott
Rothamsted Experimental Station
Harpenden Herts AL5 2 JQ

A. Armstrong
Field Drainage Experimental
Unit
Min. of Agr. Fisheries and Food
Anstey Hall, Maris Lane
Trumpington, Cambridge CB2 2LF

A. Armstrong
Institute of Irrigation Studies
Southampton University
Highfield
Southampton

K. Beven
Institute of Hydrology
Maclean Building
Crowmarsh Gifford
Wallingford OX10 8BB

P. Bullock
Rothamsted Experimental Station
Harpenden, Herts AL5 2JQ

J. Dyson
Soil Science Laboratory
Dept. of Agric. Science
University of Oxford
Parks Road OX1 3PF Oxford

P. Leeds-Harrison
Silsoe College
Silsoe, Bedford, MK 45 4DT

P. Loveland
Soil Survey of England and Wales
Rothamsted Experimental Station
Harpenden, Herts AL5 2JQ

L.A. Mackie
Soil Science Department
Mest Walk, Aberdeen

R. Parkinson
Seale-Hayne College
Newton Abbot, Devon TQ12 NQ

United Kingdom

A. Ringrose-Voase
Rothamsted Experimental Station
Harpenden, Herts AL5 2JQ

C.J.P. Shipway
Silsoe College
Silsoe, Bedford, MK 45 4DT

G. Sturdy
Soil Survey of England and Wales
East Malling Research Station
Maidstone, Kent ME 19 6BJ

G.D. Towner
Rothamsted Experimental Station
Harpenden, Herts AL5 2JQ

R.E. White
University of Oxford
Soil Science Laboratory
Dept. of Agric. Science
Parks Road, OX1 3PF Oxford

A.P. Whitmore
Rothamsted Experimental Station
Harpenden Herts AL5 2JQ

F. Wilkinson
Institute of Irrigation Studies
Southampton University
Highfield, Southampton

USA

J.W. Biggar
Department of Land, Air and Wate
Resources
University of California
Davis Cal. 95616

P. Germann
Dept. of Environmental Sciences
Clark Hall University of
Virginia
Charlottesville Virginia 22903

J.L. Nieber
Texas A and M University
College Station
Texas 77843 - 2117

R.F. Paetzold
USDA-ARS hydrology Lab.
BARC-West Bldg. 007 Room 139
Beltsville Maryland 20705

J. Skopp
Agronomy Department
University of Nebraska
Lincoln Nebr. 68583

J.W. Stucci
University of Illinois
S-510 Turner Hall
1102 South Goodwin Avenue
Urbana, Il. 61801

J. Lauren
Cornell University
Dept. of Agronomy
ITHACA, NY 14853

R.J. Wagenet
Cornell University
Dept. of Agronomy
ITHACA, NY 14853

L. Wilding
Dept. Soil & Crop Sciences
Texas A & M University
College Station
Texas 77843

P.J. Wierenga
Dept. of Crop and Soil Sciences
Box 3Q
New Mexico 88003

Vietnam

S. le Ngoc
c/o Institute of Land and Water
Management (ICW)
P.O. Box 35
6700 AA Wageningen
The Netherlands

West-Germany
H.H. Becher
Institut für Bodenkunde
Techn. Univ. München
D-8050 Freising-Weihenstephan

H.P. Blume
Institut für Pflanzenernährung
und Bodenkunde
Christian-Albrechts Universität
Olshansenstrasse 40
D-2300 Kiel

H. Diestel
Technical UniversityBraunschweig
Postfach 3329
33 Braunschweig

A. Gröngröft
Ordinariat für Bodenkunde
Allende-Platz 2
2000-Hamburg 13

K.H. Hartge
Institute für Bodenkunde
Herrenhäuserstr. 2
D-3000 Hannover 21

D. Horn
Inst. für Geowissenschaften
Abt. Bodenphysik
P580 Universitätsstrasse 30
Postfach 3008
8580 Bayreuth

U. Hornung
Hochschule der Bundeswehr München
Werner-Heisenbergweg 39
8014 Neubiberg

B. Maats
Ordinariat für Bodenkunde
Allende-Platz 2
2000 Hamburg 13

H. Rohdenburg
Abt. Physische Geographie und
Landschaftsökologie
Technische Universität
Langer Kamp 19c
D 3300 Braunschweig

V. Schweikle
Wilflingerstrasse 46/3
7464 Schönberg-2